GRASSLAND NITROGEN

GRASSLAND NITROGEN

D.C. Whitehead

Department of Soil Science
University of Reading
UK

Formerly of the Grassland
Research Institute, Hurley, UK

CAB INTERNATIONAL

CAB INTERNATIONAL
Wallingford
Oxon OX10 8DE
UK

Tel: +44 (0)1491 832111
Telex: 847964 (COMAGG G)
E-mail: cabi@cabi.org
Fax: +44 (0)1491 833508

A catalogue record for this book is available from the British Library.

ISBN 0 85198 915 2

Typeset in 10/12 Stempel Garamond
by AMA Graphics Ltd, Preston
Printed and bound in the UK by Biddles Ltd, Guildford

Contents

Preface

The aim in compiling this book has been to produce a comprehensive review of the transformations of nitrogen in temperate grassland systems. Topics include the changes occurring within the soil, the response of grass to fertilizer nitrogen, the fixation of atmospheric N_2 by legumes, and the impact of nitrogen from fertilizers and livestock excreta on the wider environment. In each chapter, a general account of the subject material is combined with a more detailed survey of published experimental work and, in a few instances, of unpublished data. The extensive list of references, which includes both research papers and reviews, provides an important source of additional information. A glossary defining the more specialist terms used in the text is also included.

For many years, the countries with the most active research programmes on grassland nitrogen were the Netherlands, New Zealand and the UK but, increasingly during the past decade, relevant research has been carried out in other countries. This pattern of research activity is reflected in the data selected for inclusion in the book though, in some chapters, emphasis is placed on work from the UK. Many of the chapters feature research carried out at Hurley Research Station (known from 1954 to 1985 as the Grassland Research Institute and then, successively, as the Animal and Grassland Research Institute and the Institute for Grassland and Animal Production, before becoming part of the Institute of Grassland and Environmental Research). An active programme of work on grassland nitrogen was maintained at Hurley throughout the period from 1954 until the closure of the research station in 1992, a programme in which I was involved from 1967 to 1971 and again from 1978 to 1992. Numerous colleagues at Hurley contributed indirectly to this book through their research and discussions over a number of years.

The book has been prepared during a period of secondment to the Department of Soil Science, University of Reading, from the Institute for Grassland and Environmental Research. The Institute has been funded largely through the Agricultural and Food Research Council, now the Biotechnology and Biological Sciences Research Council, and I would like to thank the Council for their continued financial support. I also thank the members of the Department of Soil Science at Reading for providing a welcoming environment during this period.

Several people read a draft copy of the book, in whole or in part, and I would like to express my appreciation for their suggestions. In particular, I thank Dr R.J. Stevens of the Department of Agriculture for Northern Ireland who reviewed the whole text. Professor R.J. Wilkins, Dr S.C. Jarvis and Dr D. Scholefield of the Institute for Grassland and Environmental Research also read the whole text and made a number of useful comments; Dr M. Wood of the Department of Soil Science, University of Reading, commented on Chapter 3, and Dr A.T. Chamberlain of the Department of Agriculture, University of Reading, commented on Chapter 4: I thank them for their contributions to the preparation of the final copy. In addition, I thank Mr John Cook for his help with literature searches and with the preparation of many of the figures.

The book is based, in part, on *The Role of Nitrogen in Grassland Productivity* (D.C. Whitehead), published by the Commonwealth Bureau of Pastures and Field Crops in 1970. Since that time, there has been considerable progress in our understanding of some topics but little change with others. It is hoped that the present book does justice to the advances that have been made, and that it will be of value to those engaged in research, teaching and advisory work related to grassland, and to students taking courses in the agricultural and environmental sciences.

David C. Whitehead
Reading
December 1994

Notes

Units

Data for herbage yield and rates of fertilizer application are stated in kg ha^{-1}, some having been converted from lb per acre.

For conversion:

1.0 kg = 2.21 lb
1.0 ha = 2.47 acres
1.0 kg ha^{-1} = 0.89 lb $acre^{-1}$
1.0 lb $acre^{-1}$ = 1.12 kg ha^{-1}

Terminology

Chemical symbols are used for nutrient elements.

Quantities and percentage contents of nutrient elements (P, K etc.) are stated in terms of the actual elements, not their oxides.

The figures cited for concentrations of various constituents of herbage refer to concentrations in the dry matter (DM), unless otherwise stated.

Abbreviations

cv. = cultivar
DM = dry matter
DoE = Department of the Environment (UK)
M = molar (concentration in solution)
MAFF = Ministry of Agriculture, Fisheries and Food (UK)
OM = organic matter
Pa = Pascal (unit of pressure)

1 Introduction

The Importance of Nitrogen

All living organisms require nitrogen (N) as part of the chemical structure of two of their essential constituents, proteins and nucleic acids. Very few organisms (all of which are bacteria) can utilize nitrogen gas (N_2), which is abundant in the atmosphere: all other organisms require N in a form in which it is combined with other chemical elements, e.g. hydrogen or oxygen. Nitrogen combined with other elements is often referred to as 'fixed' N; and the conversion of gaseous N_2 into combined N is known as fixation. A small amount of fixation occurs naturally in the atmosphere during thunderstorms, some occurs biologically through the action of bacteria and some is the result of industrial processes. The molecules of N_2 contain a triple bond which is extremely stable and, as a result, the industrial fixation of N_2 requires conditions of high temperature and pressure. Natural fixation in the atmosphere also requires exceptional conditions, such as those occurring during thunderstorms, and the amounts of N involved are therefore small. However, the biological fixation of N_2 occurs at ambient temperature and pressure, and those bacteria that are able to fix N_2 have specialized enzymes that convert it to ammonium-N (see Chapter 3). Other microorganisms and all higher plants absorb N in fixed forms, mainly as ammonium (NH_4^+) or nitrate (NO_3^-) ions; and animals consume the protein of plants or other animals to obtain a source of N for their own proteins and nucleic acids. The fixation of N_2 from the atmosphere is the original source of virtually all the N in soil/plant/animal systems, since igneous rocks contain only 10–50 mg N kg^{-1} (Stevenson, 1965) and their weathering provides negligible amounts. Gaseous N_2 constitutes about 78% by volume (76% by weight) of the atmosphere and amounts to

80,000,000 kg N for each hectare of the earth's surface (Jollans, 1985). This is many hundreds of times greater than the amount present in a hectare of soil, plus plant and animal biomass, in any terrestrial ecosystem.

In natural ecosystems, the amount of biomass produced by living organisms is often limited by the supply of N that is available to them. For most microorganisms and for plants, the N must be in the form of either ammonium or nitrate, and for animals it must be mainly in the form of protein. Before the development of fertilizers such as ammonium nitrate and urea, agricultural yields of both crop and livestock products were also limited by supplies of plant-available N. However, despite the widespread deficiency of N in available forms, the soils of most ecosystems, both natural and agricultural, contain large amounts of non-available N as part of their organic matter. This fraction of the soil, which ranges in different soils from almost nil to about 70% of the soil weight, accumulates slowly, often over hundreds or thousands of years, from the decomposition of plants, animals and microorganisms. For the soils of temperate regions, typical concentrations of N in the top 15 cm are 0.30–0.60% of soil dry weight under long-term grassland, and 0.05–0.20% under arable cultivation. Nitrogen added to the soil in the form of organic matter is thought to have an average residence time in the soil of about 175 years, though some may be lost from the soil within a few hours and some, which becomes incorporated into the most stable components of the soil organic matter, may have a residence time of more than 1000 years (Stevenson, 1986). However, in any one year, a small proportion of this non-available organic soil N is converted to ammonium and/or nitrate, and thus becomes available for uptake by plants. In temperate regions, the amount of N present in living plant material at any one time is typically less, and often much less, than 2% of the amount present in the soil organic matter.

Because the amount of N mineralized from organic matter, plus that deposited from the atmosphere, is usually sufficient to produce only low crop yields, intensive farming systems rely on the provision of additional N. This may be supplied either through the use of fertilizers or through growing leguminous crops that have a symbiotic relationship with N-fixing bacteria. In most agricultural cropping systems, N has a greater effect than other plant nutrients on crop yield and is taken up in larger amounts than the other nutrients, except possibly potassium.

Fertilizer N is produced industrially by a process that involves mixing atmospheric N_2 with hydrogen, and passing the mixture over a catalyst at high temperature and pressure to form ammonia:

$$N_2 + 3H_2 \xrightarrow[\text{400–540°C; 80–270 atm}]{\text{Fe catalyst}} 2NH_3$$

Most of the ammonia is then converted chemically to one of the main forms of fertilizer N, i.e. urea, ammonium nitrate or ammonium phosphate (see p. 202). Natural gas (methane) is now a major source of the hydrogen that is needed for the fixation process, though coal and oil are alternatives:

$$CH_4 + H_2O \xrightarrow[\text{75–115°C; 35 atm}]{\text{Ni catalyst}} CO + 3H_2$$

The process as a whole requires a considerable amount of energy, about 30 MJ kg^{-1} fertilizer N in recently constructed installations. However, there have been major improvements in the efficiency of energy use, and the figure of 30 MJ kg^{-1} N is about half that for the mid-1960s (France and Thompson, 1993).

Biological fixation, the other main source of N for agricultural production, avoids the use of fossil fuels though it is dependent on energy from photosynthesis. In agricultural terms, the most important N_2-fixing bacteria are those of the genus *Rhizobium*, which live symbiotically in the roots of leguminous plants such as clovers, peas and beans. For temperate grassland, white clover is generally the most important legume species, though lucerne and red clover are preferred in some areas, especially in swards intended for cutting rather than grazing. Lucerne, often in association with grass, is particularly important as a hay crop in the USA. A major advantage of growing a legume in association with a grass is that the N fixed by the legume benefits not only the legume itself but also the grass growing with it, and the mixed herbage often provides a better diet for livestock than would the legume alone.

On a worldwide basis, it is estimated that 60 Mt of fertilizer N are added each year to agricultural soils, while the estimate for biological fixation, though much less certain, is about 90 Mt N (Hauck, 1988). Of this biologically fixed N, about 50 Mt is thought to be fixed by the forage legumes, including white clover and lucerne, and about 40 Mt by grain legumes. In addition to the inputs of combined N provided by fertilizers and biological fixation, small amounts of fixed N, mainly as nitrate and ammonium, are added to soils and vegetation through rainfall and dry deposition from the atmosphere. Some of this nitrate and ammonium occurs naturally through fixation during thunderstorms, but most is a by-product of various industrial and agricultural processes. In contrast to the various fixation processes which result in the transfer of N_2 from the atmosphere to the soil, there are a number of processes that result in the transfer of N from the soil to the atmosphere. These are mainly due to the activities of soil microorganisms. Denitrification is particularly important as it results in the conversion of nitrate in the soil to gaseous N_2 and nitrous oxide (N_2O) which are released to the atmosphere (see Chapter 9).

The various transformations of N that involve the atmosphere, the soil and living organisms are often referred to collectively as the N cycle. As the range of transformations is wider for N than for other nutrient elements such as P and K, the N cycle is relatively complex. The wide range of transformations is due partly to the fact that N occurs in many different organic and inorganic compounds, some of which are gaseous and some solid, and which vary in oxidation state from –3 (in ammonia) to +5 (in nitrates). Some of the organic compounds, such as the proteins and nucleic acids, have a very large molecular size.

Agricultural Grassland in the UK and Other Temperate Regions

In the context of agricultural grassland, N is required ultimately for the production of meat, milk and wool. Of the major areas of temperate grassland in the world (Fig. 1.1), some have grassland as their natural vegetation and some are man-made. In most of the areas where grassland is the natural climax vegetation (e.g. the steppes of central Asia and the prairies of North America), the climate is semi-arid and a lack of rainfall prevents the growth of forests. The productivity of these grasslands is generally low because the growth of the grass is also restricted by the lack of rainfall. In more humid temperate areas (e.g. northwestern and central Europe, New Zealand, Japan and parts of Australia and North and South America), most grassland is man-made and would eventually revert to forest if agricultural management ceased. In general, the man-made areas of grassland are used much more intensively for livestock production than are the natural grasslands, and they provide most of the world's supplies of milk and beef. Although cattle and sheep in these areas

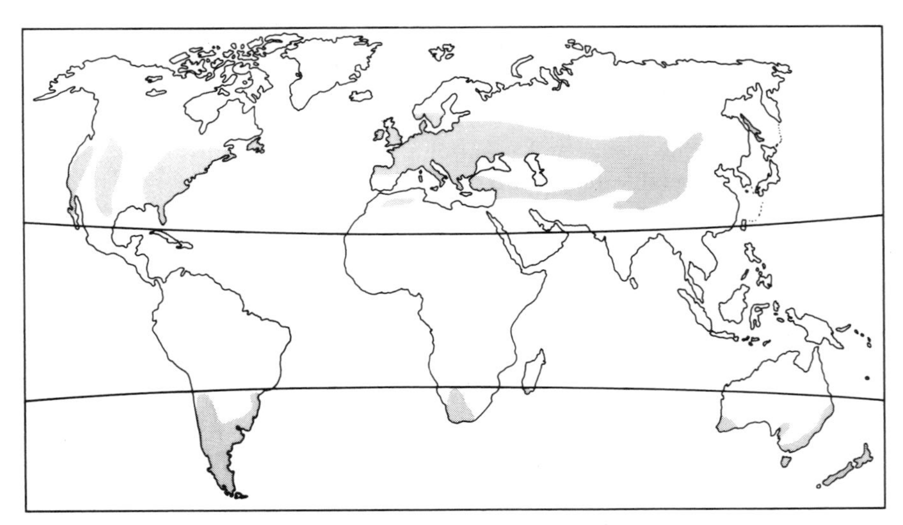

Fig. 1.1. The world's main areas of temperate grassland.

obtain the bulk of their food from grassland and forage crops, dairy cattle in particular are often provided with additional feeds, most of which are based on cereals (van der Meer and Wedin, 1989).

In western and northern Europe, and in the former USSR, grassland occupies about 50–60% of the utilized agricultural area whereas, in central and southern Europe, the proportion is about 35–40% (Weissbach and Gordon, 1992). There is a wide variation between individual countries even at similar latitudes. For example, grassland occupies about 21% of the utilized agricultural area in Denmark and about 90% in Ireland (Weissbach and Gordon, 1992). In the UK, enclosed grassland occupies more than half the agricultural area and, in addition, there is a considerable area of rough grazing which is mainly in the uplands of Scotland, Wales and northern England (Table 1.1). Regions of the UK differ widely in the proportion of the agricultural area occupied by grassland (Fig. 1.2), and these differences reflect climate and topography. In the drier lowland areas of east and southeast England, which are most suitable for arable cultivation, grassland occupies less than 25% of the agricultural land. In much of southern and central England, and eastern Scotland, grassland occupies between 25 and 75%, and in the western and northern areas, where rainfall is generally high and the topography is less suited to arable crops, grassland occupies more than 75% of the agricultural land (Holmes, 1989).

Grassland swards vary considerably in age and botanical composition. For example, in the UK most of the rough grazing and about half the enclosed grassland, has remained unploughed for more than 20 years and some, particularly rough grazing, has never been ploughed (Hopkins, 1987). However, a large proportion of the enclosed grassland was ploughed during the 1939–1945 wartime period, either to provide additional land for arable cropping or to improve its productivity by re-seeding, and little enclosed grassland has remained unploughed for more than 100 years. In 1993, the grass in about 23% of the enclosed grassland was less than 5 years old and was mainly perennial and Italian ryegrass sown, in some instances, with white clover. The rough grassland, which occurs mainly in hill and mountain areas, is often dominated by grasses of poor nutritional quality, such as *Nardus stricta* and *Molinia caerulea*, though in some areas there are also large proportions of non-graminaceous species such as heather (*Erica* spp.), ling (*Calluna* spp.) and bracken (*Pteridium aquilinum*).

Table 1.1. Areas of grassland, rough grazing and arable cropping in UK agriculture, 1992. (From Ministry of Agriculture Fisheries and Food, 1993.)

	Area ($\times 10^6$ ha)
Total area of UK	24.0
Total agricultural land	18.5
Grassland, including leys (enclosed)	6.8
Rough grazing	5.9
Arable cropping (including horticulture)	5.0

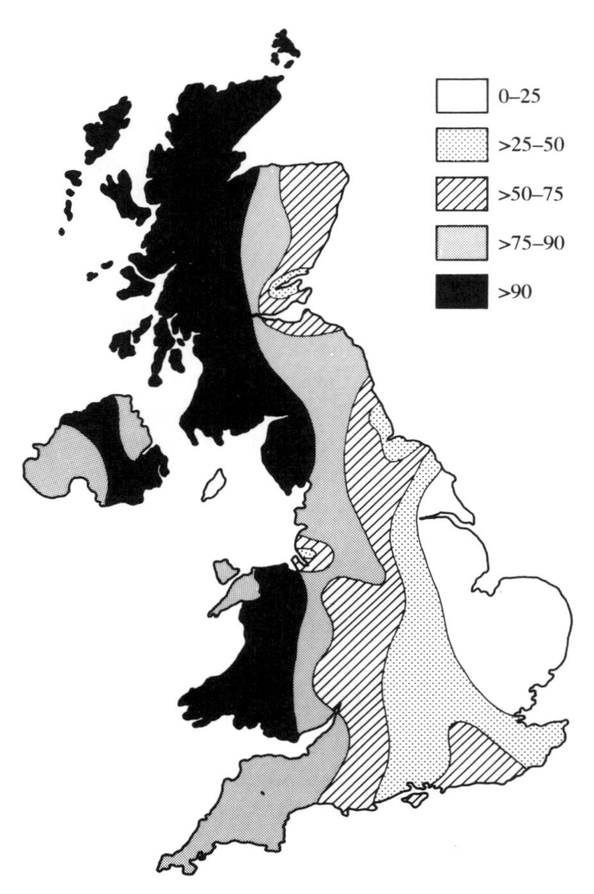

Fig. 1.2. Regional variation, in the UK, in the proportion of agricultural land occupied by grass and rough grazing. (From Holmes, 1989.)

The ploughing and re-seeding of UK grassland during the 1939–1945 period, together with the use of fertilizer-N and increased stocking rates, resulted in large increases in productivity, and these practices were therefore continued after the wartime period with further increases in productivity. For example, the output of milk and meat in the UK increased by about 1.8% per year during the period 1950–1970 (Green and Baker, 1981) and increased by about 1.4% per year during the period 1970–1985 (Wilkins, 1987). The increase during 1970–1985 occurred despite the area of enclosed grassland being reduced by 9% during the period. Similar increases in the output of milk and meat occurred in many other countries in Europe and, to a lesser extent, elsewhere. For example, milk production during the period 1960–1987 increased by about 34% in western Europe, 63% in eastern Europe and the USSR, 14% in Oceania and 13% in North America (van der Meer and Wedin,

1989). Greater productivity was promoted by various factors including research and advice programmes, guaranteed prices and subsidies. In practice, the increasing use of fertilizer N on grassland was a major factor, especially after the wartime period. In addition, especially in Europe and the USA, cereal-based feeds made a substantial contribution to the production of milk and meat.

Inputs of N to Agricultural Grassland

Grass has a large potential for the production of plant biomass, partly because the leaf area index (the ratio of the total area of leaf to the area of ground covered) is higher than for most other crops and partly because it has the capacity to re-grow after cutting or grazing. The leaf area index of grass is typically between 2 and 6. Most grass species, except those of semi-arid areas, are perennial and will grow (i.e. show net herbage accumulation) whenever the temperature is above about 5°C and when there is adequate moisture. The number of grass-growing days depends mainly on climate and, within the UK, ranges from 150–200 growing days in the dry areas of East Anglia, and in the hill areas above about 400 m, to 250–300 days in lowland areas near to the south and west coasts. Most of the other lowland areas have 200–250 growing days per year (Fig. 1.3). Partly because grass has a long growing season and partly because it is harvested (either by cutting or grazing) as vegetative growth, its response to fertilizer N is greater than that of most other crops. The large response in terms of herbage yield, and hence in the amount of milk or meat produced per unit area of land, has encouraged farmers to apply large amounts of fertilizer N. This practice has been further encouraged by the cheapness of fertilizer in relation to the value of milk and meat. In the UK, the average application of fertilizer N to enclosed grassland increased from about 10 kg N ha^{-1} per year in 1950 to a maximum of 135 kg ha^{-1} per year in 1986. Since 1986, there has been some fluctuation in use (Fig. 1.4). In 1986, 44% of the total area of grass leys (2–7 years old) in the UK received more than 200 kg N ha^{-1} $year^{-1}$, with 7% of the area receiving more than 400 kg N ha^{-1} (Chalmers and Leech, 1987) whereas in 1989 the corresponding figures were 24% and 2% (Chalmers *et al.*, 1990).

The rates of fertilizer N recommended for intensively farmed grassland in temperate regions are generally within the range of 200–400 kg N ha^{-1}, depending on soil type, weather and sward management. The influence of these factors on the response of grass to fertilizer N, and on the optimum rate of application, is described in Chapters 10–13. In the UK, the recommended rate is often between 250 and 380 kg N ha^{-1} $year^{-1}$ (see Chapter 10) whereas in countries such as the USA that have a wider range of grassland systems, there is a wider range of recommended rates. Examples include

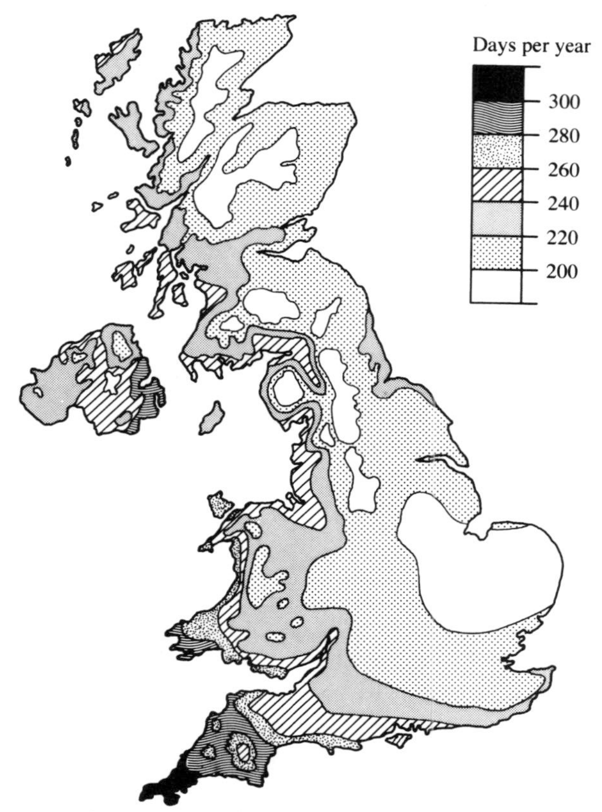

Fig. 1.3. Regional variation, in the UK, in the number of grass growing days per year. (From Lazenby, 1988.)

210 kg N ha^{-1} year^{-1} for cocksfoot in central Indiana, and 880 kg N ha^{-1} year^{-1} for irrigated Bermudagrass in Texas (Follet and Wilkinson, 1985).

Although the intensive production of milk and meat is based mainly on the use of fertilizer N, it is possible, though often more difficult, to achieve similar rates of production from grass–clover swards that receive no fertilizer N. Such swards are dependent on symbiotic N_2 fixation. One problem with grass–clover swards is that the growing season is shorter for clovers than for grasses. A second problem is that, in many areas, it is difficult to maintain an adequate proportion of clover in grass–clover swards over a number of years. Consequently, in most countries, grass swards receiving fertilizer N are much more widespread than grass–clover swards. An exception is New Zealand where well-managed grass–clover swards often produce herbage yields of 10–15 t dry matter (DM) ha^{-1} year^{-1} (Daly, 1990), similar to the yields produced by N-fertilized all-grass swards. The contrast between the UK and New

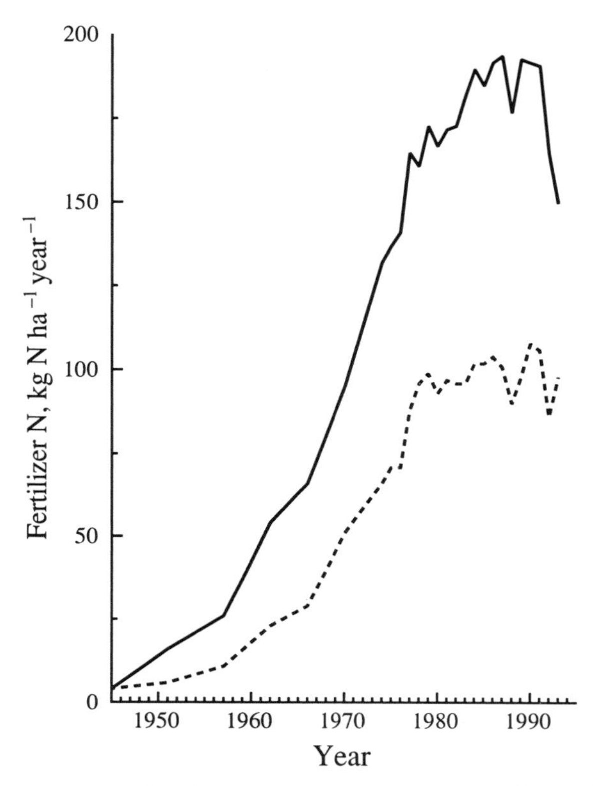

Fig. 1.4. Average rate of application of fertilizer N (kg N ha^{-1} $year^{-1}$) to 2–7 year leys (—) and permanent grassland (-----) during the period 1944–1991, and to grass < 5 (—) and > 5 years (-----) in 1992 and 1993. (Data from Cooke, 1975; and from annual Surveys of Fertilizer Practice conducted by MAFF in conjunction with the Fertilizer Manufacturers Association, Rothamsted Experimental Station (until 1991), and the University of Edinburgh (1992/93))

Zealand, which have similar areas of grassland, is illustrated by the fact that about 1,000,000 t of fertilizer N are used each year on grassland in the UK but only 10,000–20,000 t in New Zealand. Fixation by legumes has been estimated to provide about 1,000,000 t N $year^{-1}$ in New Zealand but only about 80,000 t in the UK (Ball and Crush, 1985). However, attempts are being made, particularly in Europe, to produce varieties of white clover that are more resistant to pests and diseases and to die-back in winter, and to develop management practices that will encourage clover to persist in mixed swards for a number of years.

When grassland herbage is consumed by livestock, only a small proportion of the N contained in the herbage is retained by the animal and most is excreted in dung and urine. In fields that are grazed, the excreta are returned directly to the land surface whereas, with animals that are housed and fed silage or hay, the excreta are usually stored and applied subsequently in the form of slurry or manure. Whether the excreta are returned during grazing or after storage, the N is susceptible to loss from the soil by leaching, ammonia volatilization and denitrification, as described in Chapters 7–9.

In contrast to the situation with most enclosed grassland, rough grazing receives very little fertilizer, and usually the only inputs of N from external sources are small amounts from biological fixation, and from wet and dry deposition from the atmosphere. The outputs of N in livestock products from rough grazing, and the losses through leaching, ammonia volatilization and denitrification are also low (Batey, 1982). In upland areas, the average temperature declines by about 0.6°C for every 100 m increase in altitude (Smith, 1984) and, above about 300 m, temperature is an important limiting factor in grass growth. The effect is greatest in spring and often results in about 75% of total grass growth taking place in a 6-week period during May and June (Newbould, 1981). Another factor restricting the growth of grass, and its potential to respond to fertilizer N, in upland areas is the infertility of the soils which are often shallow, acid and poor in P and K.

Transformations of N in Grassland Systems

The transformations that occur in the soil are central to the cycling of N, as illustrated by the diagram of the grassland N cycle shown in Fig. 1.5. However, although the concept of the N cycle is useful, it is important to appreciate that, at various stages of the cycle, there is not a clear-cut sequence of transformations but a number of alternatives. For example, nitrate in the soil may be taken up by plants, leached or denitrified. There are also large differences in the rates of the various transformations, which suggest that it may be more precise to regard the N cycle as a set of interlocking subcycles (Floate, 1987; Jansson and Persson, 1982). One subcycle involves N_2 in the atmosphere, and includes N_2 fixation and denitrification as major transformations. A second subcycle involves the N circulating between soils, plants and animals, and a third subcycle involves mainly the soil microorganisms and their processes of mineralization and immobilization.

Many of the important transformations of N represent inputs to, and outputs from, the soil. Thus, inputs of N occur through the application of fertilizers and manures, through the deposition of excreta from grazing animals, through the fixation of N_2 by microorganisms and through the deposition of ammonia and nitrate from the atmosphere. Outputs of N from the soil occur through the removal of grass or livestock products, and through leaching, the volatilization of ammonia and denitrification. Different types of grassland differ in both the absolute and relative amounts of N undergoing the various transformations, with climate and sward management being important

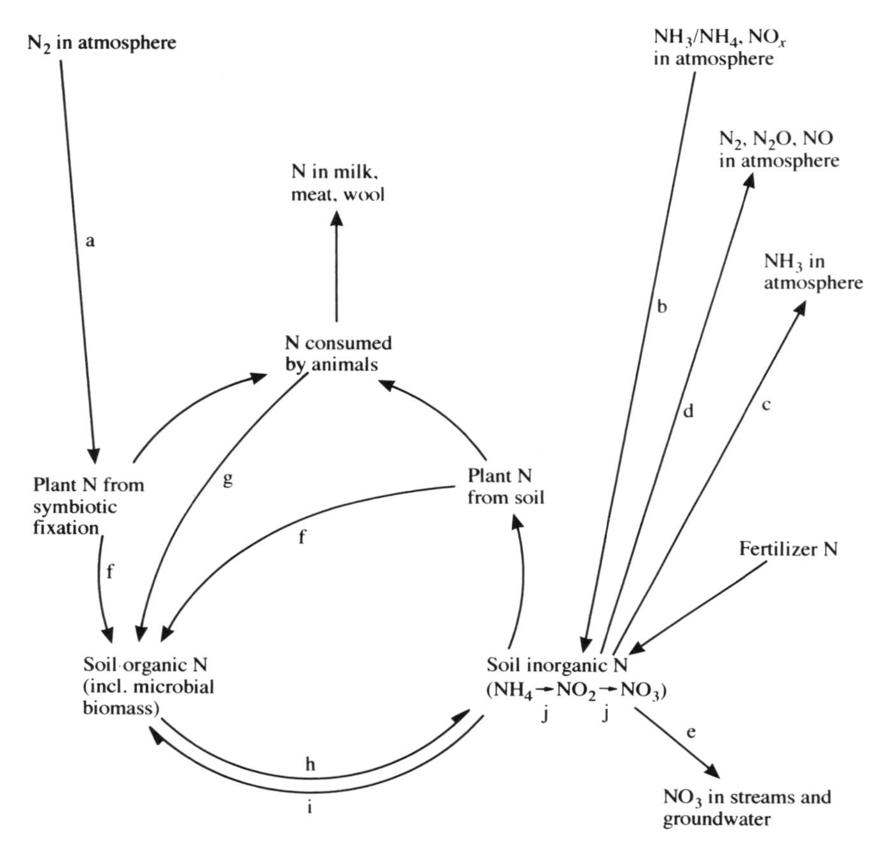

a Fixation by legumes
b Wet and dry deposition
c Volatilization of ammonia
d Denitrification/nitrification
e Leaching of nitrate
f Death and decay of plant tissue
g Excretion of dung and urine; application of manures and slurries
h Mineralization
i Immobilization
j Nitrification

Fig. 1.5. Major transformations of N in grassland.

factors. Aspects of sward management that have major effects are whether significant amounts of clover are present, whether the sward is cut or grazed, and the amount of fertilizer, if any, that is applied. Some transformations occur rapidly and some slowly. Thus, in grazed grassland, it is possible for an atom of N, initially present as nitrate or ammonium in the soil, to be taken up by the grass, to be consumed by a grazing animal, to be returned to the soil in excreta and transformed again to nitrate, all within a few days. As a result, some of the N in the system may be recycled several times during a growing season. On the other hand, a proportion of the N returned in excreta will remain in the soil for many years, immobilized in the soil organic matter. The intensity of management has an important influence on the outputs of N from the system: for example, the amounts of nitrate leached and of ammonia volatilized are likely to be at least 50 times greater from intensively fertilized and heavily grazed grassland than from unfertilized and lightly grazed grassland. Some quantitative estimates of the inputs and outputs of N in different grassland systems are presented in Chapter 15. An important feature of intensively grazed grassland is the heterogeneity in the soil that arises from the combined effects of consumption and excretion by the grazing animals (Floate, 1987; Azal and Adams, 1992). Herbage is consumed from a wide area but the N excreted in dung and urine is concentrated into small patches within the area. The nature and intensity of N transformations in the soil may therefore show large differences even within distances of a few centimetres.

Environmental Impacts of Fertilizer N Applied to Grassland

Although the widespread use of fertilizer N has greatly increased the production of food, it has incurred environmental costs. These have arisen partly from the energy requirements of fertilizer manufacture, partly from the pollution of the wider environment by compounds derived from fertilizer N, and partly from the loss of botanical diversity. The energy required for the manufacture of fertilizer N has been estimated to account for 60 to 90% of the support energy used in livestock production from British grassland (White *et al.*, 1983). Pollution of the wider environment through the use of fertilizer N arises from the leaching of nitrate to groundwater and streams (see Chapter 7), and from the volatilization of ammonia, nitrous oxide and nitric oxide to the atmosphere (see Chapters 8 and 9). Also, the botanical diversity of old grassland (permanent pasture) is inevitably reduced by the application of large amounts of fertilizer N (see p. 264). Repeated applications eventually eliminate many of the dicotyledenous species that are visually attractive during the flowering season, and that provide food for insects such as butterflies (Wells and Sheail, 1988). The disappearance of species is more significant in lowland areas, where old grassland is rare, than in upland areas where old grassland (though of different botanical composition) is widespread.

The rate of fertilizer N applied to grassland has a major impact on the transformations that are liable to result in pollution, though it is not the only factor involved. Thus the increase in the average rate of fertilizer N applied to agricultural land during the period from about 1950 to the mid 1980s was responsible, either directly or indirectly, for much of the increase in the amounts of N leached through the soil and lost in gaseous forms to the atmosphere, though factors such as the ploughing of long-term grassland and the importation of animal feeding stuffs were also partly responsible. Nitrogen that is leached from soils into groundwater is almost entirely in the form of nitrate and, if the groundwater is used for domestic supplies, the nitrate may occasionally constitute a health risk to consumers. Although nitrate itself has a low toxicity, if the acidity of the stomach is abnormally low, it is susceptible to bacterial conversion to the more toxic nitrite. The risk of nitrite poisoning to the general population is small, but it is appreciable for infants below the age of about 3–6 months who are liable to develop methaemoglobinaemia if they consume nitrate in more than trace amounts (Royal Society, 1983). Even in infants, this type of poisoning is rare, the last case in the UK being reported in 1972 but, when it does occur, it is sometimes fatal (Heathwaite *et al.*, 1993). In addition to the risk of methaemoglobinaemia, there is a possibility that stomach cancer in the general population might be increased by the consumption of nitrate. It has been suggested that nitrite produced in the stomach might react with naturally-occurring amines to form nitrosamines, some of which have been shown to cause cancer in laboratory animals. However, attempts to link the distribution of stomach cancer in different areas with the concentration of nitrate in drinking water have not been successful. For example, in the UK, the incidence of stomach cancer is highest in northern and western areas where nitrate concentrations are generally lower than in southern and eastern areas (Heathwaite *et al.*, 1993). Nevertheless, in view of the possibility that nitrate in drinking water might constitute a health risk, the European Community in 1980 adopted a directive that the concentration of nitrate in water intended for human consumption should not exceed 50 mg nitrate l^{-1} (11.3 mg NO_3^-–N l^{-1}) and this directive is in force in member countries. Subsequently, the European Commission introduced legislation requiring member countries to identify 'nitrate vulnerable zones' and to restrict the use of fertilizers and manures in these areas (see p. 303 and Heathwaite *et al.*, 1993).

In the atmosphere, the concentrations of N oxides and ammonia have also increased in recent decades. The N oxides include nitric oxide (NO) and nitrogen dioxide (NO_2), which are often referred to collectively as NO_x, and also nitrous oxide (N_2O). Nitric oxide and nitrogen dioxide react readily with water, forming nitrous acid (HNO_2) and nitric acid (HNO_3) respectively. The main sources of NO_x are combustion processes (particularly those in vehicle engines and power stations) which result in the oxidation of atmospheric N_2 to NO. However, it is possible that, on a global scale, relatively large amounts of NO are also released from the soil and by the burning of vegetation and crop residues (Warneck, 1988). Nitrous oxide (not included as a component of NO_x) is produced mainly from the soil by a combination of nitrification and denitrification, especially where the concentration of nitrate is high following the application of fertilizer or the deposition of urine. Much of the ammonia in the atmosphere is derived from the hydrolysis of livestock urine and, to a lesser extent, from the interaction of soils with fertilizers containing ammonium- or urea-N, and from the decomposition of plant material. Although some volatilization of ammonia occurs in natural ecosystems, it is greatly increased in areas of intensive livestock production. Ultimately, much of the N in NO_x and ammonia released to the atmosphere is deposited on the land or sea surface, where it may have potentially undesirable effects. For example, both NO_x and ammonia contribute to the acidifying effect of acid rain on soil (Kennedy, 1992), to the damage caused to forests by atmospheric pollution (Rosen *et al.*, 1992) and to the eutrophication of lakes and coastal waters such as the Baltic Sea (Schrøder, 1990). However, the effects of nitrous oxide (N_2O) may be more harmful than those of NO_x in the long-term, as it acts as a 'greenhouse gas' and also contributes to the destruction of ozone in the stratosphere.

Trends in Research and Agricultural Practice Since 1980

In response to concerns about increasing concentrations of nitrate in drinking water, and increasing concentrations of nitrogen oxides and ammonia in the atmosphere, the Royal Society in London set up a Study Group in 1979 to examine 'The Nitrogen Cycle of the United Kingdom'. The report published under this title (Royal Society, 1983) reviewed existing information and made a number of recommendations for further research. In relation to agriculture, the recommendations were directed mainly towards:

1. the development of varieties of crops able to utilize N more efficiently in the production of harvestable products;
2. the development of more effective biological N_2 fixation systems, e.g. by genetic engineering;
3. gaining better quantitative information on the factors influencing transformations and losses of N in agricultural systems, with the aims of making better use of waste materials and developing methods to minimize losses;
4. improving the prediction, partly by modelling, of crop responses to inputs of N; and
5. improving the prediction, again partly by modelling, of the impact of agricultural practices on nitrate in water supplies.

Following these recommendations, and similar recommendations in other countries (e.g. National Research Council, USA, 1978; Thornton, 1982), there was a shift in the emphasis of research towards studies of the factors influencing nitrate leaching, ammonia volatilization and denitrification. More attention was also given to ways of improving the efficiency of utilization of the N in both fertilizers and organic manures, and improving the effectiveness of biological N_2 fixation. As a result of these developments, there is now much more quantitative information on N transformations in various grassland systems. The impact of livestock on the transformations of N has been investigated in some detail during the period since the early 1980s, following the wider recognition that less than 25%, and often less than 15%, of the N consumed is actually converted to milk or liveweight gain, and that the remainder is therefore excreted. The quantitative importance of the N in livestock excreta is illustrated by the estimate, for UK agriculture, that 1530 kt N is present in the excreta produced each year by cattle, sheep, pigs and poultry, compared with a total of 1460 kt N applied as fertilizer (Jollans, 1985). A similar ratio between excreted N and fertilizer N was reported for the USA, with an estimated 7700 kt N being present in the excreta produced each year by cattle, pigs and poultry, and 9400 kt N being applied in fertilizers (Bouldin *et al.*, 1984).

A useful approach to quantifying N transformations in a particular field or ecosystem is to prepare a balance sheet of inputs to, and outputs from, either the total N or the plant-available N in the soil. In humid temperate regions, where winter rainfall is normally sufficient to remove nitrate from the rooting zone by a combination of leaching and denitrification, it can be assumed that there is a negligible amount of plant-available N in the soil at the beginning of each season. It then follows that the sum of the inputs of plant-available N during a year, expressed in kg ha^{-1}, is equal to the sum of the outputs. The application of the N balance approach to different types of grassland is discussed specifically in Chapter 15. The various transformations of N interact with farm management practices, so that different patterns of fertilizer application, of grazing and cutting, of irrigation and of ploughing for re-seeding all have large effects on the quantities of N involved (Dowdell, 1986). These management factors also influence the response of grassland to fertilizer N, as discussed in Chapters 10–14. In farming practice, the rate of application of fertilizer N to grassland has often been decided solely on the basis of financial considerations. With dairy farming in particular, the cost of fertilizer N has been, and still is, low in relation to the value of the milk produced. For this reason, farmers have tended to apply high rates of fertilizer to ensure that their planned milk yields are achieved, even if some fertilizer N is wasted. However, with increasing appreciation of the pollution effects of N from agriculture, this approach is now being discouraged and more emphasis is being placed on adjusting the rate of fertilizer application to individual situations (see Chapter 16).

2 Grasses: Uptake of Nitrogen and Effects on Morphology and Physiology

Introduction

Most plant tissues contain between 1% and 5% N on a dry weight basis. The N is an essential constituent of proteins, nucleic acids and chlorophyll, and the supply of plant-available N often limits the rate of growth. Enzymes comprise a large proportion of the protein in plant tissues and the photosynthetic enzyme, ribulose biphosphate carboxylase-oxygenase (Rubisco), constitutes about half of the total enzyme material (Lawlor, 1991). The concentration of N in plant tissues tends to decline as they become more mature, due mainly to the increase that occurs in the proportion of cell wall material and the corresponding decrease in cytoplasm which contains the enzyme proteins, the nucleic acids and chlorophyll.

Higher plants, except those depending on symbiotic fixation (see Chapter 3), absorb almost all their N as nitrate and ammonium ions through the roots. Both ions are absorbed readily. Nitrate is usually the major form because ammonium is converted to nitrate by the nitrifying bacteria in the soil. However, the process of nitrification occurs slowly in acid soils and at low temperatures (see p. 115) and, consequently, under such conditions, much of the total uptake of N may be in the form of ammonium (Barber, 1984). Urea and amino acids can also be absorbed by plant roots but, because these compounds are converted rapidly to ammonium by soil microorganisms, their uptake as intact molecules is normally slight. In addition to uptake through the roots, plants can absorb some forms of N (particularly gaseous ammonia and nitrogen dioxide) through their leaves but, in most situations, absorption through the leaves contributes only a small proportion (< 5%) of the total N uptake.

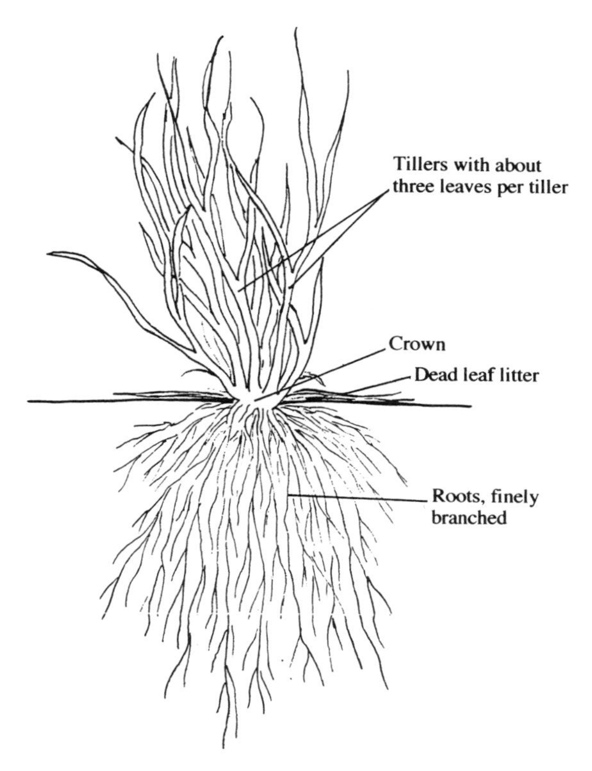

Fig. 2.1. Diagram of grass plant in the vegetative stage, showing several tillers.

With grasses, the influence of N supply on growth is apparent in several aspects of their morphology and physiology. A distinctive feature of the morphology of grasses is their development of tillers each carrying several leaves, as illustrated in Fig. 2.1. A deficiency of N restricts the number of tillers that develop and, more importantly, it also restricts the growth of individual leaves and their photosynthetic capacity. Shoot growth is restricted more than root growth by N deficiency.

Uptake of Nitrate and Ammonium by Roots

Nitrate (NO_3^-) and ammonium (NH_4^+) ions occur in soils as a result of the microbial decomposition of plant and animal residues, animal excreta and humified soil organic matter. In addition, many agricultural soils receive nitrate and/or ammonium through the application of fertilizers. The two ions differ in the reactions that they undergo in the soil, and in the mechanisms by which they are taken up by plant roots. Nitrate, an anion, is not adsorbed by the colloidal material (clay and organic matter) in the soil and it is therefore mobile in the soil solution, readily accessible to plant roots but also susceptible to leaching and denitrification. Ammonium, a cation, is retained by cation exchange on the clay and organic matter. It is therefore less mobile than nitrate, and is less accessible to roots and less susceptible to loss. Ammonium is however converted progressively to nitrate by nitrification (see p. 114). In the soil, nitrate moves mainly by mass flow with the movement of water and partly by diffusion, whereas ammonium moves mainly by diffusion and only slightly by mass flow (Barber, 1984).

When nitrate is taken up by roots, the pH of the growth medium tends to become more alkaline, with the nitrate ion in effect being exchanged for a hydroxyl ion. Conversely, when ammonium is taken up, the growth medium tends to become more acid, with the ammonium ion in effect being exchanged for a hydrogen ion (Bolan *et al.*, 1991; Hageman, 1984). Such changes in soil pH, induced by either nitrate or ammonium, may influence the availability of other nutrient elements (e.g. phosphorus) and thus have a secondary effect on growth.

The differences between ammonium and nitrate in their reactions in the soil, and variations in the extent to which ammonium is nitrified, make it difficult to compare the effects of the two ions on plant growth when the plants are grown in soil. However, comparisons with plants grown in solution culture have shown that, in general, both ammonium and nitrate ions are utilized readily but that many plants take up nitrate rather more readily than ammonium, and show greater growth responses to nitrate (Hageman, 1984). Grasses, however, often show a greater uptake of ammonium than of nitrate when the two ions are supplied in equal amounts (Clarkson *et al.*, 1986). High concentrations of ammoniacal N are toxic to all plants, and the rate of uptake of ammonium must therefore be matched by its conversion to non-toxic organic forms of N.

The rate of uptake of both nitrate and ammonium by roots is influenced by temperature and pH, but differently for the two ions. With perennial ryegrass, studies in solution culture showed that the absorption of nitrate increased with increasing temperature over the range 5–35°C and was greatest at pH 6.2, while the absorption of ammonium was greatest at a temperature of about 22°C and was only slightly influenced by pH over the range 4.0–7.5 (Lycklama, 1963). In another investigation, with perennial ryegrass grown in nutrient solution containing NH_4NO_3, the proportion taken up as ammonium was always greater than that of nitrate but was inversely related to root temperature, declining from 93% as ammonium at 3°C to 65% at 25°C (Clarkson *et al.*, 1986). Other plant species show a smaller effect of temperature on the ratio between ammonium and nitrate uptake (Wild *et al.*, 1987). Despite the difference in the uptake of the two ions when both were supplied together as ammonium nitrate, perennial ryegrass supplied with either ammonium or nitrate alone in solution showed no difference in either yield or N concentration (Jarvis, 1987). However, in another study with perennial ryegrass in solution culture, growth was maximized when the optimum concentration of nitrate was supplemented with some ammonium (Reisenauer *et al.*, 1982).

When perennial ryegrass, grown in soil under controlled environment conditions, was provided with either ^{15}N-labelled ammonium or nitrate, uptake was greater from ammonium (at least at low temperatures) but there was no difference in yield during the short experimental time period of one week (Watson, 1986). The longer term effects of N, taken up from soil as either nitrate or ammonium, were investigated with Italian ryegrass, grown in pots with six rates (0–500 mg N kg^{-1} soil) of each form, and with a chemical inhibitor to prevent nitrification (Nielsen and Cunningham, 1964). At the rate of 100 mg N kg^{-1}, the herbage yield obtained with nitrate was about 14% higher than that with ammonium but, at rates of 200 mg N kg^{-1} and above, the yield was higher with ammonium, the difference increasing from about 2% of the nitrate value at 200 mg N kg^{-1} to about 30% at 500 mg N kg^{-1}. With nitrate, the yield of herbage was highest at the rate of 100 mg N kg^{-1} whereas, with ammonium, it was highest at 200 mg N kg^{-1}.

When nitrate is absorbed by roots, part is translocated to the shoots as nitrate, and part is converted in the roots to amino acids and amides, before the N is translocated. When ammonium is absorbed, it is normally converted rapidly in the roots to amino acids and amides. The activity of the nitrate reductase enzyme in roots tends to diminish as the roots mature, and the proportion of absorbed nitrate that is transported unchanged therefore tends to increase with increasing maturity of the root system (Haynes, 1986a). The concentration of nitrate in grass roots is usually lower than the concentration in the shoots (Darwinkel, 1975).

The conversion of nitrate to ammonium in plant tissues occurs in two stages and requires a considerable amount of energy from photosynthesis:

$$NO_3^- + 2e \longrightarrow NO_2^-$$
$$NO_2^- + 6e \longrightarrow NH_4^+$$

The first stage (the conversion of nitrate to nitrite) is rate limiting and consequently there is no accumulation of nitrite as this is quickly converted to ammonium. The ammonium is also metabolized rapidly, mainly by reaction with glutamate to form glutamine, which in turn reacts with oxoglutarate to form two molecules of glutamate via the glutamate synthase cycle (Fig. 2.2). One of the molecules of glutamate then provides the starting point for the synthesis of a range of other amino acids and amides. If the total supply of N is less than that required for the potential growth rate, which is usually controlled by photosynthesis, the amino acids are converted rapidly into

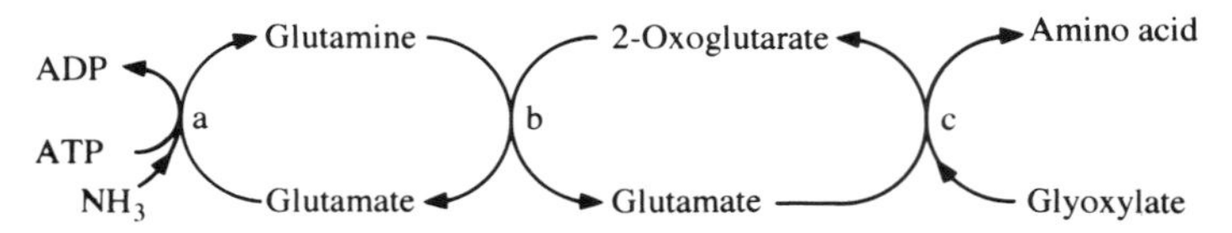

Fig. 2.2. The glutamate synthase cycle (enzymes: a, glutamine synthetase; b, glutamate synthase; c, transaminase).

proteins and nucleic acids, and the formation of amides is negligible. However, if the supply of N exceeds the amount needed to match photosynthesis, some will accumulate as nitrate (if this form is taken up) and some will accumulate in the form of amides.

Some of the N absorbed by plants may be utilized several times, and in different forms, during a growing season. This re-utilization arises partly because photorespiration releases ammonia from glycine, and partly because there is a constant turnover of proteins and nucleic acids. Particularly when the supply of external N is inadequate, some of the N from senescent leaves is translocated to young growing leaves. The translocation occurs in the phloem and is largely in the form of amino acids (Haynes, 1986a).

Amounts of N Taken up under Field Conditions

The capacity of grass swards to take up N, expressed in terms of kg ha^{-1} year^{-1}, is high in comparison with other crops and, in favourable conditions, may be more than 500 kg ha^{-1} year^{-1}. However, even with a plentiful supply of N, there is a wide variation in the daily rate of uptake, which depends on factors such as the stage of growth of the grass, and the time interval since the application of fertilizer and/or defoliation. A maximum rate of uptake of 7.5 kg N ha^{-1} day^{-1} was reported for Italian ryegrass in the period 14–21 days after

Table 2.1. Uptake of N (kg ha^{-1} day^{-1}) by established grassland supplied with five rates of fertilizer N and assessed after four periods of regrowth (mean values for seven field trials). (From Dilz, 1988.)

Rate of fertilizer N, kg ha^{-1}	Applied in April		Applied in June	
	30 days	54 days	32 days	46 days
0	0.89	1.05	1.05	1.10
30	1.41	1.37	1.47	1.47
60	1.84	1.74	2.03	1.80
90	2.42	2.05	2.44	2.09
120	2.91	2.44	2.98	2.39

the application of 140 kg N ha^{-1} as fertilizer (Wilman, 1965). A similar rate of uptake occurred with the same species in the period 16–23 days after the application of 175 kg N ha^{-1} (Hunt, 1966a). Daily rates of uptake, averaged over longer growth periods, are often in the range of 1–3 kg N ha^{-1} day^{-1} (Table 2.1). The rate of uptake varies with time of year, as illustrated by cut swards of perennial ryegrass receiving 420 kg N ha^{-1} $year^{-1}$ in which the rate of uptake increased markedly during the spring, from nil to about 3–4 kg N ha^{-1} day^{-1} in late April, before decreasing to about 0.5 kg day^{-1} in July and increasing again to 1–2 kg ha^{-1} day^{-1} in late August/September (Anslow and Robinson, 1986).

With grass that is not defoliated, the rate of uptake of N is greater during the vegetative stage of growth in spring than during the subsequent reproductive stage. The total amount of N in the herbage reaches a peak a few days before flowering stems emerge (the mean date of ear emergence) and well before maximum herbage yield (Fig. 2.3). The concentration of N in the herbage declines continually with advancing maturity and, after the time of ear emergence, the uptake of N is more than offset by various losses. These arise partly through the translocation of N from senescent leaves and partly through the death and decay of older leaves. There may also be some loss of N from senescent leaves through leaching by rainfall and the volatilization of ammonia. With grass that is defoliated at intervals during the growing season, defoliation causes a major though temporary reduction in nitrate uptake (Jarvis and Macduff, 1989). Other changes in the uptake of N during the season reflect weather conditions, such as temperature and rainfall, and fluctuations in the supply of N.

When intensively managed temperate grassland is harvested on several occasions during the growing season, either by cutting or grazing, the total annual yield of herbage generally amounts to between 8000 and 15,000 kg dry weight ha^{-1} and contains 200–550 kg N ha^{-1}. If the supply of N is not limiting, the herbage yield and its content of N at a particular location will depend largely on weather and soil conditions, and often mainly on the factors that determine the supply of water. The annual production of stubble and roots in grassland is less well defined but, with intensive management, probably amounts to about 6000–12,000 kg dry weight containing 100–250 kg N ha^{-1}. The total uptake of N into herbage, stubble and roots by intensively managed grass swards receiving fertilizer N is, therefore, often in the range of 300–700 kg N ha^{-1} $year^{-1}$.

Absorption of Ammonia by Leaves

Plant leaves have the capacity to absorb gaseous ammonia from the atmosphere and to utilize the ammonia as a source of N. Growth is increased by the

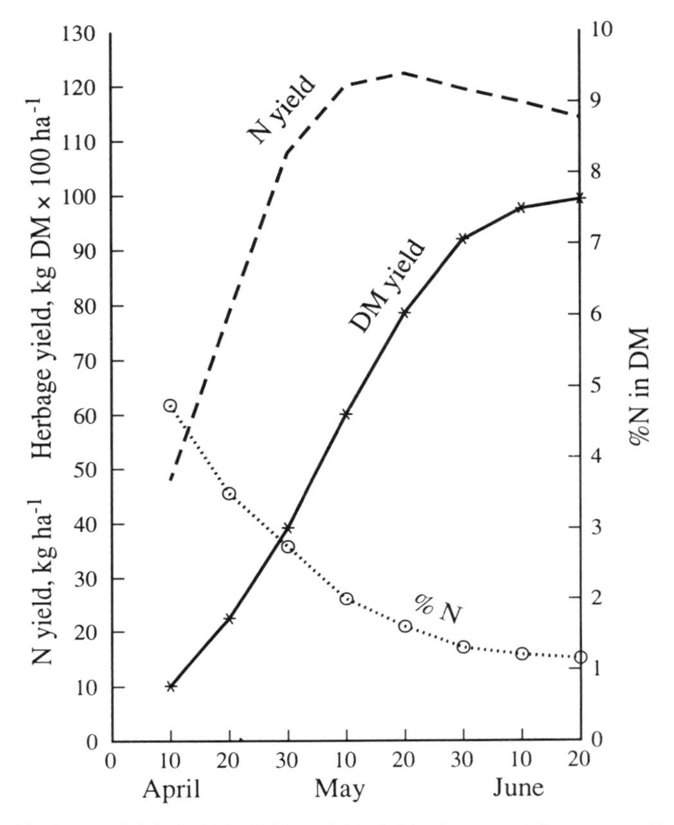

Fig. 2.3. Herbage yield, % N in DM and N yield of perennial ryegrass (S24) with advancing uninterrupted growth during April–June in southern England, following the application of 60 kg N ha^{-1} on 25 February and 78 kg N ha^{-1} on 24 March. (Data from Green and Corrall, Grassland Research Station, Hurley, 1965, unpublished.)

ammonia if the supply of N to the roots is less than optimum (Cowling and Lockyer, 1981; Whitehead and Lockyer, 1987), the ammonia being assimilated into amino acids via the glutamate synthase cycle (Fig. 2.2). At high concentrations, ammonia has toxic effects, and plant leaves may be damaged when the long-term average concentration in the atmosphere is greater than about 75 μg NH_3 m^{-3} (van der Eerden, 1982). Conversely, at the low concentrations found normally in the atmosphere (1–3 μg NH_3 m^{-3}), plant leaves with a temporary surplus of N may release ammonia to the atmosphere (Farquhar *et al.*, 1980; Schjoerring, 1991). Whether there is net absorption or net release by a plant leaf depends on the difference in ammonia concentration between the stomatal cavity of the leaf and the free atmosphere. If there is no difference, neither absorption nor release occurs and the concentration is referred to as the 'compensation point'. The concentration in the stomatal cavity reflects the balance between the reactions that produce ammonia in plant tissues (mainly photorespiration and proteolysis) and the reactions (mainly amino acid and protein synthesis) that assimilate it. The concentration at the 'compensation point' therefore varies with environmental factors, such as temperature and the supply of N to the roots, and also with plant factors such as the age of the leaf (Denmead, 1990; Fangmeier *et al.*, 1994; Parton *et al.*, 1988a; Schjoerring, 1991; Wetselaar and Farquhar, 1980). For several plant species and under normal growth conditions, the concentration at the compensation point has been reported to range between 1 and 4 μg NH_3 m^{-3} but, in some situations and particularly with senescing plants, the compensation point may be as high as 20–25 μg NH_3 m^{-3} (Denmead, 1990; Schjoerring, 1991). The concentrations of ammonia that have been reported in the atmosphere generally vary from < 3 μg m^{-3}, in rural areas unaffected by large numbers of grazing animals or by the disposal of manures or slurries, to > 25 μg m^{-3} in localized areas of slurry disposal or intensive livestock production (see p. 152). Actual concentrations in the atmosphere vary widely with time, and with height above ground level, and it is therefore difficult to assess the importance of leaf absorption of ammonia. However, there may well be times when the absorption of ammonia by leaves exceeds release, and other times when release exceeds absorption. In areas, such as most of Europe and the USA, with at least moderately intensive agriculture and/or industrialization, average concentrations of ammonia in the atmosphere often exceed 3 μg m^{-3}, suggesting that actively growing crops (except possibly those heavily fertilized with N) would absorb ammonia from the atmosphere. However, during senescence, the balance might well shift in favour of the release of ammonia (Sutton *et al.*, 1993).

The potential for plants to absorb and utilize ammonia from the atmosphere was shown in a study in which Italian ryegrass was grown, in pots, in special chambers with controlled concentrations of atmospheric ammonia and with ^{15}N-labelled nitrate incorporated into the soil (Whitehead and Lockyer, 1987). After 33 days exposure to the ammonia, the proportion of total plant N derived from the ammonia ranged from 4% in plants grown at a concentration of 14 μg NH_3 m^{-3}, and with a high rate of nitrate addition, to 78% in plants grown at a concentration of 700 μg NH_3 m^{-3} and with a low rate of nitrate addition. Some of the N derived from the NH_3 was transported to the roots. However, although the proportion of the total plant N derived from atmospheric ammonia is influenced by the supply of N to the roots, there may be little effect on the actual amount of ammonia absorbed by the leaves (Rogers and Aneja, 1980; Whitehead and Lockyer, 1987). In addition to the absorption of gaseous ammonia from the atmosphere, plants are also able to absorb aerosols, such as those of ammonium sulphate, through their stomata (Gmur *et al.*, 1983) and aerosol N in water-soluble forms can presumably be utilized.

Absorption of NO_2 and NO by Leaves

The nitrogen dioxide (NO_2) and nitric oxide (NO) that are present in the atmosphere result partly from reactions between O_2 and N_2 during the combustion of fuels, and partly from nitrification and denitrification in soils (see Chapter 9). The NO_2 and NO are interconvertible in the atmosphere and a wide range of ratios is possible, NO_2 being favoured by the presence of ozone, and NO by UV light (Wellburn, 1990). Although both compounds can be absorbed by plant leaves, NO_2 is absorbed more readily, causes less damage at high concentrations and, at low concentrations, may promote plant growth. For example, ryegrass grown with NO_2, added to the atmosphere at a concentration of 200 $\mu l\ m^{-3}$ showed increased growth, a higher content of chlorophyll and delayed senescence (Taylor and Bell, 1988). At high concentrations, NO_2 tends to inhibit plant growth, possibly due to the acidity that accompanies the assimilation of NO_2 to nitrate (Wellburn, 1990). When NO is absorbed, either alone or in a mixture with NO_2, the effect on plant growth is always negative, probably due to the formation of nitrite which has a number of toxic effects (Wellburn, 1990). In general, normal concentrations of NO_2 and NO in the atmosphere make only a slight contribution to the N supply of grass or other crops.

Herbage Analysis in the Diagnosis of Plant N Status

The N in plant tissues can be categorized into structural, metabolically active and storage components. In plants that are severely deficient, most of the N is in the structural category, associated with cell walls and nucleic acids, some is metabolically active in the form of enzymes, and a negligible amount is present in the storage category. When the supply of N is increased, the greatest response is in the metabolically active category and this promotes photosynthesis and growth. If the supply of N exceeds the short-term requirement, the excess is stored, usually as nitrate or amides. In the context of diagnosing plant N status by tissue analysis, reference is often made to the 'critical' concentration, which is the concentration of N in a specified portion of the plant (e.g. leaf material of a certain age) associated with a yield of plant material that is 5 or 10% below the potential maximum (Fig. 2.4). In terms of the categories described above, the critical concentration is likely to reflect a normal amount of structural N, a slightly deficient amount of metabolically active N and very little storage N. The critical concentration of N in any specific plant tissue declines with age, reflecting the gradual increase that occurs in cell wall material and the corresponding decrease in the proportion of cytoplasm. When the concentration of N is used to assess the adequacy of supply, it is therefore important to standardize the portion of the plant selected

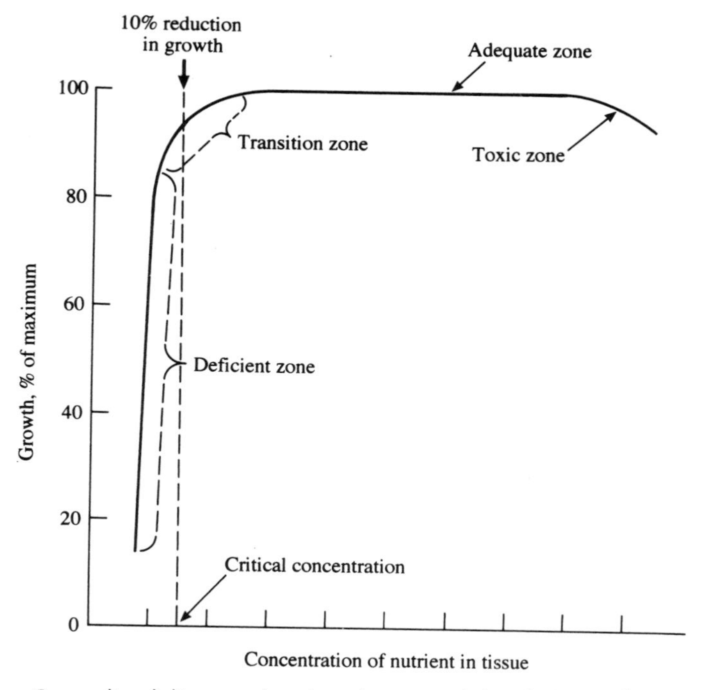

Fig. 2.4. Generalized diagram showing plant growth in relation to the concentration of an individual nutrient in plant tissue. (Redrawn from Epstein, 1972.)

for analysis and its age at the time of sampling. For grasses, it is usual to select either the total herbage, cut at a specified height after a certain period of regrowth, or individual leaves at a certain stage of growth. The critical concentration in grass, assessed with 4-week-old regrowth cut at a height of 2.5 cm, is about 3.2–3.5% N (de Wit *et al.*, 1963; Smith *et al.*, 1985); with 6-week-old regrowth it is about 2.5%. At the critical concentration, almost all the N is in organic forms, and little nitrate is present. However, when nitrate is the main source of N, the concentration of nitrate-N, though low, can be used to indicate the adequacy of supply. Estimates of the critical concentration in young grass leaves have ranged from about 0.05% to 0.15% nitrate-N on a dry-matter basis (George *et al.*, 1973; Hylton *et al.*, 1964; Reid and Strachan, 1974; Smith *et al.*, 1985; van Burg, 1966).

Effect of Nitrogen on the Rate of Tiller Production

In grasses, each leaf (except that immediately below the inflorescence) has a tiller bud in its axil and, if supplies of nutrients are adequate and if competition for space is not severe, most of these buds develop to form tillers. When N is deficient, the development of tillers is inhibited, and increasing the supply of N to plants that are grown individually increases the number of tillers per plant (Langer, 1963). However, when grasses are grown as swards, many tillers are short-lived due to competition, and fertilizer N generally has less effect on the number of tillers per unit area than with plants grown individually (Wilman and Pearse, 1984). With well-established swards, the number of tillers is influenced mainly by sward management: ryegrass swards that have been subject to continuous grazing by sheep may well have 40,000 tillers m^{-2}, whereas the more open swards that result from rotational grazing usually have 10,000–15,000 tillers m^{-2} and those harvested by cutting may have only 4000–7000 m^{-2}. The tillers developed under continuous sheep grazing, though numerous, tend to be small and to grow rather slowly (Robson *et al.*, 1989).

The effect of N supply on the number of tillers of individual grass plants was shown in a study with cocksfoot grown in soil/sand mixtures: the number of tillers was three times greater when fertilizer was applied at a rate equivalent to 224 kg ha^{-1} than when no N was applied (Auda *et al.*, 1966). Similar results were obtained with individual seedlings of perennial ryegrass and cocksfoot grown in small field plots (O'Brien, 1960), and with cocksfoot grown as simulated swards on vermiculite (Ryle, 1970). With young plants of perennial ryegrass subject to cutting, fertilizer N increased the number of tillers in the regrowth (Davies, 1971).

Under sward conditions, the number of tillers produced by four varieties of perennial ryegrass was increased by fertilizer N when the swards were harvested at intervals of 3–5 weeks but was decreased by high rates of fertilizer N when harvested at longer intervals (Table 2.2). With swards of Italian ryegrass that were cut regularly, fertilizer N and water supply appeared to interact in their effects on tillering (D'Aoust and Tayler, 1968). Under dry conditions the number of tillers per unit area was substantially higher with an application rate of 250 kg N ha^{-1} than with 125 kg N ha^{-1}, but the number with 500 kg N ha^{-1} was similar to that with 250 kg N ha^{-1}. With irrigation there was little or no difference between the 125 and 250 kg N ha^{-1} treatments, but the number of tillers with 500 kg was appreciably higher. In another study,

Table 2.2. Influence of the rate of fertilizer N, and the interval between successive harvests, on the number of tillers of ryegrass per m^2, assessed after 30 weeks (mean of four varieties). (From Wilman *et al.*, 1976b.)

Fertilizer N	Interval between harvests (weeks)			
(kg ha^{-1} $year^{-1}$)	4	6	8	10
0	4380	4360	3880	3770
263	5250	5130	4620	3410
525	5850	5100	4570	3020

the number of tillers per unit area of timothy was increased substantially by ammonium nitrate when conditions were cool, but much less when conditions were warm (Smith and Jewiss, 1966).

In eight field experiments in Wales, involving perennial ryegrass cut at intervals of 4–5 weeks, the average number of tillers was increased by 660 m^{-2} per 100 kg fertilizer N ha^{-1} $year^{-1}$, for rates up to about 200 kg N ha^{-1} $year^{-1}$: at higher rates of application, the effect was less and, with more than 500 kg N ha^{-1} $year^{-1}$, the number of tillers tended to decline (Wilman and Wright, 1983). When the harvesting interval was longer than 5 weeks, fertilizer N caused a decline in tiller numbers at a lower rate of application than when the harvesting interval was 3–4 weeks. The results of a subsequent investigation showed that, when the number of tillers present without fertilizer N was less than about 4000 m^{-2} (possibly due to unfavourable conditions when the sward was sown), the response could be 1000–1500 tillers m^{-2} per 100 kg ha^{-1} $year^{-1}$ up to a rate of 200 kg N ha^{-1} (Hollington and Wilman, 1985). When swards of perennial ryegrass that were continuously grazed by dairy cattle received fertilizer N at rates ranging from 250 to 700 kg N ha^{-1} $year^{-1}$, the rate had no consistent effect on the number of tillers per unit area though, at certain times, the numbers were least at the highest rates of fertilizer N. On average, the number of tillers varied during a 2-year period between about 8000 and 14,000 m^{-2} (Deenen and Lantinga, 1993).

Effect of Nitrogen on the Rate of Leaf Production per Tiller

The vegetative tillers of grasses continually produce new leaves at a rate that depends largely on temperature. In general, few tillers have more than three green leaves at any one time, and the production of new leaves is balanced by death and decay. During the active growing season, a grass plant is likely to produce a new leaf at intervals of about 11 days and, if the plant is not defoliated, leaves senesce and die at a similar rate (Parsons, 1988). Fertilizer N therefore has little effect on the rate of leaf production per tiller, irrespective of whether the plants are grown as spaced plants or in swards (Wilman and Mohamed, 1980; Wilman and Wright, 1983). For example, no difference was found in the number of leaves per tiller produced by plants of ryegrass and cocksfoot grown from seed and receiving either no fertilizer N or 135 kg N ha^{-1} (O'Brien, 1960). In an investigation with cocksfoot grown as a simulated sward in a constant environment, plants receiving 150 mg N l^{-1} in nutrient solution had slightly fewer leaves per tiller than did plants receiving 15 mg N l^{-1} (Ryle, 1970). Similarly with Italian ryegrass, although the application of fertilizer N increased the number of leaf primordia per tiller it did not affect the number of leaves which emerged, and there was no consistent effect on the number of leaves per tiller (Wilman *et al.*, 1977).

Effect of Nitrogen on Leaf Size and the Rate of Leaf Extension

The supply of N influences herbage yield mainly through its effect on leaf size, an effect that occurs with both individual plants and swards. With plants of timothy grown individually, the mean leaf area of 6-month-old plants increased from 0.94 to 2.30 cm^{-2} as the concentration of nitrate-N in solution was increased from 6 to 150 mg N l^{-1} (Langer, 1959). And, with simulated swards of cocksfoot in constant-environment conditions, increasing the concentration of nitrate-N in the nutrient solution from 15 to 150 mg N l^{-1} increased the average area of individual leaves from 8.5 cm^2 to 13.5 cm^2, mainly by increasing leaf length (Ryle, 1970).

In the field, the effect of fertilizer N on the length, width and weight of leaf blades in ryegrass swards was examined in several investigations carried out in Wales (Wilman and Wright, 1983). On average, the length of leaf blade was increased by 2.2 cm (from 13.0 cm) by applying 100 kg N ha^{-1} $year^{-1}$ compared with no N, but the incremental response diminished with increasing rate of application. The width of leaf blade was increased by 0.024 cm (from 0.323 cm) by applying 100 kg N ha^{-1} compared with no N but, as with length, the incremental response diminished with higher rates of application. Corresponding effects were noted for the area and dry weight per leaf blade. In percentage terms, N application had its largest effect on area per blade (86% increase due to applying 500 kg N ha^{-1} compared with no N), followed by weight per blade (59%), length (52%) and width (23%). The dry weight per unit area of leaf blade declined slightly as the rate of N application was increased (Wilman and Wright, 1983). However, in these investigations, the effect of fertilizer N on leaf size, although substantial, was limited by the relatively large supply of N from the soil: it would have been greater on soils with less plant-available N. The effect that N supply has on leaf size is influenced by the time interval between successive harvests, as illustrated in Table 2.3.

Table 2.3. Influence of the rate of fertilizer N on the length of leaf blade of perennial ryegrass harvested at various time intervals (cm of leaf blade per tiller in the 6th week after initial defoliation, with fertilizer applied in equal amounts for each harvest). (From Wilman and Pearse, 1984.)

Fertilizer N (kg ha^{-1} per 6-week period)	Interval between harvests (weeks)			
	1	2	3	6
0	5.6	6.2	6.5	8.2
66	7.5	11.3	12.1	17.5
132	8.0	15.6	16.7	24.9

Table 2.4. Influence of the rate of fertilizer N on the leaf extension of perennial ryegrass (mm per tiller per day, with fertilizer applied in equal amounts for each harvest). (From Pearse and Wilman, 1984.)

Fertilizer N (kg ha^{-1} per 6-week period)	Interval between harvests (weeks)			
	1	2	3	6
0	3.8	3.6	4.4	4.8
66	7.1	7.5	8.4	8.9
132	9.1	12.3	14.3	14.3

The influence of N supply on grass growth has also been examined in terms of the rate of leaf extension. For example, in two experiments, the mean rate of extension in mm per day was approximately doubled by applying 500 kg N ha^{-1} year^{-1} compared with nil, and there was little evidence of diminishing response over this range of fertilizer application (Wilman and Wright, 1983). Other detailed measurements on ryegrass swards at various times of year showed that the mean rate of leaf extension during a 6-week period was increased from 4.8 to 14.3 mm per tiller per day by a single application of 132 kg N ha^{-1} during the summer, but that the increase was less during the spring and autumn (Table 2.4). The application of only 33 kg N ha^{-1} had a substantial effect, compared with no fertilizer N, and nearly doubled the rate of leaf extension during the first week of summer growth periods (Pearse and Wilman, 1984). With two genotypes of tall fescue, the application of 336 kg N ha^{-1} more than doubled the rate of leaf extension compared with 22 kg ha^{-1}, the mean rates being 4.4 and 10.5 mm per tiller per day (Volenec and Nelson, 1984). In agricultural practice, the enhanced growth rate brought about by fertilizer N results in a yield of herbage suitable for cutting or rotational grazing being achieved more quickly, thus increasing the number of harvests during a growing season.

Effect of Nitrogen on the Metabolic Activity of Leaves

Increasing the supply of N increases the metabolic activity of leaves when a serious deficiency is being overcome, but otherwise the effect is small. The effect of N supply on net assimilation rate (i.e. the increase in dry weight per unit leaf area per unit of time) interacts with cutting frequency: with timothy and meadow fescue, there was no difference in net assimilation rate between swards receiving fertilizer N at either 58 or 175 kg N ha^{-1} year^{-1} when they were cut only three times during the season, but the net assimilation rate was higher with 175 kg N ha^{-1} when the swards were cut at monthly intervals (Lambert, 1964).

The rate of photosynthesis per unit leaf area responds in a similar way to net assimilation rate. For example, with simulated swards of perennial ryegrass, severe N deficiency reduced the rate of photosynthesis by about one-third compared with a plentiful supply of N (Robson and Parsons, 1978). Supplying fertilizer N immediately after cutting a ryegrass sward substantially increased the rate of photosynthesis (per unit area of leaf) in the next leaf to expand, but had little effect later in a 4-week growth period (Woledge and Pearse, 1985). Photosynthesis appears to be reduced when the organic N concentration of leaves is less than about 3% (Woledge and Pearse, 1985).

Nitrogen in Leaf Senescence

In all grass species, both annuals and perennials, there is a continual turnover of leaf, tiller and root material. The metabolic changes involved in senescence are controlled genetically but the timing of senescence, and its rate of progress, are influenced by factors such as nutrient supply, light intensity and water stress. The changes include a net breakdown of protein and chlorophyll, and the eventual cessation of metabolic activity. Senescence has been studied to a greater extent in leaves than in stems and roots, and an important feature of leaf senescence is the change in enzyme activity. Enzymes that are involved in assimilation (e.g. nitrate reductase) lose their activity while those involved in degradation (e.g. peptide hydrolases) gain in activity (Feller and Keist, 1986). Mobile nutrient elements, including N, are partially remobilized from senescing leaves and translocated in the phloem to other parts of the plant, with the result that the concentration of N in leaf material declines during senescence. However, during the later stages of senescence there is a disintegration of membrane tissue and this restricts further translocation of N in the phloem. In grasses grown in the field, the concentration of N in dead brown leaves is often about half that in green leaves, but there is some evidence that when mature leaves senesce rapidly (e.g. under severe water stress) they retain a greater proportion of their N content (Hill, 1980). Remobilization by the plant is likely to be the major factor in the reduction of leaf N concentration (Hunt, 1983; Robson and Deacon, 1978) but senescent leaves may also lose N through the leaching of N by rainfall and possibly through the volatilization of gaseous compounds such as ammonia (Sutton *et al.*, 1993; Wetselaar and Farquhar, 1980). Remobilization from senescing leaves increases as the season progresses (Thomas *et al.*, 1990) and is proportionately greater when the supply of N is poor (Hunt, 1983; Robson and Deacon, 1978).

In grassland swards, all herbage that is not harvested by cutting or grazing undergoes senescence, and the amount of senescing herbage is therefore influenced by the intensity and frequency of harvesting. Senescence was estimated to account for about 30% of the leaf lamina material in a grass sward receiving 360 kg N ha^{-1} $year^{-1}$ and grazed intensively by steers, compared with about 60% in a grass–clover sward receiving only 60 kg N ha^{-1} $year^{-1}$ and grazed less intensively (Laidlaw and Steen, 1989). The effect of the length of the time interval since the previous defoliation on the accumulation of dead material in a sward is illustrated in Fig. 2.5. However, these data were obtained during dry conditions, and less dead material would have accumulated in moist conditions as more would have decomposed (Hunt, 1965).

Fertilizer N has only a small effect on the longevity of tillers, except at high rates when the death of tillers may be increased (Wilman *et al.*, 1977; Wilman and Wright, 1983). In general, fertilizer N appears to retard the death of leaves in the first 2–3 weeks after application, but to accelerate leaf death subsequently (Table 2.5). The N transferred from herbage to soil as a result of the senescence and death of leaves, in a grazed grass sward receiving no

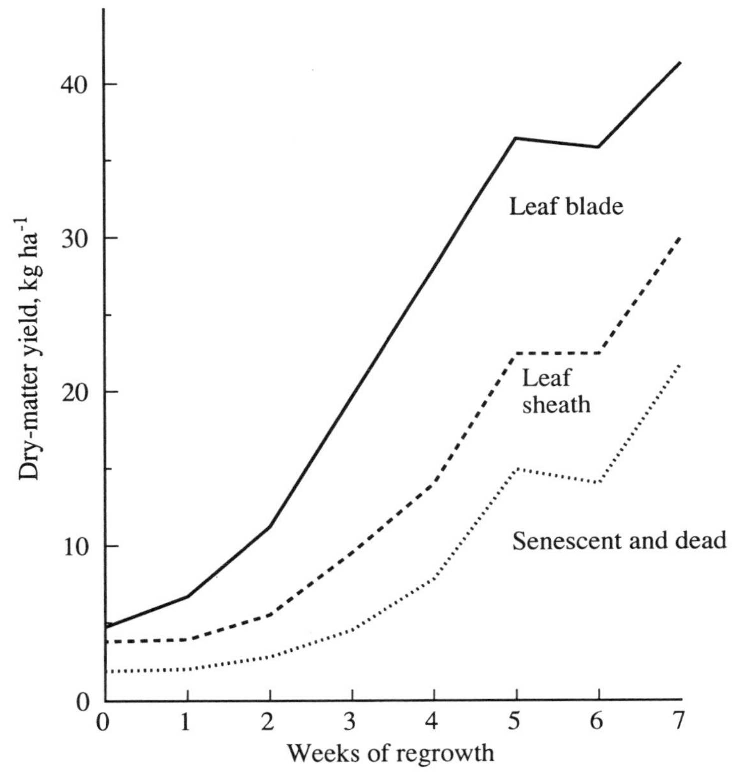

Fig. 2.5. Changes in the amounts of leaf blade, leaf sheath and senescent and dead material of perennial ryegrass following defoliation in summer. (Data from Hunt, 1965.)

Table 2.5. Influence of the rate of fertilizer N on leaf death in perennial ryegrass (mm per tiller per day) at various time intervals after defoliation in summer. (From Pearse and Wilman, 1984.)

Fertilizer N (kg ha^{-1} per 6-week period)	Interval after defoliation (weeks)					
	1	2	3	4	5	6
0	1.9	2.7	2.2	2.3	3.2	2.9
66	1.5	2.2	2.8	3.7	4.7	4.9
132	1.3	1.3	4.0	3.9	6.4	6.0

fertilizer N, amounted to between 0.06 and 0.42 kg N ha^{-1} day^{-1} (Thomas *et al*., 1990). A similar range was reported for rotationally-grazed ryegrass grown with fertilizer N at a rate of 155 kg N ha^{-1} (Hunt, 1983).

In areas of prairie grassland in USA that are subject to drought in summer and sub-zero temperatures in winter, substantial amounts of N are transferred from the leaves to the roots during the autumn, and this N is used for new growth in the following spring (Clark, 1977). There is a lack of information on the extent to which this type of transfer occurs in areas of grassland where the seasonal variation in temperature is less pronounced, and where a higher proportion of grass leaves survive the winter.

Effect of Nitrogen Supply on Root Growth

Although grass roots often amount to much more than half the total plant weight at any time, information on their growth and senescence is limited. It is, of course, difficult to measure the length or weight of roots grown in the field, and it is also difficult to separate living from dead roots. An additional problem is that roots are subject to attack by invertebrates in the soil. In measurements that have been made of root production in grass swards, the area sampled has usually been small in relation to the variation in root density, and the data therefore have a high degree of variability. Nevertheless it is clear that root density is higher in well-established grassland than under arable crops (Evans, 1978). Most of the roots are in the top 10 cm of soil (Table 2.6). Root densities of between 60 and 140 cm of root per cm^3 of soil were reported for the top 10 cm of soil under perennial ryegrass, cocksfoot and timothy, with densities of only 5–15 cm per cm^3 at a depth of 10–20 cm, and 1–4 cm per cm^3 at a depth of 50–60 cm (Garwood and Sinclair, 1979). In prairie grassland in Canada, 35% of the root weight was in the top 10 cm of the soil and 9% below a depth of 90 cm (Sims and Coupland 1979).

Increasing the supply of N generally increases shoot growth more than root growth and therefore increases the ratio of shoot to root. The shoot : root

Table 2.6. Distribution, in successive depth increments, of the root mass of a grass–clover sward on a silt loam soil in New Zealand. (From Williams *et al.*, 1989.)

Depth (cm)	Root DM (kg ha^{-1})	Proportion of total root (%)
0–5	4340	53.4
5–10	1850	22.8
10–15	820	10.1
15–20	460	5.7
20–25	410	5.0
25–30	240	3.0
Total	8120	100.0

ratio is also influenced by supplies of other nutrients and of water: the ratio is greatest when all nutrients and water are non-limiting, and is least when there is a major imbalance in supplies of nutrients and water (Davidson, 1969). When plants are severely deficient in N, the application of fertilizer N may result in some increase in root weight but, once a moderate supply of N has been attained, further increases tend to reduce root weight (Boote, 1976; Cunningham, 1968; Toomre, 1966). It is likely that, when N is deficient, plants maximize their exploration of the soil by allocating a relatively large proportion of their photosynthate to root growth; and changes in growth rate following changes in N supply have been demonstrated. For example, when the supply of nitrate to ryegrass grown in solution culture was stopped for 11 days, there was an increase in the rate of root growth (Jarvis and Macduff, 1989). Conversely, when fertilizer N at a rate of 336 kg N ha^{-1} was supplied to bromegrass and cocksfoot grown in soil in a glass-sided box, the rate of root growth was reduced by about 18% (Oswalt *et al.*, 1959). The growth of grass roots into mesh bags filled with soil and inserted into a sward was also reduced progressively as the rate of fertilizer N was increased from 50 to 150 to 300 kg N ha^{-1} $year^{-1}$ (Steen, 1991). Decreases in the weight of grass root material (which do not necessarily indicate a change in growth rate) following the application of fertilizer N have been reported in several field experiments (Ennik *et al.*, 1980; Garwood, 1967a; Oswalt *et al.*, 1959) and pot experiments (Auda *et al.*, 1966; Nielsen and Cunningham, 1964). However, increases have also been found in a pot experiment (Schuurman and Knot, 1974) and in a field investigation (Power, 1985). Despite this variation in results, it is clear that the N supply at which root growth reaches a maximum is considerably less than that producing maximum herbage growth.

It is possible that the form in which N is supplied may influence the extent of root growth. With perennial ryegrass grown in pots of soil and receiving fertilizer N at several rates as either ammonium or nitrate, ammonium produced significantly more root weight than did nitrate, although the yield of herbage was similar from the two sources (Herlihy, 1978), but no such effect was found in solution culture (Jarvis, 1987).

In addition to influencing the growth of roots, the supply of N may affect their morphology and physiological activity. An increase in root diameter with increasing rate of N application has been reported in some investigations (Oswalt *et al.*, 1959; Schuurman and Knot, 1974) though not in all (e.g. Boot and Mensink, 1990). For a number of plants including cereals and grasses, positive correlations have been found between N supply, the concentration of N in the roots and their cation exchange capacity (McLean *et al.*, 1956), a factor that may influence the uptake of other nutrients.

Nitrogen Metabolism following Defoliation

The grasses grown for agricultural purposes are able to tolerate frequent cutting or grazing because most of their growing points remain below the height of defoliation. When such grasses are defoliated, the vigour of the regrowth depends mainly on the amount and composition of the stubble and roots, and particularly on the contents of available carbohydrate and nitrogen. The amount of photosynthetic tissue that survives the defoliation is also important. In the first few days after defoliation, most of the N utilized in regrowth is obtained from the roots and stubble. This was shown in a study of perennial ryegrass grown in nutrient solution and cut at a height of 4 cm above the root system: the N in the leaves that grew during the first six days after defoliation was derived mainly from the organic N of roots and stubble, and only to a small extent from the nutrient solution (Ourry *et al.*, 1988). However, after the sixth day, most of the N appearing in the regrowth was obtained from the nutrient solution. In a subsequent study, with ryegrass grown in nutrient solution containing ^{15}N-labelled ammonium nitrate, increasing the supply of N increased the proportion of remobilized N obtained from the roots rather than the stubble (Ourry *et al.*, 1990). Calculations, based on the uptake of ^{15}N by cut plants of ryegrass grown in nutrient solution, indicated that the time taken for half the transfer of N from roots plus stubble to new leaf lamina was about 11 days with a poor supply of N (as ammonium) and about 4.5 days (followed by a slower phase of transfer) with a plentiful supply of N (Millard *et al.*, 1990).

Effects of Nitrogen Supply on Hardiness and Disease Resistance

The application of fertilizer N, especially at high rates, may lower the resistance of grasses to adverse weather conditions and to attack by pathogens. An increased susceptibility to drought, cold and heat was found in several grass species when ammonium sulphate was applied at 154 kg N ha^{-1} $year^{-1}$ to swards from which cores were removed and then subjected to a range of controlled-environment conditions (Carroll, 1943). During the severe winter of 1962–1963, an increased susceptibility to herbage death during the winter was noted in grass plots that received more than about 300 kg N ha^{-1} $year^{-1}$ at three locations in the UK (Baker and David, 1963; Miles and Williams, 1964; Widdowson *et al.*, 1966). At one location, Rothamsted, the effect was associated with an increased attack by snow-mould fungi (Widdowson *et al.*, 1966). Grass varieties differ in their susceptibility to death during the winter but, with four varieties of Italian ryegrass, susceptibility was increased by increasing rate of fertilizer N (Hides, 1978). Increased susceptibility to death during the winter following the application of high rates of fertilizer N has also been reported with grass swards in the USA (Adams and Twersky, 1960; Kresge and Decker, 1966; Wilkinson and Duff, 1972).

It is well known that high rates of fertilizer N increase the susceptibility of many crop plants to attack by pathogens, and there is some evidence of this effect in grasses. Thus, the incidence of *Helminthosporium* leaf-spot disease in Midland Bermudagrass was greatest on plots receiving high rates of fertilizer N (450–670 kg ha^{-1} $year^{-1}$) with little or no K (Kresge and Decker, 1966). Also, with grassland in Wales, the application of fertilizer N increased the incidence of fungal infection of grass by 50%, though the loss of yield due to infection was outweighed by the increased growth (Carr and Catherall, 1964).

3 Legumes: Biological Nitrogen Fixation and Interaction with Grasses

Introduction

Of all living organisms, only a few genera of bacteria are able to fix atmospheric N_2, and these produce the nitrogenase enzyme system that converts N_2 to ammonia. The ammonia that results from the fixation process is readily metabolized by the bacteria themselves, and also by higher plants if it is available to them through a symbiotic relationship with the bacteria. Some N_2-fixing bacteria (e.g. *Azotobacter*) live freely in the soil, but those that fix N_2 most effectively are the rhizobia which live symbiotically in the roots of leguminous plants. Although a few plant families other than the *Leguminosae* have symbiotic relationships with N_2-fixing bacteria, their symbioses are much less widespread, and less effective, than those involving legumes and rhizobia. In the grasslands of Europe and New Zealand, the most important legume species is white clover but, in North America, lucerne, red clover and birdsfoot trefoil are also widely grown, the trefoil especially on acid soils (Brophy *et al.*, 1987). In temperate areas with a pronounced dry season, such as parts of Australia, annual clovers (especially subterranean clover, *Trifolium subterraneum*) and medics are the forage legumes that grow most successfully (Brougham *et al.*, 1978). All of these species form symbiotic associations with root nodule bacteria which belong to the genera *Rhizobium* and *Bradyrhizobium*.

White clover is particularly important because it grows well in association with grasses and is tolerant of grazing; it also grows over a fairly wide range of climatic conditions and its herbage has a high nutritional quality for livestock. By producing laterally-growing stolons, white clover is able to colonize gaps in the sward and to maintain most of its growing points below grazing height. Of the other cultivated legumes, lucerne, red clover and sainfoin are not stoloniferous, and they are often grown alone rather than in mixtures with grass. They are usually cut for hay or silage rather than grazed.

The amount of N fixed by a legume depends partly on the inherent genetic characteristics of both components of the legume–rhizobium association, and partly on the environment in which it is grown. Attempts to increase fixation therefore involve one of two possible approaches. In the long term, it may be possible to increase fixation by modifying the genetic characteristics of either the legume or the rhizobium but, in the short term, the only approach that is feasible is to ensure, so far as possible, that the environment is optimal for the growth of the particular legume and its associated rhizobia.

Growth and Morphology of White Clover

Initially, a white clover seedling consists of a short primary stem and a tap root but, as the plant develops, stolons grow from the leaf axils of the stem and the tap root forms a number of branches. Subsequently, adventitious roots grow from the nodes of the stolons and the stolons themselves produce branches (Fig. 3.1). The adventitious roots gradually replace the tap root and, as the original centre of the plant dies back, some branch stolons become independent plants (Caradus, 1990; Frame and Newbould, 1986). Because the senescence of stolon tissue increases with age, and stolons become rejuvenated in the direction of growth, it is particularly important for new adventitious

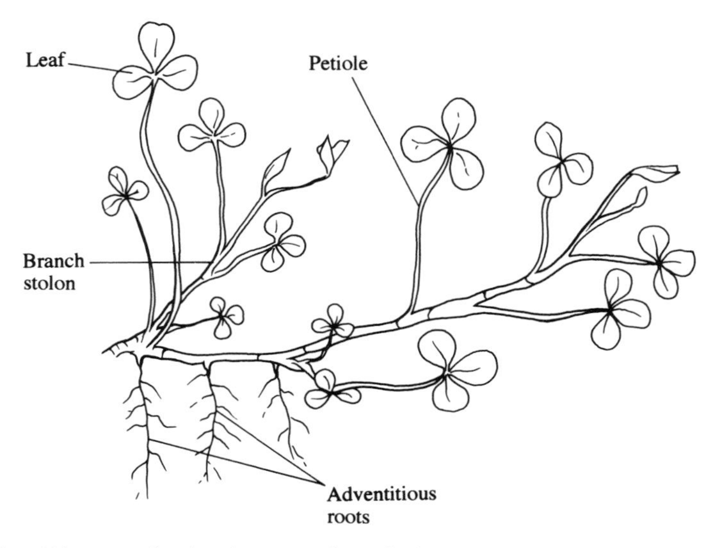

Fig. 3.1. Diagram of a developing stolon of white clover.

roots to become established in the soil. Although some roots of white clover may reach the same depth as those of perennial ryegrass and cocksfoot, most (like those of grasses) are concentrated in the top 10 cm (Caradus, 1990). When clover is grown in association with grass, the amount of root produced is much less than that produced by the grass (Young, 1958). The persistence of white clover in grass–clover swards usually depends much more on the growth and branching of stolons than on the germination of seeds, and branching is encouraged by intermittent defoliation (Harris, 1978). With intensive grazing, the number of viable clover seeds in the soil is often low, probably because most of the flowering heads are consumed before the seeds develop (Murphy, 1987). Varieties and cultivars of white clover differ in their growth characteristics, and those varieties that are strongly stoloniferous (e.g. wild white clover) tend to survive better in grazed grass–clover swards than do the less stoloniferous, large-leaved types (Frame and Newbould, 1986).

Despite the advantages of including white clover as a component of intensively-managed grassland, it does have a number of disadvantages. First, in comparison with grass, it grows slowly and produces little herbage in the cool conditions of early spring. Second, it has a relatively small root system with fewer and shorter root hairs, and so competes poorly with grass for nutrients other than N (Brougham *et al.*, 1978). Third, white clover is more susceptible than grass to cold wet conditions, and so its proportion in grass–clover swards tends to decline during the winter (Woledge *et al.*, 1990), the decline being most severe with large-leaved varieties (Eagles and Othman, 1989). Fourth, in some grass–clover combinations, the clover may be shaded by the grass to an extent that reduces photosynthesis (Donald, 1963; Schwank *et al.*, 1986) or the branching of stolons (Caradus and Chapman, 1991). However, some clover varieties are able to overcome the effect on photosynthesis by increasing the length of developing petioles in response to the height of the grass (Dennis and Woledge, 1982, 1983). A fifth potential disadvantage of white clover is that it is more susceptible than grass to attack by pests such as nematodes. These various factors provide some explanation for the decline in clover growth that sometimes occurs unexpectedly in well-managed grass–clover swards (Frame and Newbould, 1986). Several of the factors are being addressed in plant breeding programmes, in which important objectives for white clover are to improve tolerance to low temperatures, to improve compatibility with grasses (e.g. by increasing the potential for stolon branching) and to improve pest and disease resistance.

Physiology of Symbiotic N_2 Fixation by Clover

The basis of the symbiotic relationship between a legume and rhizobial bacteria is that the legume provides the rhizobium with an energy supply through photosynthesis, and the rhizobium provides the plant with ammonia, or a related compound, derived from N_2 fixation. The symbiosis is dependent on the legume developing root nodules which contain colonies of rhizobia. Initially, rhizobia present in the soil colonize a root by causing a root hair to invaginate and then forming an infection thread through which the rhizobia migrate to the root cortex (Crush, 1987). Colonies of rhizobia eventually produce root nodules which, in white clover, are characteristically about 1.5 mm in diameter. In total, the root nodules may comprise up to 20–35% of the total root weight when the plant is relying entirely on fixation for its N supply (Arnott, 1984; Ryle *et al.*, 1979). Rhizobia are able to exist independently in soil for several years, though the numbers decline in the continued absence of an appropriate legume (Masterton and Sherwood, 1970). Numbers increase when a legume is present and, in the surface soil of a grazed grass–clover sward, there may be as many as 100,000 rhizobial cells per gram of soil (Murphy, 1987).

The nitrogenase enzyme system produced by rhizobia consists of two enzymes which have molecular weights of 220,000 and 55,000. The larger enzyme, which contains both Fe and Mo, is the actual N_2 fixation enzyme and the smaller enzyme, which also contains Fe, provides the necessary electrons. Considerable energy is required to split the triple bond of the N_2 molecule, and this is supplied by ATP from the respiration of the bacterial mitochondria.

$$N{\equiv}N + 16ATP + 8e^- + 10H^+ \rightarrow 2NH_4^+ + 16ADP + H_2$$

A study with red clover, involving the use of ^{14}C-labelled CO_2 and ^{15}N-labelled N_2, indicated that the respiratory cost of N_2 fixation was equivalent to 4 mg C per mg N fixed (Warembourg and Roumet, 1989). However, the total energy requirement of fixation is greater than the respiratory cost, and has been estimated as being equivalent to 15 units of plant dry matter (about 7.5 units of C) per unit of N fixed (Heichel, 1985). Despite this energy requirement, there is no direct evidence that N_2 fixation restricts the growth of the legume, and N_2-fixing legumes produce yields similar to those of the same variety well supplied with fertilizer N. In an investigation with the legume *Pisum sativum* (L), it was calculated that 32% of the carbon derived from photosynthesis was diverted to the nodules, with 5% being used for nodule growth, 12% being used for nodule respiration and 15% being incorporated into amino acids that were translocated to other plant organs (Minchin and Pate, 1973).

Nitrogenase is destroyed by oxygen at the concentrations that occur normally in the atmosphere. However, some oxygen is required for the production of ATP; and the root nodules contain a biochemical mechanism that controls the supply of oxygen to the rhizobia. The mechanism depends on the protein leghaemoglobin, which forms an addition product with oxygen and provides a low equilibrium concentration of free oxygen in the nodule. Root nodules that are actively fixing N_2 have a distinctive pink coloration due to the presence of leghaemoglobin. A large proportion of the N_2 that is fixed is released to the plant cells of the nodule in the form of ammonia, and much of the ammonia is then converted to glutamine by reaction with glutamate. Legume species differ to some extent in their subsequent metabolism of N but, in white clover, there is a transfer of N from glutamine to asparagine which is then transported in the xylem (Dunlop and Hart, 1987).

Rhizobia exist as various strains which show some specificity for particular legume species and, with any one legume species, the effectiveness of N_2 fixation varies considerably among different strains of rhizobium (Date and Brockwell, 1978). Varieties of individual legume species also differ in their effectiveness of N_2 fixation with a particular strain of rhizobium and, with white clover, the interaction between variety of clover and strain of rhizobium is thought to be important (Mytton, 1975).

When legumes depend mainly on N_2 fixation for their supply of N, their growth rate may be limited either because fixation is insufficient to match photosynthesis, or because photosynthesis is insufficient to match the potential fixation. When N_2 fixation itself is restricted, the concentration of N in the plant tissue is likely to be less than a critical value, whereas when photosynthesis is restricted (e.g. by light or nutrient supply) the concentration of N is likely to be higher than the critical value. With white clover, the critical concentration of N in the leaf + petiole material after 4 weeks regrowth appears to be about 3.5% (Whitehead, 1982) and in whole-plant 15-week-old material (including roots) about 3.2% (Haydock and Norris, 1966). In practice, the concentration of N in the herbage (leaf + petiole) of white clover, grown in the field and defoliated at intervals of a few weeks, is often well above 3.5% and may be as high as 5.8% (Metson and Saunders, 1978; Munro and Davies, 1974; Simpson *et al.*, 1988). When white clover is defoliated, N is remobilized from roots and stolon material to supply new leaves, and turnover may amount to about 2% of the plant N per day in the 7-day period following defoliation (Marriott and Haystead, 1990).

Methods for the Quantitative Assessment of N_2 Fixation

The amount of N_2 fixed by an individual legume plant that is entirely dependent on fixation can be measured simply by analysing the whole plant for total N. However, for plants grown in soil containing nitrate and ammonium, more detailed methods are necessary. Three main types of method have been developed, and these are based on (i) the difference in total N between fixing and non-fixing plants, (ii) acetylene reduction and (iii) the use of the ^{15}N isotope. In the first method, based on total N, the amount of N_2 fixed is estimated by subtracting from the total amount of N in the legume the amount of N present in a corresponding control plant not fixing N_2. The control plant is usually a non-legume but, with some species, it may be a non-nodulating isoline of the same species. The method is based on the assumption that both fixing and non-fixing plants take up the same amount of N from the soil (including inputs of inorganic N from the atmosphere), an assumption that may not be wholly correct, especially if the plants differ in their seasonal pattern of growth and in root morphology (Ledgard and Peoples, 1988). Non-nodulating plants, which are entirely dependent on soil N, are likely to have larger root systems and hence explore a larger volume of soil. However, when grass–clover swards are compared with all-grass swards receiving no fertilizer N, the difference in uptake from the soil is likely to be small because both will take up virtually all the available N. In most of the assessments of N_2 fixation that have been published for white clover, the amount of N_2 fixed has been based on the amount of N present in harvested herbage, taking no account of any additional N that may have accumulated in the stolons, stubble, roots and soil of the grass–clover as compared with the all-grass sward. This approach will tend to underestimate the actual amount of N_2 fixed.

The acetylene reduction technique for measuring N_2 fixation is based on the fact that the nitrogenase enzyme system, in addition to catalysing the chemical reduction of gaseous N_2 to ammonia, also catalyses the reduction of acetylene to ethylene. When this method is used in controlled environment conditions, whole plants or detached root systems are enclosed in a vessel containing acetylene, usually at a concentration of 10%, and the subsequent accumulation of ethylene is determined by gas chromatography. The technique involves the assumption that nitrogenase activity is not affected by the substitution of acetylene for N_2 and, while this may be valid initially, activity has been found to decline in many legume species after a few minutes exposure to acetylene (Witty and Minchin, 1988). However, it may be feasible to obtain a value for a longer period by making repeated short-term measurements of nitrogenase activity to allow for diurnal and seasonal fluctuations. An additional problem with this method is that, in calculating results, it is necessary to assume a specific ratio between acetylene reduction and N_2 fixation, and yet the ratio varies considerably. The chemical reduction of N_2 to ammonia requires eight electrons (including two electrons required for the associated reduction of two protons to H_2) while the reduction of ethylene requires two, giving a theoretical ratio of 4 : 1. However, measured values for the ratio are often greater than 4 : 1 and, ideally, appropriate conversion ratios for a particular situation should be determined using ^{15}N (Witty and Minchin, 1988). Although the method has been used for assessing N_2 fixation in the field and has provided useful data on comparative rates of fixation under a range of environmental or management conditions (e.g. Masterton and Murphy, 1976), the problems outlined above make it unreliable for assessing the absolute amounts of N_2 fixed (Witty and Minchin, 1988).

In the third type of method, based on the use of the ^{15}N isotope, the legume and a non-fixing control plant are supplied, via the soil, with a small amount of nitrate or ammonium labelled with ^{15}N (Ledgard and Peoples, 1988; Wood and McNeill, 1993). The two sources of N for the legume (the inorganic N from the soil and the N_2 in the atmosphere) then differ in their abundance of the two isotopes, and the proportion (P) of the plant N obtained from N_2 in the atmosphere can be calculated:

$$P = 1 - \frac{[\text{atom \% } ^{15}\text{N excess (compared with the air) in N}_2\text{-fixing legume}]}{[\text{atom \% } ^{15}\text{N excess (compared with the air) in non-fixing control plant}]}$$

When the ^{15}N abundance of plant-available N in the soil is higher than that of the atmospheric N_2, the fixation of N_2 will result in the ^{15}N excess being lower in the legume than in the control plant. A major assumption in the method is that both the N_2-fixing legume and the reference plant absorb, through their roots, N with the same ratio between added ^{15}N and native soil N. This assumption may not be valid if the two types of plant take up N from different depths in the soil and/or differ in their pattern of uptake with time (Ledgard and Peoples, 1988). The potential error is reduced if the ^{15}N-labelled nitrate or ammonium is supplied at frequent intervals, but in small amounts (< 5 kg N ha^{-1}), to avoid inhibiting N_2 fixation. Any transfer of fixed N_2 from legume to reference plant will result in an underestimate of fixation, but transfer is likely to be negligible during short growth periods (several weeks) if the clover is growing actively. The method is expensive but has the major advantage of providing a 'time-averaged' estimate of fixation for the measurement period. More details of the procedure, and potential sources of error, are available elsewhere (e.g. Ledgard and Peoples, 1988; Wood and McNeill, 1993). In addition to the use of supplementary ^{15}N, it is possible with sensitive mass spectrometry to utilize the small difference that occurs in the natural ^{15}N abundance between soil and atmosphere. The abundance of ^{15}N, which in the atmosphere is 0.3663 atom %, is often between 4% and 17% greater in agricultural soils, and this difference can be measured and used in assessing N_2 fixation by legumes (e.g. Unkovich *et al.*, 1994).

Amounts of N_2 Fixed by Legumes in the Field

There are wide variations in the amount of N_2 fixed by all legume species. For example with white clover, the amount fixed may range from nil to more than 500 kg N ha^{-1} $year^{-1}$. In New Zealand, where conditions are particularly favourable for white clover, a maximum of 670 kg N ha^{-1} $year^{-1}$ was reported during the establishment of a grass–clover pasture on a subsoil supplied with additional P and K (Sears *et al.*, 1965). However, the usual range for grass–clover swards in New Zealand is between 100 and 350 kg N ha^{-1} (Caradus, 1990; Hoglund *et al.*, 1979; Ledgard *et al.*, 1990) with less than 100 kg N ha^{-1} in unimproved pastures and in dry areas. High rates of fixation by white clover

Table 3.1. Examples of the amounts of N fixed by white clover in the UK (mean and range values) assessed from the N in the harvested herbage (kg ha^{-1}) of grass–clover swards compared with grass-only swards: the swards were harvested 4–7 times per year.

Location(s)	Sward type(s)*	No. of years	N fixed (kg ha^{-1} year^{-1})	Reference
S. England	8 grass spp./wc (S100)	3	174	Cowling and Lockyer, 1967
Wales (2)	prg/wc (wild white)	4	90 (0–165)	Munro and Davies, 1974
Wales	prg/wc (8 cvs)	3	115 (21–192)	Evans *et al.*, 1990
Scotland	prg/wc (S100)	3	168 (148–195)[†]	Reid, 1970
Scotland (3)	8 grass spp./wc (S100)	5	' 200 (147–255)	Chestnutt, 1972
N. Ireland	prg/wc (4 cvs)	5	142 (74–237)[†]	Laidlaw, 1988
UK (22)	prg/wc (Blanca)	4	176 (5–445)	Morrison *et al.*, unpublished[‡]

*prg = perennial ryegrass, wc = white clover.
[†]Range for combined species/varieties from year to year.
[‡]Data from experiment GM 23; see Annual Report of the Grassland Research Institute, Hurley, 1982, 76–77.

have also been reported from the Netherlands; exceptionally up to 565 kg N ha^{-1} year^{-1} (on a previously arable field on a clay soil) though usually much less ('t Mannetje, 1994). In Britain, the amount of N_2 fixed by white clover in a mixed grass–clover sward can vary from nil to about 400 kg N ha^{-1} year^{-1}, and productive swards often fix between 100 and 200 kg N ha^{-1} year^{-1} (Table 3.1). In Switzerland, between 270 and 370 kg N ha^{-1} were reported as being fixed by ryegrass with white clover swards in the second year after sowing (Boller and Nösberger, 1987).

The amount of N_2 fixed in a grass–clover sward often reflects the vigour of clover growth, and this is usually limited by factors such as temperature, water or nutrient supply and by competition from the grass. Irrigation sometimes has a substantial effect in increasing N_2 fixation, as illustrated in Table 3.2. The fixation of N_2 is curtailed by plant-available inorganic N in the soil as this substitutes for fixation if it is taken up by the clover. Inorganic N suppresses N_2 fixation by reducing nitrogenase activity, by decreasing the formation of nodules and, sometimes, by increasing the rate at which nodules senesce. These effects are less marked when the clover is growing rapidly and, if the exposure to increased inorganic N continues for no more than 7–9 days, there is often a rapid recovery of nitrogenase activity (Hoglund and Brock, 1987). With grass–clover swards, inorganic N also increases competition from the grass and this, in turn, may reduce N_2 fixation per unit area.

Table 3.2. Influence of irrigation on the amounts of N fixed (kg ha^{-1} $year^{-1}$) by white clover (cv. Blanca) in mixtures with perennial ryegrass in four successive years at three sites in southern or eastern England (assessed from N in the harvested herbage (kg ha^{-1}) of grass–clover and grass-only swards harvested 6–7 times per year). (From Morrison *et al.*, unpublished*.)

	Year 1 Irrigation		Year 2 Irrigation		Year 3 Irrigation		Year 4 Irrigation	
Location	+	–	+	–	+	–	+	–
Hurley	341	298	399	292	410	363	359	249
Rothamsted	370	340	312	286	365	334	275	218
Gleadthorpe	350	263	240	201	110	27	298	148

*Data from experiment GM 23; see Annual Report of the Grassland Research Institute, Hurley, 1982, 76–77.

When clover is grown with grass, more of its N is derived from fixation than when it is grown alone, because the grass reduces the amount of soil N available. The proportion of the clover N derived from fixation tends to increase as the grass : clover ratio is increased (Nesheim and Boller, 1991). In cut grass–clover plots, the proportion of the total N in the clover herbage that was attributed to fixation was, on average, 75% for white clover and 86% for red clover; and fertilizer N applied at 30 kg N ha^{-1} per cut had little effect on these proportions (Boller and Nösberger, 1987). With well-managed grazed grass–clover in New Zealand, N_2 fixation often contributes about 40% of the total N in the herbage (Hoglund and Brock, 1987). However, the total amount of N_2 fixed per unit land area is generally greater with clover grown alone than with grass–clover swards. It has been suggested that, if more than 200 kg N ha^{-1} per year is to be fixed by the white clover in grass–clover swards, three conditions should be met. First, soil and weather factors should allow a total herbage yield of more than 10,000 kg DM ha^{-1} $year^{-1}$. Second, the proportion of clover in the mixed herbage should be at least 50% and, third, the clover should obtain at least 70% of its total N from fixation (Boller and Nösberger, 1987).

The main varieties of white clover are generally similar in the amount of N fixed, as shown in a 5-year comparison of four varieties that differed widely in leaf size, all grown with perennial ryegrass (Laidlaw, 1988). In many studies of grass–clover swards, the amount of N fixed has been greatest in the second or third harvest year, and has declined thereafter. One factor in this decline is that, after the initial development of a grass–clover sward, there is an increase in the supply of available N from the soil, particularly if the sward is grazed. This increase in soil N tends to reduce fixation while at the same time increasing grass growth and hence competition for the clover (Ball and Crush, 1985).

Under good conditions for growth, legumes other than white clover are also able to fix more than 100 kg N ha^{-1} $year^{-1}$. For example, in Minnesota,

USA, lucerne grown with reed canarygrass (*Phalaris arundinacea* L.) fixed about 330 kg N ha^{-1} in the second year after sowing, while the corresponding figure for red clover was about 170 kg N ha^{-1} (Heichel, 1989). In a study with lucerne, assessments based on ^{15}N dilution during 4 years indicated that fixation ranged from 160 to 220 kg N ha^{-1} $year^{-1}$ (Heichel *et al.*, 1984).

Influence of Companion Grass on N_2 Fixation by Clover

When a legume is grown in association with a grass, the grass inevitably influences the growth of the legume and its fixation of N_2. Although the grass reduces the amount of inorganic N available to the legume and therefore tends to increase N_2 fixation per unit weight of legume, it also tends to reduce the growth of the legume due to competition. Grass species and varieties differ in the extent to which they compete with clover, and therefore in the proportions of grass and clover that develop in grass–clover swards. For example, clover was found to be more compatible with red fescue, crested dogstail and smooth meadowgrass than with perennial ryegrass, whereas it was less compatible with Yorkshire fog and *Agrostis* (Frame, 1990). With perennial ryegrass, different varieties differed in the yield of associated white clover (Collins and Rhodes, 1989; Davies and Fothergill, 1990), possibly due partly to differences in the seasonal growth pattern of the grasses and partly to differences in the spatial distribution of roots and shoots. Tetraploid varieties of ryegrass may be more compatible than diploid varieties with white clover (Davies and Fothergill, 1990; Swift *et al.*, 1993). Tiller density per unit area is less with tetraploid than with diploid varieties, thus allowing more space for the stolons of clover to develop. Studies in a grass–clover sward showed that N_2-fixing activity in early spring was positively correlated with the number of rooted nodes, and with stolon and leaf dry weight, and that subsequent activity (except during drought) was positively correlated with the amount of clover leaf material (Marriott, 1988).

Influence of Light and Temperature on N_2 Fixation by Clover

Light intensity influences N_2 fixation through its effect on photosynthesis and hence on the supply of carbohydrate to the rhizobia. In addition, prolonged shading for more than 2–3 weeks reduces the number and weight of nodules per plant (Butler *et al.*, 1959; Chu and Robertson, 1974). Although diurnal changes in light intensity have little effect on N_2 fixation, differences maintained for several days can have large effects (Halliday and Pate, 1976; Haystead *et al.*, 1979). With some combinations of varieties of grass and white clover, and in the absence of frequent defoliation, the shading of clover by the grass may restrict clover growth and N_2 fixation (Donald, 1963; Schwank *et al.*, 1986) but, with other combinations, the effect is small because the clover petioles elongate in response to the growth of the grass (Dennis and Woledge, 1982; Woledge, 1988).

Temperature is a major factor in the increase in the rate of fixation that normally occurs during spring (Marriott, 1988). Active N_2 fixation requires a minimum soil temperature of about 9°C (Frame and Newbould, 1986; Leconte, 1987) and nitrogenase activity has been shown to have a broad optimum range of 13–26°C with a sharp decline below 13°C and above 26°C (Halliday and Pate, 1976). The fixation of N_2 is more susceptible to low temperature than is clover growth *per se* and, on a diurnal basis, the change in temperature often has a greater effect than the change in light intensity (Haystead *et al.*, 1979; Masterton and Murphy, 1976; Ryle *et al.*, 1979). The effect of temperature relative to light intensity appears to be greater when the proportion of stolon material is high and there is a plentiful reserve of carbohydrate (Haystead and Marriott, 1978). At continuous low temperatures, clovers are in more competition with grasses for plant-available N, and lose some of the competitive advantage that they would have if they were actively fixing N_2 (Nesheim and Boller, 1991).

The nodulation process is also influenced by temperature, and poor nodulation may sometimes be responsible for the failure of clover to establish under cool conditions. In general, nodulation increases with increasing temperature over the range 10–35°C but there are differences between varieties of white clover, and between strains of rhizobium, in their ability to nodulate and fix N_2 at low temperatures (Richardson and Syers, 1985).

In conditions such as those in the UK, white clover is more susceptible than grass to the death of tissue during the winter, and this appears to be a major factor in the decline that often occurs from year to year in the proportion of clover in grass–clover swards. Thus in a 3-year study, the proportion of clover in the herbage of a ryegrass with white clover sward decreased during each winter, though there was some recovery during spring and summer (Woledge *et al.*, 1990). Clover lost about two-thirds of its maximum weight during the most severe of the three winters, and it lost some weight even in the mildest winter when the grass actually gained in weight. The loss in weight was greater in leaf tissue than in stolon material and amounted to between 60 and 95% of the leaf lamina present in the autumn (Woledge *et al.*, 1990). Differences between varieties of white clover in winter hardiness are not well established, but the vigour of clover growth in early spring has been related to the amount of stolon that survives the winter (Collins *et al.*, 1991; Frame and Newbould, 1986). The concentration of non-structural carbohydrate in the stolon material in early spring is also important (Collins and Rhodes, 1995).

This concentration is related to carbohydrate accumulation during the previous autumn and hence to light intensity and shading during that period (Lüscher and Nösberger, 1992).

Influence of Water Supply on N_2 Fixation by Clover

In general, legumes are more affected than grasses by a shortage of water, though lucerne which is deep-rooting is relatively drought-resistant. With white clover, both N_2 fixation and herbage production tend to be greater when summer rainfall is high, an effect noted in both the UK (Cowling, 1961a) and New Zealand (Crouchley, 1979). In most years in the UK, irrigation increases the proportion of clover in grass–clover swards and the amount of N_2 fixed. Examples of the effect of irrigating a grass–clover sward on the amount of N_2 fixed by white clover include an increase from about 170 to 280 kg N ha^{-1} $year^{-1}$ (Stiles, 1966), and the increases shown in Table 3.2. However, even in a high rainfall area, the growth of white clover may be reduced by a shortage of water if the soil is shallow (Simpson *et al.*, 1987). In New Zealand, in a pasture under uniform management, N_2 fixation amounted to 90 kg N ha^{-1} in a year with a dry summer but 240 kg N ha^{-1} in the following year in which the summer was moist (Crouchley, 1979). However, in contrast, no correlation was detected between the clover content of Dutch permanent grassland and summer rainfall over a 6-year period (Kleter, 1968). In studies in controlled environment conditions, white clover fixed less N_2 per plant, and per unit weight of nodule, in conditions of water stress, but there was a complete recovery in the rate if the stress had not been excessive (Engin and Sprent, 1973). Some varieties of clover also respond quickly in terms of growth to a renewed supply of water after a period of drought: the response is linked to the amount of stolon material maintained through the drought period (Thomas, 1984).

Influence of Soil pH and Nutrient Supply on N_2 Fixation by Clover

Legumes are generally more sensitive to soil acidity and to nutrient deficiencies than are grasses grown in association with them. In addition to the usual requirements for growth, several nutrient elements (Ca, Fe, Cu, Co and Mo) are thought to be involved either in nodulation by rhizobia or in the N_2 fixation process. Mo in particular is an essential constituent of one of the nitrogenase enzymes. The harmful effects of soil acidity on legume growth are usually due to the toxic effects of soluble aluminium or manganese, or to a lack of calcium (Dunlop and Hart, 1987; Jarvis and Hatch, 1985). With clovers in temperate grassland, aluminium toxicity probably has the most widespread effect. Liming to a pH of at least 5.5 normally ensures that the soil concentrations of soluble Al and Mn are too low to be toxic, and that the supply of Ca is adequate (Dunlop and Hart, 1987; Frame and Newbould, 1986). Nodulation is more sensitive to low pH than is the growth of the clover itself (Crush, 1987).

The greater susceptibility of clover than of grass to nutrient deficiency is due, at least partly, to the grass having a more extensive and finely branched root system that enables it to compete more effectively (Dunlop and Hart, 1987; Evans, 1977; Richardson and Syers, 1985). The difference in root morphology between grasses and clovers is particularly important in the uptake of P, which is not readily mobile in soils: it results in fertilizer requirements for P (and to a lesser extent for other nutrients e.g. K and S) being greater for grass–clover swards than for comparable all-grass swards. There is also some evidence that the requirement for P is greater for active N_2 fixation than for legume growth *per se*: for example, the application of fertilizer P (and S) to white clover was followed by an increase in the concentration of N in the herbage (McNaught and Christoffels, 1961).

Although the application of fertilizer P or K sometimes increases the proportion of clover in a mixed sward (Richardson and Syers, 1985; Simpson *et al.*, 1988), the effect of nutrient supply is not always apparent. Thus, the proportion of white clover in several grass–clover swards in Wales was not related to the soil status of either P or K, possibly due to natural selection favouring ecotypes that were adapted to such conditions (Simpson *et al.*, 1987).

Influence of N Supply from Fertilizer, Manure and Soil on N_2 Fixation by Clover

When fertilizer N is applied to a legume growing alone, some of the N is absorbed and there is a corresponding decrease in N_2 fixation (Allos and Bartholomew, 1959; McAuliffe *et al.*, 1958; Walker *et al.*, 1956). Fertilizer N also reduces the number and size of root nodules: the application of 112 kg N ha^{-1} to white clover almost halved the weight of nodules, though there was no effect on herbage yield (Cowling, 1961b). With white clover in grass–clover swards, fertilizer N has been shown to decrease the diameter of stolons and to increase the length of petioles (Wilman and Asiegbu, 1982). It also tends to inhibit the branching of clover stolons (Caradus and Chapman, 1991) and to increase the mortality of growing points (Davies and Evans, 1990; Dennis and Woledge, 1985). However, with grass–clover swards receiving fertilizer N at normal rates, most of the N is absorbed by the grass and only a small proportion is taken up by the clover. The growth of the grass is increased by the fertilizer N and, as a consequence, the amount of clover in the sward is often reduced (e.g. Davies and Evans, 1990; Frame and Newbould, 1986;

Table 3.3. Influence of the rate of fertilizer N on herbage yield, and the proportion of clover in the herbage, of perennial ryegrass with white clover mixtures (means of three years, two cutting heights and four clover varieties). (From Frame and Newbould, 1986.)

Fertilizer N (kg ha^{-1} $year^{-1}$)	Total herbage yield (kg DM ha^{-1} $year^{-1}$)	Proportion of clover in herbage (%)
0	7,830	53
120	8,710	28
240	9,980	11
360	11,700	4

Laidlaw, 1984; Reid, 1970). An example is shown in Table 3.3. The reduction in clover growth is due mainly to increased competition for light, water and/or nutrients (Donald, 1963; Frame and Newbould, 1986; Mouat and Walker, 1959; Richards, 1976). In addition, long-petioled varieties of clover in mixed swards supplied with fertilizer N may be affected more than grass by defoliation, losing a greater proportion of their photosynthetic tissue and therefore losing competitiveness (Woledge, 1988). However, although varieties of white clover differ in their tolerance of fertilizer N, it has not been possible to relate tolerance to well-defined morphological characteristics (Caradus *et al.*, 1993).

Although fertilizer N does not necessarily impair the ability of the remaining clover to fix N_2 (Boller and Nösberger, 1987; Brockman and Wolton, 1963; Cowling, 1961a; Holmes and MacLusky, 1954), the overall effect of an application of fertilizer N to a well-managed grass–clover sward is often to reduce N_2 fixation by about half the fertilizer rate (Cowling, 1982; Crush *et al.*, 1982). The size of the effect is influenced by such factors as the time of year at which the fertilizer is applied, the frequency of defoliation, the adequacy of moisture and nutrients, and the form of the fertilizer N. In some situations, fertilizer N has less effect when the sward is defoliated frequently, presumably because competition from the grass is reduced (Blackman and Templeman, 1938; Holliday and Wilman, 1965; Sprague and Garber, 1950). However, with medium to large-leaved varieties, the frequency of defoliation appears to have little effect on clover growth (Frame and Newbould, 1986). Irrigation also tends to offset the effect that fertilizer N has on clover growth, especially in relatively dry areas such as southeast England (Low and Armitage, 1959; Penman, 1962; Stiles, 1966) and parts of the USA (Prine *et al.*, 1963; Robinson and Sprague, 1947), though not in wetter areas such as southwest Scotland (Reid and Castle, 1965a). The application of fertilizer K has sometimes been found to reduce the effect of fertilizer N on clover (Blaser and Brady, 1950; Lowe, 1966), though this has not occurred in all instances, even with a soil deficient in K (Lowe, 1967).

Not all forms of fertilizer and manure, applied at equivalent rates of N, have the same effect on the clover content of mixed swards. For example, clover was found to be suppressed more by ammonium sulphate than by calcium ammonium nitrate, probably due to the soil acidity induced by the ammonium sulphate (Moloney and Murphy, 1963). Calcium ammonium nitrate and urea, each applied at 50 kg N ha^{-1} in spring, had similar effects on the proportion of white clover in a grass–clover sward, but the urea appeared to have less effect on N_2 fixation (Murphy *et al.*, 1986). And urine and liquid manure, despite supplying substantial amounts of N, have been reported to increase the proportion of clover in some instances (Castle and Drysdale, 1962; Drysdale, 1966; Gisiger, 1950) though not in others (e.g. Herriott and Wells, 1962; Wheeler, 1958). In a comparison of the effects of liquid manure, solid fertilizer and an aqueous fertilizer solution, all containing the same amounts of N and K, the three sources produced similar herbage yields, but the liquid manure resulted in consistently higher clover contents in the sward at rates of N up to 300 kg N ha^{-1} $year^{-1}$ (Drysdale, 1966). Clover contents were also high where both liquid manure and fertilizer N were applied together. Thus, at a rate of 224 kg N ha^{-1}, swards receiving liquid manure, fertilizer, or a mixture providing half the N in each form, contained 20%, 8% and 18% clover, respectively. The effects did not appear to be due to either K or water and were unexplained (Drysdale, 1966). In a later comparison of liquid manure (urine plus water), urea and ammonium nitrate applied to a grass–clover sward, the liquid manure again produced a greater proportion of clover, and this effect was apparently not due to any change in soil pH (Castle and Reid, 1987). A positive but unexplained effect of diluted urine on legumes in a mixed sward was also observed in a study in Austria (Schechtner *et al.*, 1980).

The influence of N supply from the soil, as distinct from fertilizer application, on N_2 fixation by clover has been shown most clearly in New Zealand. Thus, on soils that were low in organic matter and had not previously been grazed by cattle or sheep, N_2 fixation by newly-sown grass–clover, supplied with adequate P and K, amounted to more than 500 kg N ha^{-1} $year^{-1}$ (Sears *et al.*, 1965). However, subsequent studies on soils where the supply of mineral N had increased as a result of the decomposition of clover residues, and the excretion of N by grazing animals, showed much less N_2 fixation. Estimates for productive long-established pastures at nine sites in New Zealand ranged from about 85 to about 265 kg N ha^{-1} $year^{-1}$, with a mean for the nine sites of 184 kg (Hoglund *et al.*, 1979). Initially when a grass–clover mixture is sown on an infertile soil, and nutrients other than N are supplied, the clover is dominant and there is little if any transfer of N to the grass. After a few months, increasing amounts of N are transferred (see p. 54) with the result that grass growth is promoted. The grass then tends to compete more strongly, and a smaller proportion of clover results in less fixation. Continued grass growth eventually results in a depletion of available soil N which curtails grass growth, enabling the clover to increase again. Clover grows well only when the supply of N is insufficient for maximum grass growth (Ball and Crush, 1985; Crush, 1987). Some evidence of alternating grass or clover dominance has been reported from New Zealand, with the proportion of clover reaching a peak at approximately 4-year intervals (Steele and Shannon, 1982). Nevertheless, once an approximate equilibrium between grass and clover has been reached, the balance at any one time is influenced to a large extent by seasonal and year-to-year differences in weather. In ley-farming systems, N_2 fixation by a grass–clover ley is promoted by the previous arable phase because the supply of readily mineralizable N tends to be depleted by the arable cropping (Ball and Crush, 1985).

Influence of Defoliation and the Presence of Grazing Animals on N_2 Fixation by Clover

When clover is severely defoliated, N_2 fixation declines rapidly, reflecting a decrease in the number of root nodules and the degradation of leghaemoglobin (Chu and Robertson, 1974). In grass–clover swards, defoliation also reduces the uptake of mineral N by the grass thus allowing more to be taken up by the clover (Crush, 1987). Fixation is usually at a minimum 1–2 days after defoliation, but often recovers during the next 1–2 weeks depending on the severity of the defoliation (Chu and Robertson, 1974). The effect of defoliation is relatively small when the clover has well developed stolons and these remain intact. When Ladino white clover was cut to a height of 6 cm, the rate of N_2 fixation was at a maximum after 16 days regrowth whereas, in plants cut to a height of 1 cm, 26 days were required for full recovery (Yoshida and Yatazawa, 1977, cited by Crush, 1987). In a study of the effects of defoliation on white clover grown in sand, the removal of all fully expanded leaves (but no others) caused a decrease in the number of root nodules, assessed after 10 days, but had no effect on stolon growing points: N was remobilized from root and stolon tissue to supply new leaves, which were developed more rapidly than in plants that were not defoliated (Marriott and Haystead, 1990). The speed at which clover recovers after defoliation also depends on the amount of carbohydrate available for new shoot growth. White clover appears to be more vulnerable to defoliation in early spring when carbohydrate reserves are low and when the leaves are close together on the stolons (Wilman and Simpson, 1988).

In grass–clover swards, competition from the grass tends to be curtailed by frequent defoliation. Thus, with a mixture of ryegrass and white clover, the amount of N_2 fixed was equivalent to 168 kg fertilizer N ha^{-1} when the sward was cut three times per year, to 202 kg when cut six to eight times and to 275 kg when cut twice weekly throughout the season (Cowling, 1966). Defoliation in early spring was found to increase the subsequent proportion of clover in the herbage of a grass–clover sward (Davies and Evans, 1990).

The effect of a sequence of rotational grazings on N_2 fixation by a grass–clover sward in Ireland is shown in Fig. 3.2. A rapid decline in fixation

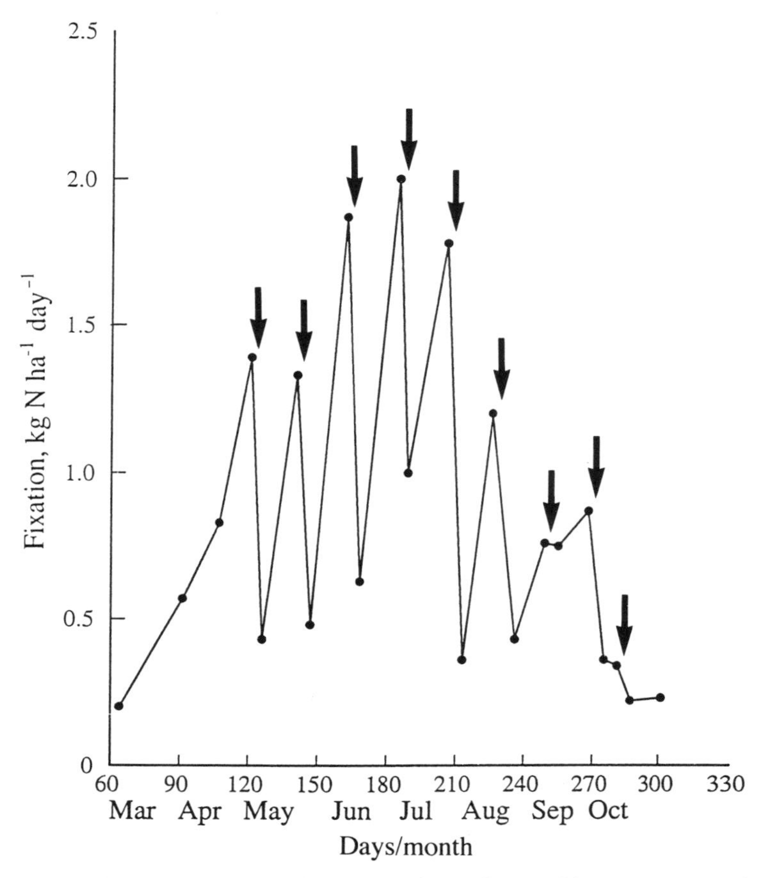

Fig. 3.2. N_2 fixation in a grass–clover sward as influenced by a sequence of grazings: arrows indicate times of grazing. (Redrawn from Murphy *et al.*, 1986.)

occurred after each defoliation but regrowth was accompanied by a recovery in fixation (Murphy, 1987).

Defoliation by grazing is often less severe than with cutting because there is less tendency for stolons to grow upwards in swards that are continuously or regularly grazed, and fewer growing points are removed. The more prostrate short-petioled forms of white clover in particular are favoured by a grazing regime. However, both the timing and intensity of grazing are important and, in order to encourage white clover, both overgrazing in spring and undergrazing in summer should be avoided (Wilman and Simpson, 1988). On the other hand, hard grazing by sheep in late autumn has been found to promote the growth of clover in the following spring (Laidlaw and Stewart, 1987). With continuous grazing, both under- and overgrazing can be prevented by adjusting the area available to a given number of animals and cutting any surplus.

The influence of grazing animals on clover growth and N_2 fixation is not confined to defoliation but includes the effects of treading and the excretion of dung and urine. One effect of treading is that some stolons are covered by soil and, while this may provide some protection against frost damage in winter, it tends to increase the death of stolons during the growing season, especially if combined with defoliation (Grant *et al.*, 1991). The effects of excreta on clover are due more to urine than to dung. Urine supplies a large amount of soluble N that is readily converted to ammonium and nitrate, and therefore suppresses N_2 fixation in the urine patch (Ball *et al.*, 1979; Carran *et al.*, 1982; Ledgard *et al.*, 1982; Marriott *et al.*, 1987). The initial decrease in N_2 fixation is due to the uptake of ammonium and nitrate by the clover (Ledgard *et al.*, 1982) but, in the longer term, urine also causes the death of nodules and reduces the length and weight of stolons (Marriott *et al.*, 1987). The substitution effect of urine may last for 6–8 weeks and, although N_2 fixation per unit of clover then tends to recover, the decrease in clover growth may persist for much longer (Crush and Lowther, 1985; Marriott *et al.*, 1987). The effect is more severe and protracted in winter than in spring (Crush and Lowther, 1985). With dung, particularly cattle dung, there is often some smothering of both grass and clover, but this effect is sometimes followed by a stimulation of clover growth (Weeda, 1967) especially when the soil is deficient in P (Brougham *et al.*, 1978).

Influence of Inoculation with Rhizobia on Clover Growth

The results of inoculating clover seed with rhizobia have been inconsistent: in some instances there has been a substantial increase in growth and in others no response. Where the soil contains no effective strains of rhizobium and only a poor supply of mineral N, inoculation is essential if white clover is to be established (Crush, 1987). Inoculation was necessary for the initial introduction of white clover into New Zealand (Crush, 1987) but, in the UK, inoculation has had much less effect. A large scale trial in upland areas of the UK, involving two varieties of white clover and inoculation with three strains of rhizobium, showed a positive response in only 18 of 139 comparisons made at 26 sites, with most of the 18 positive responses being on peaty soils (Newbould *et al.*, 1982). The lack of response to inoculation in the UK was considered to be due to competition from less effective strains already present in the soil (Frame and Newbould, 1986). In a survey of 192 grassland soils in Northern Ireland, the numbers of clover rhizobia were positively correlated (though not closely) with soil pH, base saturation and exchangeable calcium, and were negatively correlated with the degree of aluminium saturation of the cation exchange capacity (Wood *et al.*, 1985). The data indicated that grassland soils with a pH greater than 5.5 generally contained large populations of clover rhizobia, of which about 80% were effective on white clover, cv. Grasslands Huia.

Influence of Pests and Diseases on N_2 Fixation by Clovers

White clover is susceptible to attack by various pests and diseases, as discussed in more detail elsewhere (Baker and Williams, 1987; Caradus, 1990; Frame and Newbould, 1986). Pests and diseases are responsible for some of the unpredictability of clover growth, and nematodes are sometimes especially damaging to young roots (Caradus, 1990).

In New Zealand, the control of invertebrate pests by the application of insecticides has been shown to increase the growth of white clover and the amount of N_2 fixed (Steele *et al.*, 1985). At five sites, the increase in herbage N resulting from the application of insecticide to grass–clover swards ranged from 31 to 118 kg N ha^{-1} $year^{-1}$. Although the general treatment of grass–clover swards with insecticide is both uneconomic and environmentally undesirable, the selection and breeding of cultivars of clover with increased tolerance or resistance is thought to have potential for increasing grassland productivity (Steele *et al.*, 1985).

Transfer of Fixed N from Clover to Grass

When no fertilizer N is applied, grass–clover swards not only produce a higher yield of herbage than do grass swards but the yield of the grass component of the mixed sward often exceeds the yield of the grass grown alone (Herriott and Wells, 1960; Sears, 1953; Shaw *et al.*, 1966). The increased growth of grass in the mixed sward is due to the transfer of N from clover to grass, and occurs despite the grass in the mixed sward being subject to some competition from the clover. The proportion of the N_2 fixed by the clover that is transferred to the grass varies from nil to about 75%, depending partly on the time interval since sowing, and partly on sward management. The N transferred from clover may account for a large proportion of the N contained in the grass component, possibly up to 80% (Boller and Nösberger, 1987; Broadbent *et al.*, 1982). Usually, there is little transfer during the first few months after sowing a grass–clover sward (Boller and Nösberger, 1987; Broadbent *et al.*, 1982; Haystead and Marriott, 1979; Holliday and Wilman, 1962; Walker *et al.*, 1954), presumably because the clover is itself using all the N_2 fixed and there is little decomposition of dead material. The transfer of N increases after the first year and, in some investigations, has been found to increase over a period of three or four years (Bland, 1967; Cowling *et al.*, 1964; Herriott and Wells, 1963; Mallarino *et al.*, 1990). In an investigation in Wales, most transfer appeared to occur during the spring (Evans *et al.*, 1990), probably reflecting the death of clover material during the winter. The proportion of grass N represented by N transferred from clover is greater when the ratio of grass to clover is low (Boller and Nösberger, 1987).

The amount of N transferred in established grass–clover swards is influenced by the severity and frequency of defoliation, by the presence of grazing animals, and by factors such as light intensity, which govern the balance between N_2 fixation and photosynthesis. When swards are grazed, much of the N in the herbage is returned in excreta and some of this excreted N is subsequently absorbed by the grass. However, this mode of transfer, although substantial, is restricted by the dung and urine being deposited on only a small proportion of the total area (see Chapter 5) and because much of the N is lost through a combination of ammonia volatilization, denitrification and leaching. A major route for the transfer of N in both cut and grazed swards is through the decomposition of dead clover material, including roots, stolons and leaves. The stolons of white clover are particularly susceptible to death and decomposition during the winter (Collins *et al.*, 1991; Eagles and Othman, 1989; Woledge *et al.*, 1990). In addition, small amounts of N may be transferred by the excretion of organic N from living roots (Wacquant *et al.*, 1989), by the leaching of N from clover leaves (Tukey and Morgan, 1964) and by the volatilization of gaseous ammonia from clover and its subsequent absorption by grass (Denmead *et al.*, 1976).

A review of data from the UK indicated that non-excretal transfer (in cut swards) amounted, on an annual basis, to about half the amount of N in the clover component of the herbage. The average amount transferred was 30 kg N ha^{-1} $year^{-1}$ and the maximum was 110 kg N ha^{-1} $year^{-1}$ (Hobson and Richards, 1978). Defoliation tends to increase the transfer of N from clover to grass and the effect is particularly marked if all the clover leaf material is removed (Dilz and Mulder, 1962). Defoliation causes a rapid turnover of clover root and nodule tissue, involving the death of older material and its replacement by new roots and nodules (Butler *et al.*, 1959; Chu and Robertson, 1974). The decomposition of root nodules is a potentially significant route of transfer, as their concentration of N is high. Concentrations that have been reported for the root nodules of white clover include 4.8% (Wardle and Greenfield, 1991), 5.6% (Jensen, 1947), 6.3% (Butler and Bathurst, 1957) and 7–9% N (Chu and Robertson, 1974). Assuming for a grass clover sward that the clover root weight is 400 kg ha^{-1} (Young, 1958), that nodules comprise 25% of the weight (Arnott, 1984) and that their concentration of N is 6%, then the amount of N in root nodules at that time would be 6 kg ha^{-1}. However, when root nodules decompose as a result of natural senescence, their concentration of N may be less than those of active nodules due to remobilization. In addition to the turnover of roots and nodules, the decomposition of clover leaves and stolons is probably a major route for the transfer of N to grass (Brougham *et al.*, 1978; Carran, 1983; Field and Ball, 1982). The weights of stolon material in perennial ryegrass with white clover swards have been reported to range from 400 to 1200 kg dry weight ha^{-1} during three successive September–March periods in England (Woledge *et al.*, 1990) and from 300 to 500 kg ha^{-1} during a March–April period in Scotland (Marriott, 1988). Reported mean concentrations of N in white clover stolons have ranged from 2.2 to 2.8% (Cowling, 1961b; Marriott, 1988; Parsons *et al.*, 1991). In grazed grass–clover swards most clover stolons survive for less than one year and there is often a large decrease in the amount of clover stolon during the winter (Vez, 1961; Woledge *et al.*, 1990), as well as after severe defoliation. Trampling by livestock tends to increase the incorporation of readily degradable clover residues into the soil; and stolons buried by soil decompose more rapidly than those on the soil surface (Marriott and Smith, 1992).

In two investigations involving grass–clover swards grazed by dairy cattle, approximately equal amounts of N were considered to be transferred through livestock excreta and through the decomposition of dead clover material (Ledgard, 1991; Shaw *et al.*, 1966). In an investigation in the UK, grazed grass–clover swards contained less clover than did cut swards, at rates of fertilizer N from nil to 200 kg N ha^{-1} $year^{-1}$, but the amount of N transferred per kg clover N was approximately doubled by grazing (Table 3.4). In an investigation involving a grazed grass–clover pasture in New Zealand, the amounts of N transferred by the two main routes were assessed using a combination of ^{15}N-labelling and comparison with a cut sward. Slightly more N was transferred through the decomposition of dead clover material than through excreta (Table 3.5) and about half the N in the grass component of the sward was provided by the clover.

There has been some controversy regarding the extent to which N compounds are excreted from living legume roots, since an early report from Finland indicated that significant transfer from lucerne could occur by this mechanism. Subsequent investigations elsewhere failed to show any appreciable excretion of N from legume roots (Butler and Bathurst, 1957), and it was

Table 3.4. Amounts of N (estimated by herbage analysis for total N) transferred from clover to grass in cut and grazed ryegrass with white clover swards receiving no fertilizer N). (From Shaw *et al.*, 1966.)

Year after sowing	N transferred (kg ha^{-1} $year^{-1}$)		N transferred (kg per kg clover N)	
	Cut	Grazed	Cut	Grazed
3	43	67	0.20	0.45
4	79	89	0.37	0.56

Table 3.5. Amounts of N (estimated by ^{15}N-labelling) fixed by clover and transferred to grass in a grazed grass–clover sward. (From Ledgard, 1991.)

	Amount of N (kg ha^{-1} year^{-1})
Clover N derived from fixation	269
Clover N derived from root uptake	100*
Clover N transferred to grass via excreta	60
Clover N transferred to grass via decomposition	70
Grass N derived from clover	130
Grass N derived from soil	126

*Includes 23 kg N re-cycled via excreta.

suggested that excretion might require unusual circumstances (e.g. a combination of long days with relatively low temperatures). In general, the excretion of N by living legume roots is considered to be of minor importance (Vallis, 1978). However, more recent work with red clover and ryegrass grown in sand culture (during the spring–summer period in southern France) has demonstrated the excretion of soluble N by the clover during the period of darkness, the N being released as nitrate, or a form readily converted to nitrate (Wacquant *et al.*, 1989). There is also some evidence that transfer may be increased when the clover is infected by mycorrhizal fungi (Haystead *et al.*, 1988).

In a study of the transfer of N from lucerne and birdsfoot trefoil to reed canarygrass, calculations based on ^{15}N indicated that the grass obtained a maximum of 68% of its N from lucerne and 79% from trefoil, and that these amounts represented 17% of the N fixed by the lucerne and 13% of that fixed by the trefoil (Brophy *et al.*, 1987).

Fixation by Free-living Non-symbiotic Bacteria

A number of long-term experiments have shown increases in soil N that could not be attributed to symbiotic fixation and, in some instances, these have exceeded inputs through rainfall (e.g. Karraker *et al.*, 1950). There are two possible sources: dry deposition of N from the atmosphere (see p. 87) and non-symbiotic fixation.

Non-symbiotic fixation by soil bacteria (e.g. *Azotobacter*) requires a readily available source of energy and is encouraged by a restricted oxygen supply. It is therefore likely to occur to a greater extent in grassland than in arable soils (Jensen, 1965; Parker, 1957). The energy requirement is high, the fixation of one unit of N requiring at least 50–100 units of organic matter having the same energy value as glucose (Jensen, 1965). Although as much as 10,000 kg organic matter per ha may well be decomposed each year under a grass sward, the quantity available to support non-symbiotic fixation would be much smaller. The conversion of the N in plant and animal residues to microbial and humified forms of soil organic N would itself involve the consumption of much of the carbonaceous material present. Only materials with C : N ratios wider than about 30 : 1 are likely to provide appreciable energy for non-symbiotic fixation. This probably explains, at least in part, the finding of Delwiche and Wijler (1956) that the incorporation of dried grass material, including roots, at rates up to 40 mg g^{-1} soil (equivalent to about 36 t ha^{-1} to a depth of 15 cm) resulted in very small increases in N during 40 days incubation at 21°C, whereas the incorporation of 10 mg glucose per g soil resulted in fixation equivalent to 40 kg N ha^{-1}. There is considerable evidence that conditions of limited aeration increase the efficiency of N_2 fixation by free-living microorganisms (Jensen, 1965), probably through an increased production of organic metabolites from materials such as cellulose, and their more efficient utilization by N_2-fixing organisms.

Non-symbiotic fixation is reduced by the presence of ammonium and nitrate (Virtanen and Miettinen, 1963) and consequently by the application of fertilizer N. Decreases in the numbers of *Azotobacter* in soils receiving fertilizer N have also been reported (see Jensen, 1965). Little non-symbiotic fixation is therefore likely to occur in swards receiving moderate to heavy applications of fertilizer N. Non-symbiotic fixation was assessed by acetylene reduction in cores of soil from 56 grassland fields in England and Wales, and it was concluded that fixation rarely exceeded 5 kg N ha^{-1} $year^{-1}$ although, in parts of the soil where cyanobacteria were visible, rates were equivalent to 18 kg ha^{-1} (Lockyer and Cowling, 1977). Fixation was much less in areas that had been treated previously with fertilizer N; and where 100 kg N ha^{-1} had been applied six weeks previously, there was little or no fixation. Non-symbiotic fixation was estimated to amount to about 8 kg N ha^{-1} $year^{-1}$ in an area of prairie in Ohio (DuBois and Kapustka, 1983).

Among the free-living N_2-fixing bacteria are some (e.g. *Azospirillum*) that develop in the rhizosphere of various plants, including grasses. Fixation by this type of association is generally small but appears to be of most significance with tropical grasses, as reported for example in Texas (Morris *et al.*, 1985). Inoculation of the seeds of perennial and Italian ryegrass varieties with *Azospirillum* spp. appeared to give small increases in herbage yield, and amount of N in the herbage, when the rate of fertilizer N was 200 kg but not 400 kg ha^{-1} $year^{-1}$ (Carlier and Verbruggen, 1994).

4 Consumption, Digestion and Excretion of Nitrogen by Ruminant Livestock

Grassland Herbage as a Feed for Ruminant Livestock

The main function of agricultural grassland is to provide a source of feed for cattle and sheep. When grassland is managed extensively, the herbage is consumed mainly by grazing, though some herbage may be conserved as hay for use during the winter. When the management is more intensive, and particularly with dairy farming, herbage is usually consumed by grazing during the growing season, and as silage or hay during the winter. Grassland herbage, including hay and silage, supplies about 60% of the feed requirement of dairy cattle in the UK (Lazenby, 1981). Similar proportions are provided by grassland in the Netherlands, and by grass and other forage crops in the USA (van der Meer and Wedin, 1989). The remaining 40% of the requirement of dairy cattle is provided mainly by cereals but with some protein-rich supplements such as soyabean meal. In comparison with dairy cattle, beef cattle and sheep generally obtain a higher proportion of their total feed intake from grass and forage crops, often more than 80% with beef cattle and more than 90% with sheep (Frame, 1992).

In order to maximize the utilization of grassland herbage, it is necessary to take account of the variations in grass growth that occur due to season of the year and weather. For example, in many temperate regions, grass grows at its maximum rate in late spring, and often has a second period of quite rapid growth in late summer (Anslow and Green, 1967). The period of maximum growth rate is associated with the development of flowering stems, and this factor is reinforced by the favourable growing conditions that usually occur in late spring. Subsequently, the rate of growth depends to a large extent on the amount of water available to the grass which, in the absence of irrigation,

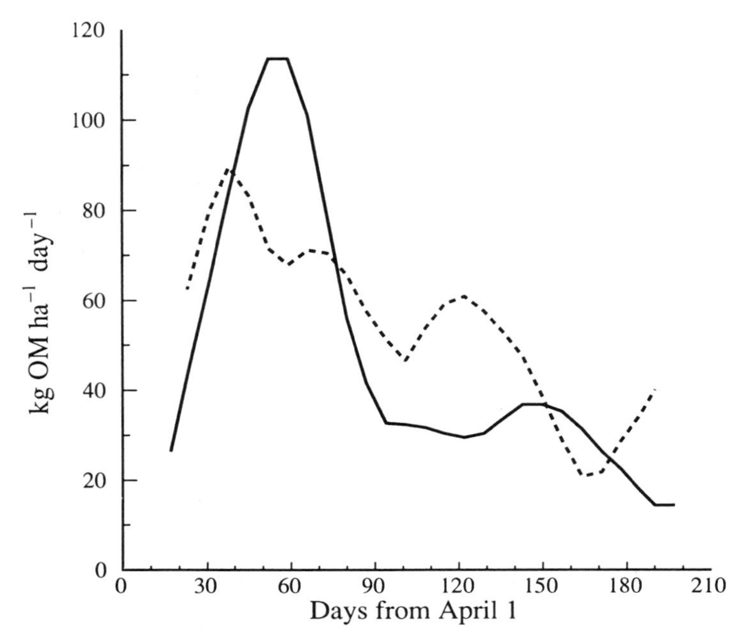

Fig. 4.1. Changes in the rate of growth during the growing season of grass swards harvested either by cutting (—) or by continuous grazing by sheep (-----). (Data from Orr *et al.*, 1988.)

is influenced both by current rainfall and by the water storage capacity of the soil. If the grass is irrigated, both the mid-season depression in growth and the late summer peak are smaller than without irrigation (Corrall and Fenlon, 1978). The seasonal variation in growth is less with continuously grazed swards than with cut swards, due mainly to the flowering process being prevented by grazing. The variation in growth that does occur reflects more closely the seasonal changes in light, temperature and water supply (Robson *et al.*, 1989). Examples of the seasonal variation in the growth of both cut and continuously grazed swards in the UK are shown in Fig. 4.1.

The seasonal variation in grass growth makes it difficult for farmers to ensure that the supply of herbage matches the needs of livestock. Although herbage from the periods of most rapid growth can be converted into silage or hay, additional costs are incurred: the total cost of producing grass silage is about twice that of grass for grazing (Doyle and Elliott, 1983). For a farm as a whole, the grassland area is utilized most efficiently when there is maximum flexibility in deciding, at intervals during the growing season, the proportion of the area to be grazed and the proportion to be allocated for silage or hay. However, it is sometimes not possible to utilize previously grazed areas for silage or hay because the herbage is fouled by dung.

Nitrogenous Constituents in Grass and Legume Herbage

Numerous factors influence the concentration of total N in grassland herbage which is usually within the range of 1–5%. The factors include the species and variety of grass (and possibly other plants), the stage of maturity, the ratio of leaf to stem, weather factors and, of course, the supply of N. Differences between grass species and varieties are usually small when comparisons are made on regularly cut swards, but may be appreciable during primary (uncut) growth in spring due to differences in the date of flowering. For example, the concentration of N at any particular date during spring growth is higher in the late-flowering S23 perennial ryegrass than the earlier flowering S24 ryegrass or S37 cocksfoot (Minson, 1990). The decline in concentration that occurs with increasing maturity during primary growth is illustrated in Fig. 2.3 (p. 22) and is due mainly to the relative increase in cell wall material and the relative decrease in cytoplasm. The concentration of N in clover herbage (usually between 3.5 and 5.8%) is generally higher than that in grass except, possibly, grass receiving high rates of fertilizer N.

Nitrogen is not distributed uniformly through grass herbage material. The concentration in leaf laminae is often at least twice that in the leaf sheath and stem (Minson, 1990), and is greater in young leaves than in older leaves (Parsons *et al.*, 1991a; Simpson and Stobbs, 1981). With grass separated from grass–clover swards (grazed by sheep to a height of about 6 cm), the mean concentrations of N in the three progressively older leaf laminae (separated from individual tillers) were 4.1%, 3.8% and 3.4%; the mean concentration in the leaf sheath was 1.4% and the mean concentration in the stem was 1.5% (Parsons *et al.*, 1991a). With white clover, the concentration is much higher in the leaf laminae than in the petioles and stolons. Mean values for clover from a grazed grass–clover sward were 5.6% N in the laminae, 2.9% N in the petioles and 2.7% N in the stolons (Parsons *et al.*, 1991a). The influence of increasing rate of fertilizer N on the overall concentration of N in herbage is described in Chapter 14 (p. 265).

In the context of the nutritional value of herbage for livestock, the concentration of total N is often expressed as 'crude protein'. This is the concentration of total N multiplied by 6.25, a factor derived from the average concentration of N in plant proteins, viz. 16%. Because it is based on total N, the 'crude protein' includes both genuine protein and other nitrogenous constituents. Usually in grass and legume herbage, about 70–90% of the total N is present as genuine protein, and 10–30% is present in forms such as amino acids, peptides and amides (van Straalen and Tamminga, 1990). Occasionally, substantial amounts of nitrate are also present in grass herbage, and the non-protein N may then exceed 30%. About half of the actual protein in grass and clover herbage is water-soluble and, of this, about half is the enzyme primarily responsible for photosynthesis, ribulose biphosphate carboxylase-oxygenase (Lawlor, 1991; Mangan, 1982). Most of the insoluble protein is combined with lipid in the membrane of the chloroplasts but a small amount is associated with the cell walls (Sanderson and Wedin, 1989).

Proteins are polymers of amino acids, with up to 22 amino acids contributing to each protein. Although different proteins differ in their proportions of the individual amino acids, the proteins of grasses and legumes are similar to one another in amino acid composition (Lyttleton, 1973). However, the grass and legume proteins generally contain less of the sulphur-containing amino acids than do proteins of animal origin. Glutamic acid, aspartic acid and arginine are major amino acids in the proteins of grasses and legumes, while lysine, alanine and glycine are also present in substantial amounts (Lyttleton, 1973). The amino acid composition of the protein is not appreciably influenced by either the stage of growth or the application of fertilizer N.

Some non-protein N (mainly amino acid and peptide) occurs normally in plant material, and amides tend to accumulate when the growth of plants well supplied with N is restricted by a factor such as K deficiency (Griffith *et al.*, 1964). The dominant amides are glutamine and asparagine, glutamine being particularly important in grasses, and asparagine in legumes (Mangan, 1982). The amount of ammoniacal N in herbage is usually negligible: less than 2% of the total N in Italian ryegrass was in the form of ammonium even when large amounts of ammonium, together with a nitrification inhibitor, were added to the soil (Nowakowski *et al.*, 1965). However, nitrate is sometimes a major constituent, especially within 3–4 weeks of fertilizer application, and occasionally it may amount to more than half of the total N (see p. 270).

When herbage is cut for hay or silage, proteolysis (the hydrolysis of proteins to amino acids) begins almost immediately due to the action of plant enzymes. However, the rate of the process is greatly curtailed if and when the moisture content falls to less than 40–45% (Woolford, 1984). Little proteolysis occurs when herbage is dried rapidly, as in hay-making under ideal conditions. On the other hand, when herbage is cut for silage under moist conditions and is then allowed to wilt, as much as 20% of the protein may be hydrolysed within a few days (Woolford, 1984). Proteolysis continues after the herbage has been ensiled, and there is often a large increase in the proportion of water-soluble N during the first 10 days. At the same time, the fermentation of carbohydrate results in lactic acid which lowers the silage pH. The acidity reduces proteolysis and inhibits it completely when the pH is below 4.3 (Woolford, 1984). The protein in silage is therefore more stable when there is a high ratio of carbohydrate to protein in the herbage ensiled, and the pH drops rapidly. Microbial enzymes in silage contribute to both proteolysis and carbohydrate fermentation, and their importance, relative to plant enzymes, increases with time. When appreciable proteolysis does occur, the amino acids, particularly lysine, arginine and histidine, are susceptible to further degradation by other microbial enzymes, especially those produced by *Clostridia*. The degradation products include aminobutyric acid, ornithine, cadaverine and putrescine (Woolford, 1984), and these compounds may reduce the palatability of the silage to livestock. The extent of proteolysis and the degradation of amino acids can be restricted by the addition of formic acid and/or formaldehyde during the preparation of the silage. Formic acid rapidly lowers the pH and thus inhibits proteolysis, whereas formaldehyde decreases the degradation of protein through the formation of chemical bonds between formaldehyde and proteins (van Straalen and Tamminga, 1990). Formaldehyde also inhibits the growth of *Clostridia* in the silage.

Consumption of Herbage by Cattle and Sheep

The amount of food consumed on a daily basis by ruminant animals depends to a large extent on the amount and type of food available, and on the physiological state of the animal (e.g. whether or not it is gaining weight, or lactating). Animal feeds are often compared in terms of their voluntary intake i.e. the weight of dry matter (DM) consumed by animals offered that feed *ad lib*. When comparisons are being made between different species or breeds of animal, or between animals of the same species but of different physiological states, intake is often expressed in terms of g DM per kg liveweight (or liveweight raised to a power, usually 0.75). With cattle and sheep given herbage, hay or silage of 70% digestibility (see p. 66), a typical intake would be about 30 g DM per kg liveweight per day, with more during late pregnancy and early lactation.

In situations where cattle or sheep obtain all their food by grazing, the supply of herbage restricts intake only when the average height of the herbage is less than about 7–10 cm (Minson, 1990) or 4–6 cm for sheep (Parsons *et al.*, 1991a). When herbage is grazed, the amount actually consumed reflects the time spent grazing, the average weight of herbage per bite and the rate of biting. Typical values for these characteristics for cattle and sheep are shown in Table 4.1. When a grass sward is grazed intensively, the average height of the sward decreases and consequently the intake per bite decreases and the rate of biting increases. In time, the increase in the rate of biting fails to compensate for the reduced intake per bite and the rate of consumption declines (Mayne and

Table 4.1. Some grazing characteristics of cattle and sheep. (From Thomas and Chamberlain, 1990.)

	Cattle	Sheep
Grazing time (h day^{-1})	5.8–10.8	6.5–13.5
Biting rate (bites min^{-1})	20–66	22–94
Intake per bite (mg OM kg^{-1} LW)	0.3–4.1	0.4–2.6
Rate of intake (mg OM kg^{-1} LW min^{-1})	13–200	22–80

Wright, 1988; Penning *et al.*, 1991). Consumption per unit time is then related to the height of the sward, though the relationship differs between swards grazed continuously and those grazed rotationally, due to differences in sward morphology. With good management, there is little difference in animal output between swards grazed continuously and those grazed rotationally, but rotational grazing has two advantages. First, it encourages the regular application of fertilizer and, second, it enables any surplus herbage to be seen more easily and then harvested by cutting (Leaver, 1976). Rotational grazing is, however, more labour intensive and is therefore confined to intensively managed grassland.

As intake is related to sward height, it is also related to grazing pressure, i.e. the quantity of herbage available per day in relation to the number of grazing animals. When grazing pressure is low, intake per animal is at a maximum but the proportion of the herbage consumed is low. When grazing pressure is increased, intake per animal declines but the proportion of the available herbage that is consumed increases (Leaver, 1976). Intensive grazing pressure results in less opportunity for the animals to select young leaf material, which has the highest digestibility and the highest concentration of N. However, at any grazing pressure, the herbage that is actually consumed usually has a higher concentration of N than does a bulk sample of cut herbage, because the grazed herbage contains a relatively high proportion of young leaves from the upper layer of the sward. If other factors are non-limiting, increasing digestibility, at least up to 80% digestibility, increases the intake of herbage by grazing animals (Thomas and Chamberlain, 1990).

For high-yielding dairy cattle, and for actively growing beef cattle and sheep, the optimum concentration of N in the diet is normally between 2.4 and 3.2% of dry weight (15–20% crude protein). Although animal production may increase as dietary N concentration increases above 2.4%, the efficiency of utilization of the N tends to decrease (see p. 281). The concentration of N in herbage varies with factors such as the rate of fertilizer application, the time of year and the botanical composition of the sward, and is often greater than 2.4% when swards are managed intensively. High concentrations, up to about 4.5%, occur with heavily fertilized grass, and concentrations up to about 4.0% occur with grass–clover swards containing a substantial proportion of clover. When swards are grazed continuously, concentrations of N in the herbage often increase towards the end of the growing season. However, concentrations of N in the herbage of unfertilized rough grassland are often in the range 1.0–2.0% N.

Herbage that is affected by dung, especially by cattle dung, tends to be avoided by grazing animals for a year or more (Forbes and Hodgson, 1985; Marsh and Campling, 1970) whereas herbage affected by urine is avoided for no more than a few days (Wolton, 1979). With cattle dung, an area four to ten times greater than that actually covered may be rejected, often resulting in 15–35% of the total area not being utilized at any one time (Marsh and

Campling, 1970). When slurry is applied, the rejection of contaminated herbage is likely to reduce intake during the next grazing period, unless grazing pressure is low (Leaver, 1976).

Digestion of Herbage by Cattle and Sheep

Digestion in ruminant animals has two main stages. The first stage occurs in the rumen and is brought about by the action of enzymes that are produced by the rumen microorganisms. These microorganisms are mostly bacteria, but there are also significant numbers of protozoa and anaerobic fungi (Gill *et al.*, 1989; Ørskov, 1992). Typical numbers of bacteria in the rumen contents of cattle and sheep are $5–50 \times 10^9$ ml^{-1}, and typical numbers of protozoa are $0.5–20 \times 10^5$ ml^{-1} (Hungate, 1966). The second stage of digestion occurs in the intestines through the action of enzymes produced by the animal itself: these enzymes act on residual dietary material and on the microbial material that is synthesized during the first stage. An outline of the process is shown in Fig. 4.2.

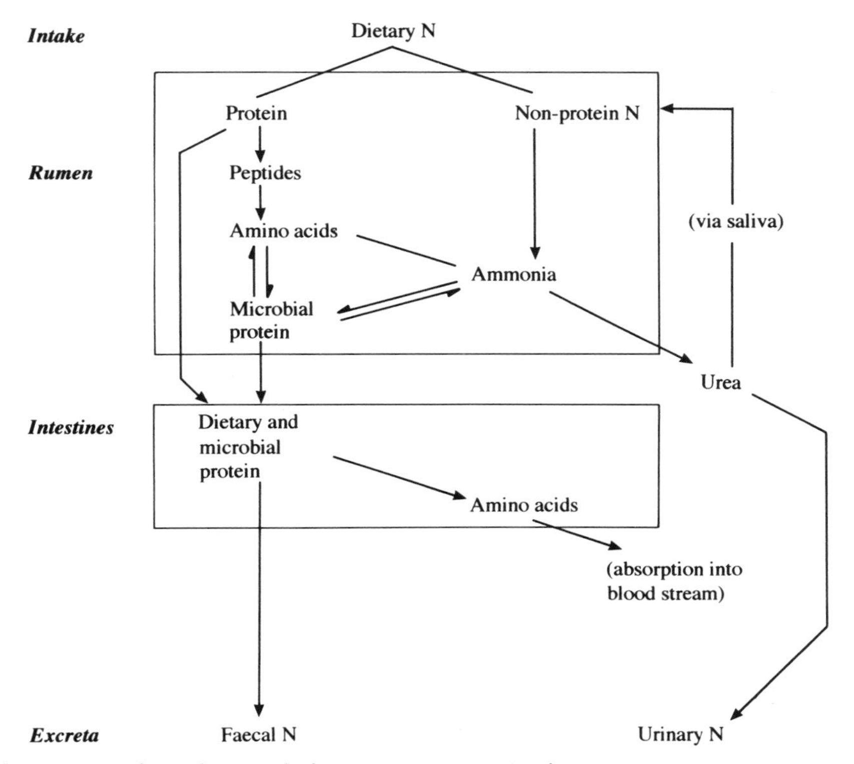

Fig. 4.2. Outline of N metabolism in ruminant animals.

The digestibility of a diet can be expressed in terms of either total dry matter or organic matter. On the basis of organic matter:

$$\text{OM digestibility} = \frac{(\text{OM consumed} - \text{OM in faeces})}{\text{OM consumed}} \times 100\%.$$

With herbage diets, digestibility is inversely related to the content of cell wall material, since cell wall is less digestible than cytoplasm and it also reduces the accessibility of the cytoplasm to digestive enzymes. Cell walls consist of cellulose, hemicellulose, pectic substances and lignin, whereas the cytoplasm contains proteins and related compounds, nucleic acids, lipids, soluble carbohydrates and chlorophyll. Digestibility decreases as herbage matures, due partly to an increase in cell wall material and partly to an increase in the lignification of cell walls. There is also an increase in the ratio of stem to leaf which tends to increase the proportion of cell wall material.

As indicated above, part of the dietary protein is digested in the rumen by microbial enzymes, and part remains intact until it reaches the small intestine. Protein digestion within the rumen involves (i) the hydrolysis of the peptide bond by protease and peptidase enzymes and (ii) the deamination and decarboxylation of the resulting peptides and amino acids with the release of ammonia (van Straalen and Tamminga, 1990). More than half the protein entering the rumen is hydrolysed to peptides and amino acids, and most of these are subsequently deaminated (Ørskov, 1992). Much of the ammonia, together with some free amino acid, is then assimilated into microbial protein. With most diets, about two-thirds of the protein reaching the duodenum is synthesized in this way in the rumen and about one-third is intact dietary protein (Buttery, 1976). Ammonia that is not synthesized into microbial protein diffuses through the rumen wall and is converted in the liver to urea, much of which is excreted in the urine. However, some is returned to the rumen in saliva or via the blood stream. The growth of microbial biomass in the rumen is influenced by the amounts of energy and nitrogen made available in the rumen, and by the ratio between them, and these factors therefore influence the amount of microbial protein that is available for digestion in the small intestine. On average, the amount of microbial protein synthesized is equivalent to 81 g crude protein kg^{-1} of forage dry matter, though values range from about 34 to 162 g kg^{-1} dry matter (Minson, 1990). The synthesis of microbial protein is at a maximum with fresh forages of high digestibility.

The digestion of microbial protein, plus the residual dietary protein, in the small intestine liberates amino acids which are absorbed into the blood stream before being further metabolized. With silage, the proportion of the original dietary protein that persists to the second stage of digestion is increased when the silage is treated with formaldehyde, as this not only prevents proteolysis in the silo (see p. 62) but also curtails microbial digestion in the rumen (van Straalen and Tamminga, 1990). The formaldehyde treatment of silage may therefore have the beneficial effect of reducing the proportion of dietary N excreted as urea in the urine.

Following the two main stages of digestion, a small amount of additional digestion may take place in the large intestine as a result of bacterial activity there. The large intestine has little if any capacity to absorb amino acids but ammonia released by deamination can be absorbed and used by the animal (van Straalen and Tamminga, 1990). Dietary constituents that are undigested on leaving the large intestine, together with undigested microbial cells and their residues, plus cells and enzyme residues from the animal's digestive system, are excreted as faeces. In general, less than 25% of the total faecal N is residual dietary N; most is microbial material (Ørskov, 1992).

The apparent digestibility of dietary protein can be assessed from the proportion of the crude protein not appearing in the faeces and, on this basis, the digestibility of herbage protein (and of the organic matter as a whole) is increased when grass receives a high rate of fertilizer N (van Vuuren *et al.*, 1992). However, the metabolic utilization of the herbage N may decrease when high rates of fertilizer N are applied (see below).

Utilization of Dietary N by Cattle and Sheep

Of the dietary N consumed by ruminant animals, less than 30% is actually utilized for the production of liveweight gain, milk or wool, and the remainder is excreted. The proportion that is utilized is usually higher in lactating than in growing animals, and is higher in growing animals than in those that are neither lactating nor growing. Utilization is usually about 20–25% of the N consumed for dairy cows, 5–10% for beef cattle and 3–15% for sheep and lambs (Henzell and Ross, 1973; Holmes, 1970).

The physiological state of the animal has a considerable influence on the requirement for protein in terms of its concentration in the diet, the requirement being greatest during lactation and during rapid growth. For animals on a maintenance-only diet, the requirement for protein is about 7% of the dry matter intake (viz. 1.1% N). For pregnant animals the protein requirement is 10–12% of dry matter intake (1.6–1.9% N) and for lactating animals, and also young animals that are actively growing, it is 15–20% (2.4–3.2% N) assuming that no amino acid is limiting (Thompson and Poppi, 1990).

The utilization of dietary N is as high as possible when the ratio between the available nitrogen and available energy in the diet is optimal for the amount of milk or liveweight gain being produced. The excretion of N is then at a minimum. With young grass herbage, the ratio between available nitrogen and energy is often higher than optimum and the excess protein is then used as a source of energy. As a result, a relatively high proportion of the ammonia produced in the rumen is excreted as urea. On the other hand, with mature grass herbage receiving little or no fertilizer N, the protein content may be insufficient to ensure that the available energy is utilized completely, especially by lactating or young growing animals which have relatively high protein requirements (Simpson and Stobbs, 1981). In this situation, increases in production can be obtained by providing supplementary protein, for example as soyabean meal or a legume hay.

Although the utilization of dietary N can be improved by adjusting the concentration of N, the extent of the improvement is limited. Thus, with dairy cattle consuming a herbage diet, the maximum production of milk by cows not receiving supplementary concentrate feeds is thought to be 20–25 kg day^{-1}, with this limit being imposed partly by the total intake of energy and partly by the amount of amino acid N available for absorption in the small intestine. The amount of amino acid N is limited by the fact that much of the herbage protein is hydrolysed in the rumen, but only part of the resultant ammoniacal N is synthesized into microbial protein (van Vuuren and Meijs, 1987).

When herbage contains excessive N (more than about 3% N), it is often possible to improve the utilization of the N by supplementing the diet with a component low in protein. Thus, by partially substituting grass herbage containing 3.9% N by maize silage, the utilization of the dietary N was increased from 17 to 24% (Table 4.2). Alternatively, it is possible to reduce the dietary concentration of N by feeding more mature herbage or by reducing the rate of fertilizer N. However, beyond a certain point, the effect of these measures is limited by a simultaneous reduction in metabolizable energy. The utilization of N might also be improved if the diet were modified in a way that would increase the supply of intact protein to the small intestine. This is more difficult but several possibilities have been suggested. One is to include a type of herbage containing tannin (e.g. sainfoin) to reduce the rate of proteolysis in the rumen. A second is to treat the herbage with formaldehyde to stabilize the protein, and a third involves feeding Monensin to modify the microbial population and metabolism in the rumen. However, none of these techniques has been consistently successful (van Vuuren and Meijs, 1987). With silage diets, the proteolysis that occurs during production of the silage tends to

Table 4.2. Intake and output of N in cows fed (1) grass herbage and (2) a combination of herbage and maize silage in approximately equal amounts. (From van Vuuren and Meijs, 1987.)

Diet	% N in diet	N intake (g day^{-1})	N output (g day^{-1})			Proportion of dietary N used in milk production (%)
			Milk	Urine	Faeces	
Grass herbage	3.9	626	107	361	158	17
Grass herbage + maize silage	2.8	494	118	198	178	24

curtail the supply of amino acids to the small intestine, and protein supplements have been found to improve both milk production by dairy cattle, and liveweight gain and the ratio of protein to fat in beef cattle (Gill *et al.*, 1989).

Several different attempts have been made to develop a means of assessing the protein requirement of different classes of livestock, fed various types of basic diet (Jarrige and Alderman, 1987). One approach, examined in detail in the UK, has been the development of a model based on knowledge of the digestion process (Agricultural Research Council, 1980, 1984; Greenhalgh, 1981). Three major assumptions are made in this approach. First, the proteins in different constituents of the diet are assumed to have a specific degradability which determines the proportion of each that is hydrolysed in the rumen. Second, it is assumed that the rates of microbial assimilation of amino acid and ammoniacal N in the rumen are determined by the rate of organic matter digestion. (The proportion of the digestible organic matter that is actually digested in the rumen is assumed to be constant, and the production of microbial protein is therefore related to the intake of metabolizable energy: if the diet contains excess digestible protein, the surplus ammoniacal N is assumed to be absorbed into the blood stream and excreted.) Third, both the unhydrolysed dietary proteins and the microbial proteins synthesized in the rumen are assumed to be subject to a constant proportional loss during their digestion in the small intestine and subsequent metabolism (Greenhalgh, 1981). In a more recent development, emphasis is placed on the 'metabolizable protein', defined as the total digestible true protein available to the animal for metabolism, after digestion and absorption of the feed in the animal's digestive tract (Alderman and Cottrill, 1993). Metabolizable protein has two components. One is the 'digestible microbial true protein' which is calculated by assuming that 75% of the microbial crude protein is true protein (nucleic acids are excluded) and that 85% of the microbial crude protein is digestible in the intestines. The second component is the 'digestible undegraded feed protein' which is the fraction of the feed not degraded in the rumen but sufficiently digestible to be absorbed in the intestines. The proportion of the undegraded feed protein that is digestible varies from nil to 90% depending on the type of feed, its composition and pretreatment, but it can be predicted from the 'acid detergent insoluble N' content, or the 'modified acid detergent fibre' content, of the feed (Alderman and Cottrill, 1993).

An indication of the amounts of N utilized for production on a unit land area basis can be calculated from the amounts and N concentrations of the various products. For example, the concentration of N in the body tissue (excluding digesta) of mature livestock is about 2.4% (Agricultural Research Council, 1980) and therefore a production of 1000 kg liveweight gain ha^{-1} $year^{-1}$, a typical value for moderately intensive beef systems, would remove about 24 kg N ha^{-1} $year^{-1}$. The concentration of N in milk is more variable but typically about 0.53% for Friesian and about 0.61% for Jersey milk (Simpson and Stobbs, 1981): the production of 12,000 l of Friesian milk ha^{-1}

year^{-1} would therefore remove 64 kg N ha^{-1} year^{-1} in the milk though this would often include some N from supplementarty feeds.

Nutritional Value of Legumes

In the past, clovers were valued, in comparison with grass, for their relatively high concentrations of protein and of several mineral elements, particularly Ca. When fertilizer N and mineral supplements became widely available, these features were of less importance but other nutritional qualities of clovers were recognized. For example, the content of cell wall material is lower in clovers than in grasses, and there are differences in the composition of the cell wall (Table 4.3), with the result that patterns of digestion are also somewhat different. The protein of clover is utilized more efficiently by the microorganisms in the rumen and consequently, at an equivalent intake of organic matter, the amount of protein entering the duodenum is considerably greater with clover than with grass. This results in a greater absorption of amino acids by the animal from clover than from grass (Beever *et al.*, 1980). The voluntary food intake by animals is also greater with clovers, as illustrated by the finding that intake by sheep was 9–28% higher with white clover than with ryegrass of the same digestibility (Clark and Ulyatt, 1985). The higher intake arises partly because a shorter time is required for chewing, and partly because the herbage spends less time in the rumen (Clark and Ulyatt, 1985). White clover also has the advantage that the decline in digestibility that occurs with increasing maturity is less than with grasses.

In terms of animal production, the advantage of clover compared with grass is shown clearly by sheep and especially by lambs. The difference is less with cattle, and the risk of bloat is also a serious disadvantage when cattle are fed diets with a high proportion of clover (Greenhalgh, 1981). It is difficult to compare the overall efficiencies for livestock production of legume herbage and N-fertilized grass herbage, because the two types of sward differ in their seasonal growth pattern and in the management required for optimum

Table 4.3. Components (% in dry matter) of the herbage of perennial ryegrass (cv. Melle) and white clover (cv. Blanca) harvested at similar digestibility. (From Thomson, 1984.)

	Ryegrass	Clover
Nitrogen	2.8	4.4
Cell wall material	42.7	21.6
Cellulose	24.0	17.3
Hemicellulose	16.1	0.8
Lignin	2.7	3.8
Pectin	0.8	4.0

production (Simpson and Stobbs, 1981). However, in a study involving continuous grazing by sheep and lambs, it was concluded that grass–clover swards containing more than 20% clover could produce an output, in terms of lamb liveweight gain per ha, similar to that from grass swards receiving fertilizer N at 150–180 kg N ha^{-1} $year^{-1}$ (Vipond *et al.*, 1993).

Partition of Consumed N between Utilization and Excretion

The proportion of the dietary N that is not utilized for animal production in liveweight gain, milk or wool is excreted and, as implied above, the proportion is influenced by the type of livestock, and by the concentration and form of N in the diet. Dairy cattle generally excrete 75–80% of the N that they consume, beef cattle 90–95% and sheep 85–95%. Some of the N is excreted in the dung and some in the urine, and the distribution between dung and urine depends mainly on the concentration of N in the diet. Excretion in the dung of both cattle and sheep is approximately 0.8 g N per 100 g DM consumed, though it may increase slightly with increasing N concentration in the diet (Blaxter *et al.*, 1971). Because the excretion of N in the dung is fairly constant per unit of DM consumed, changes in dietary N concentration are reflected in the amount of N excreted in the urine (Barrow and Lambourne, 1962; Betteridge *et al.*, 1986; Henzell and Ross, 1973; Lantinga *et al.*, 1987). In general with cattle, the proportion of the excreted N occurring in the urine increases from about 45% when the diet contains 1.5% N on a dry matter basis to about 80% when the diet contains 4.0% N (Fig. 4.3). Data obtained with sheep are consistent with this general picture (Blaxter *et al.*, 1971; Parsons *et al.*, 1991a).

The relative amounts of N utilized for milk production and excreted in dung and urine, when four samples of fresh grass herbage ranging in N concentration from 2.4% to 4.1%, were fed to dairy cows, are shown in Table 4.4. The requirement of N for milk production was met by the herbage containing 2.4% N so that excretion, particularly in the urine, increased at higher concentrations. This pattern is consistent with other observations that a larger proportion of the dietary N is excreted by cattle when they consume

Table 4.4. Influence of herbage N concentration on the output of N in milk, urine and faeces of dairy cows fed grass herbage only. (From van der Meer, 1983.)

% N in diet	N intake (g day^{-1})	N output (g day^{-1})			Proportion of excreted N in urine (%)
		Milk	Urine	Faeces	
2.40	360	106	153	101	60
2.96	444	106	235	103	69
3.52	528	106	314	108	74
4.08	612	106	396	110	78

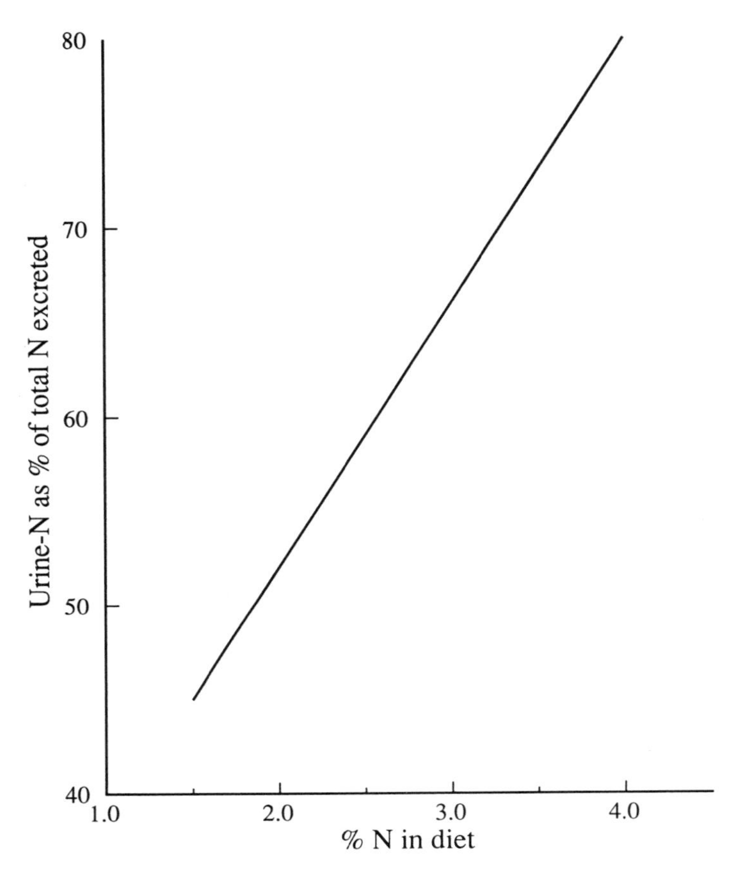

Fig. 4.3. Generalized relationship between urine N, as a proportion of the total N excreted, and the concentration of N in the diet of cattle or sheep.

grass heavily fertilized with N than when they consume grass–clover not receiving fertilizer N (Jarvis *et al.*, 1989b; Kemp *et al.*, 1979). However, whatever the composition of the diet, a substantial amount of N is excreted. For a cow weighing about 600 kg, that is neither pregnant nor lactating, the minimum amount of N excreted is about 60–80 g day^{-1} (van Vuuren and Meijs, 1987), of which about 40–65 g is faecal N, 10–15 g is urinary N and 2 g is lost via skin, hair and sweat. The amount of N excreted is greater when the animals are producing milk or gaining weight: with an output of 25 kg milk day^{-1}, and assuming an optimum balance between energy and protein in the diet, the total amount of N excreted is likely to be at least 170 g N day^{-1} (van Vuuren and Meijs, 1987).

Excretion of N in Urine

Cattle typically produce between 10 and 40 l urine day^{-1} (Betteridge *et al.*, 1986; Haynes and Williams, 1993; Holmes, 1989; Spedding, 1971). There is a wide variation in volume between breeds, between individual animals and, for an individual animal, from day to day. Daily urine production is influenced considerably by diet. For example, cows fed mainly on hay produced about 10 l urine day^{-1} whereas those fed on mown fresh grass produced about 18.5 l day^{-1} (Schechtner *et al.*, 1980). The frequency of urination in cattle is generally about 8–12 times per day (Church, 1976; Hancock, 1953; Lantinga *et al.*, 1987; MacLusky, 1960; Petersen *et al.*, 1956; Richards and Wolton, 1976a) and average volumes per urination range from about 1.5 to 3.5 l (Wolton, 1979). The smaller volumes are typical of small breeds of cattle, such as Jersey, and the larger volumes of breeds such as Friesian.

Sheep normally produce between 1 and 7 l urine day^{-1} (Frame, 1971; Herriott and Wells, 1963; Orr *et al.* 1995; Parsons *et al.*, 1991a; Spedding, 1971). In a New Zealand investigation, the average urine production was about 3 l day^{-1} (Doak, 1952). There is considerable variation in urine production depending on weather conditions. For example, in Australia the daily urine volume for sheep varied from 0.5 l in mid-summer to 2.8 l in mid-winter (Vercoe, 1962). In the UK, daily volumes varied from 1.2 l in dry weather to 6.2 l in wet weather (Herriott and Wells, 1963). Such differences are influenced by the amount of water consumed and the rate of evaporation from the skin. The average frequency of urination in sheep has been estimated at about 15–20 times per day (Doak, 1952; Frame, 1971; Haynes and Williams, 1993) and the average volume per urination is about 150 ml (Doak, 1952; Haynes and Williams, 1993; Skrijka, 1987).

Urine normally contains 4–12% of dissolved solid material (Frame, 1971; Safley *et al.*, 1984; Schechtner *et al.*, 1980), much of which consists of nitrogenous compounds. The concentration of total N in urine varies widely with factors such as the N content of the diet and the consumption of water, but reported ranges of concentration agree quite closely. For cattle, the concentration is usually between 2 and 20 g N l^{-1} (Betteridge *et al.*, 1986; Bristow *et al.*, 1992; Holmes, 1989; Lantinga *et al.*, 1987; Petersen *et al.*, 1956; Schechtner *et al.*, 1980; Wolton, 1979), with an average of 8–10 g l^{-1}. For sheep, the total concentration of N in the urine is normally between about 5 and 15 g l^{-1}, also with an average of about 8–10 g l^{-1} (Bristow *et al.*, 1992; Dale, 1961; Doak, 1952; Herriot and Wells, 1962; Sears and Newbould, 1942).

The daily excretion of N in urine varies widely with variation in both volume and N concentration, and a typical amount for dairy cows would be between 100 and 350 g N per cow per day. The amount is influenced by the physiological state of the animal and by the concentration of N in the diet. For example, a spring-calving dairy cow fed entirely on grass fertilized with N (at a rate of 300 kg N ha^{-1} $year^{-1}$) was estimated to excrete in the urine 350 g N day^{-1} in early May and about 230 g N day^{-1} in September (Kemp *et al.*, 1979). With sheep, the excretion of N in the urine of ewes grazing grass receiving 420 kg fertilizer N ha^{-1} $year^{-1}$ was about 82 g N per ewe day^{-1}, whereas the amount from ewes grazing grass–clover swards was 55–65 g N per ewe day^{-1}: the lambs of both groups excreted an average of about 20–26 g N day^{-1} during the period to 17 weeks after birth (Parsons *et al.*, 1991a). However, in subsequent studies, the average excretion of urine-N was less than 55 g per ewe day^{-1} on both sward types (Orr *et al.*, 1995). Some typical values for the

Table 4.5. Typical values for the excretion of N in urine by dairy cows, steers and sheep. (From various sources.)

	Dairy cows	Steers	Sheep
Urinations per day	8–12	8–12	15–20
Volume per urination (l)	1.5–3.5	1.0–3.0	0.1–0.2
Urine volume per day (l)	10–40	10–30	1–7
Dry matter in urine (g l^{-1})	60–120	60–120	–
N concentration (g l^{-1})	2–20	2–20	5–15
N excreted in urine (g N day^{-1})	80–320	80–240	10–70
N excreted in urine (kg N $year^{-1}$)	30–120	30–90	5–25

Table 4.6. Distribution of the N in cattle and sheep urine among various constituents, as reported in two investigations (total N expressed as g N l^{-1}; other constituents as proportion (%) of total N). (Data for diets A and C from Nehring *et al.*, 1965, and for diets B and D from Bristow *et al.*, 1992.)

	Cattle		Sheep	
Diet	A	B	C	D
Total N	0.9–18.1	6.8–21.6	2.0–12.4	3.0–12.5
Urea	70–91	59–89	45–76	75–93
Allantoin	1.7–22.2	2.4–11.7	1.2–47.0	3.6–5.2
Hippuric acid	2.3–7.3	3.2–8.0	4.0–28.5	2.6–7.1
Creatinine	0.0–8.1	1.7–5.5	1.1–6.3	0.4–0.6
Creatine	0.0–2.9	1.3–3.7	0.2–2.0	4.3–6.3
Uric acid	0.2–2.6	0.7–1.9	0.9–3.7	0.2–0.4
Amino acid-N	–	0.3–3.7	–	0.1–0.4
Ammoniacal-N	0.6–6.8*	0.3–9.1*	1.5–16.1*	0.1–0.2

Diet A: hay supplemented with three or four additional components (from barley straw, crushed barley, soya extract and peanuts) to provide dietary N concentrations of 2.4–2.9% of dry weight.
Diet B: grass, or grass or maize silage + protein supplement.
Diet C: hay.
Diet D: ryegrass and ryegrass–white clover, both grazed.
*Concentrations of ammoniacal N exceeding 1% of total N are likely to be due to the hydrolysis of urea before analysis.

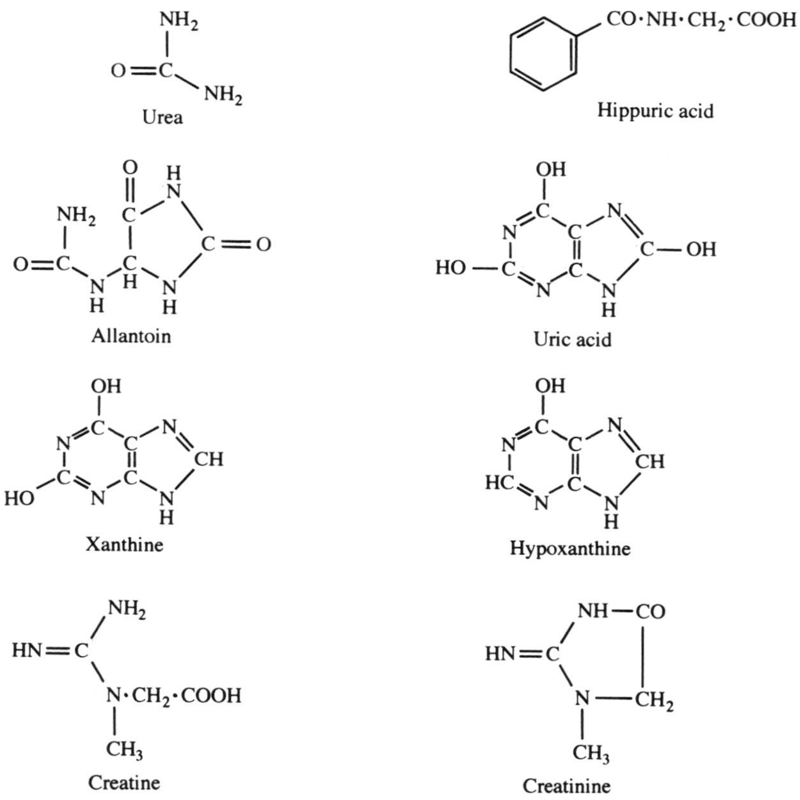

Urea

Hippuric acid

Allantoin

Uric acid

Xanthine

Hypoxanthine

Creatine

Creatinine

Fig. 4.4. Molecular structures of urinary N compounds.

amounts of urine, and of N in the urine, produced by cattle and sheep are shown in Table 4.5.

Urea generally accounts for between 60 and 90% of the total N in the urine of both cattle and sheep (Bristow *et al.*, 1992; Lantinga *et al.*, 1987; Sherlock and Goh, 1984; Thomas *et al.*, 1988), though as little as 24–25% has been reported for sheep fed on extremely low-protein diets (Field *et al.*, 1974; Topps and Elliott, 1966). The proportion of the total urinary N in the form of urea is usually > 80% when the diet is high in N, as with heavily fertilized grass (Lantinga *et al.*, 1987) but is smaller when the diet is low in N. Sheep on rough grazing produced urine containing a higher proportion of the total N as urea in summer than in winter (69% in August; 24% in February), apparently reflecting a dietary deficiency of protein during the winter (Field *et al.*, 1974).

Urine includes, in addition to urea, a number of other nitrogenous compounds, mainly hippuric acid, allantoin, uric acid, xanthine, hypoxanthine, creatine and creatinine (Fig. 4.4). The proportions of the total urinary N accounted for by these compounds vary considerably (Table 4.6). Hippuric acid normally constitutes between 2 and 7% of the total N in the urine of both cattle and sheep (Bristow *et al.*, 1992; Doak, 1952; Lantinga *et al.*, 1987) though proportions of > 20% been reported (Kreula *et al.*, 1978; Nehring *et al.*, 1965). Hippuric acid is a non-toxic condensation product of glycine (an amino acid) with benzoic acid, the latter being a potentially toxic compound derived from plant material. Allantoin, uric acid, xanthine and hypoxanthine are all metabolic products of purine metabolism and, of these compounds, allantoin is normally present in the greatest concentration in urine. It usually amounts to between 2 and 12% of the total N, though sometimes > 20% (Nehring *et al.*, 1965). Most of the allantoin appears to be derived from the nucleic acids of rumen microorganisms, and urinary allantoin can be used to indicate the extent of microbial protein synthesis in the rumen (Verbic *et al.*, 1990). Creatine is a product of the metabolism of glycine, arginine and methionine and, before excretion in the urine, is partly converted to its cyclic anhydride, creatinine. In total, these two constituents usually account for between 1% and 10% of the total N (Table 4.6).

Excretion of N in Faeces

The production of faeces varies with the amount and digestibility of the feed consumed, and typically ranges from about 2.5 to 5.0 kg DM day^{-1} for a dairy cow and from 1.2 to 2.0 kg DM day^{-1} for young cattle (Church, 1976; Haynes and Williams, 1993; Holmes, 1989; Marsh and Campling, 1970). Typically sheep produce about 0.3–0.6 kg of faecal dry matter per day (Holmes, 1989; Parsons *et al.*, 1991a). The dry matter content of faeces varies with factors such as the nature of the diet and the weather but, for cattle, is normally about 8–16% of fresh weight and, for sheep, is usually between 20 and 50% (Frame, 1971; Haynes and Williams, 1993; Parsons *et al.*, 1991a; Schechtner *et al.*, 1980). Of the total dry matter in the faeces of grazing cattle, organic matter usually amounts to between 60 and 85%, and ash to between 15 and 40% (Haynes and Williams, 1993; Kirchman and Witter, 1992). Large amounts of ash in the faeces are usually the result of the ingestion of soil, for example by animals grazing herbage splashed by soil during wet weather.

The concentration of N in the faeces of cattle and sheep fed on grass or grass–clover is usually in the range 1.2–4.0% of dry matter (Dickinson *et al.*, 1981; Kirchman and Witter, 1992; Parsons *et al.*, 1991a; Petersen *et al.*, 1956; Safley *et al.*, 1984), equivalent to about 0.2–0.5% on a fresh weight basis. With sheep fed on herbage from unfertilized hill pasture, the faecal concentration of N is at the lower end of this range, viz. 1.1–2.1% of dry weight (Floate and Torrance, 1970). The daily output of faecal N would typically be between 100 and 150 g for a dairy cow of 500 kg, and between 10 and 25 g for a sheep of

Table 4.7. Typical values for excretion of N in faeces by dairy cows, steers and sheep. (From various sources.)

	Dairy cows	Steers	Sheep
Defaecations per day	7–15	7–15	6–26
Dry weight of faeces per day (kg)	2.5–5.0	1.2–2.0	0.2–0.6
Dry matter in faeces (%)	8–16	8–16	20–50
N concentration in dry wt (%)	1.5–4.0	1.5–4.0	1.0–4.0
N excreted in faeces (g N day^{-1})	50–200	20–80	5–25
N excreted in faeces (kg N $year^{-1}$)	20–70	10–30	2–9

Table 4.8. Total N, C : N ratio and major forms of N in the faeces of cattle and pigs. (From Kirchmann and Witter, 1992.)

	Diet	Total N (% in DM)	C : N ratio	Proportion (%) of total N as		
				Organic N	Water-soluble N	NH_4-N
Cattle	Grazed grass–clover	2.33	18.6	99	22	0.9
Pigs	Barley, soya, oil seed rape	3.08	15.8	92	28	8.3

70 kg. Ewes grazing grass–clover swards receiving no fertilizer N, or a grass sward receiving 420 kg N ha^{-1} $year^{-1}$, excreted on average over the season between 13 and 25 g faecal N per ewe day^{-1} (between 400 and 630 g faeces dry weight), while their lambs excreted between 5 and 7 g N day^{-1} (Parsons *et al.*, 1991a).

The frequency of defaecation by cattle is generally 7–15 times day^{-1} (Afzal and Adams, 1992; Church, 1976; Lantinga *et al.*, 1987; MacDiarmid and Watkin, 1972b; Richards and Wolton, 1976b; Wilkinson and Lowrey, 1973). There is little information for sheep but a frequency of 26 times per day has been reported from New Zealand (Morton and Baird, 1990) and 6–8 times per day from Scotland (Frame, 1971). Some typical values for the amounts of faeces, and the concentrations of N in the faeces, produced by cattle and sheep are shown in Table 4.7.

Most of the N in cattle, sheep and pig faeces is present in insoluble organic forms (Table 4.8). In general, about 45–65% of the faecal N is amino N, about 5% occurs in nucleic acids, and 3% as ammonia, with the remainder consisting of partially degraded nucleic acids, bacterial cell walls and glycoprotein, and N bound to fibre (National Research Council, 1985). With sheep, about 25% of the faecal N is generally present as water-soluble products of animal and microbial metabolism, about 10–20% is undigested dietary N and the remainder, about 60%, is present in bacterial cells (Mason *et al.*, 1981).

Table 4.9. Typical composition of farmyard manures and fresh, undiluted slurries (on fresh weight or volume basis). (From Archer, 1988.)

Type	Approximate dry matter (%)	Nitrogen (%)	Available N* (%)
Farmyard manure			
Cattle	25	0.6	0.15
Pigs	25	0.6	0.15
Poultry:			
Deep litter	70	1.7	1.0
Broiler litter	70	2.4	1.4
In-house, air-dried droppings	70	4.2	2.5
Slurry (fresh and undiluted)			
Cattle	10	0.5	0.15
Pigs	10	0.6	0.4
Poultry	25	1.4	0.9

*Available in season of application, assuming some loss of potentially available N through the volatilization of ammonia.

Storage of Excreta from Housed Livestock

When livestock are housed, their excreta are normally collected and stored as slurries or manures before being applied to the land. The slurries consist of both dung and urine, mixed with a variable amount of water, derived mainly from washing the floors of the housing. The manures consist mainly of dung plus a variable proportion of urine, together with bedding material, often straw or wood shavings. In the UK, the total amounts of excreta produced by housed livestock have been estimated at 300,000 t for cattle (70% slurry, 30% manure), 70,000 t for pigs (50% slurry, 50% manure) and 85,000 t for poultry (10% slurry, 90% manure) (Archer, 1988). Typical values for the contents of dry matter and N in a range of slurries and manures are shown in Table 4.9.

Increases in livestock numbers and changes in housing systems during recent decades have resulted in greater amounts of excreta being stored as slurry. For example, there has been an increase in those types of housing, such as cubicles for dairy cattle and slatted floor systems for beef cattle and pigs, that do not require straw bedding. Partly as a result of increased production, the monetary value of slurry (and that of solid manures) has declined. Slurries in particular are often regarded as a disposal problem rather than an asset (Smith and Chambers, 1993; Wadman *et al.*, 1987). Other factors that have contributed to the decline in the value of manures and slurries are the relative cheapness of fertilizer N and the increasing cost of labour.

Slurries are often spread on the land, either grassland or arable, without any treatment other than storage, but they are sometimes treated to improve their ease of disposal or fertilizer value. The treatment may involve separation into mainly solid and mainly liquid fractions, or digestion for a period of weeks or months under either aerobic or anaerobic conditions. Such treatments will modify the concentration and forms of N, as described below.

In the USA, and to a lesser extent elsewhere, intensive beef production involves many animals being kept in feedlots at a high density, equivalent to a space per animal of between 5 and 40 m^2. At a density of 20 m^2 per animal, about 5500 kg N ha^{-1} are deposited as excreta during a 150-day feeding period (National Research Council, 1978). Some of the excreta are removed at intervals and used as manure, but the excessively high concentrations of N in the topsoil within the feedlot result in large losses of N by ammonia volatilization and leaching (National Research Council, 1978).

Nitrogen in Slurries

In a dairy herd, the amount of slurry produced per cow per year depends mainly on the average body weight, and on the extent to which the slurry is diluted, as illustrated in Table 4.10. The concentration of N in slurry is also influenced by the degree of dilution but, for undiluted cattle slurry is usually in the range 0.3–0.6% (Archer, 1988; Schechtner *et al.*, 1980). The influence of

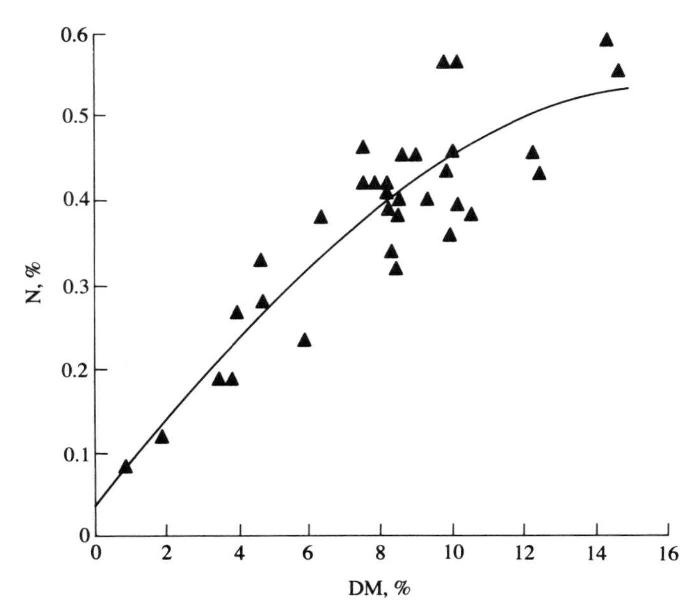

Fig. 4.5. Relationship between concentrations of dry matter and nitrogen in cattle slurry. (From Schechtner *et al.*, 1980.)

dilution is illustrated by the relationship shown between the concentration of dry matter and the concentration of N in 33 slurries (Fig. 4.5).

The urea in the urine component of slurries is hydrolysed rapidly to ammonium, the hydrolysis being complete within a few days at temperatures of > 10°C (Muck, 1981; Whitehead and Raistrick, 1993a). Considerable urease activity is present in cattle faeces (Muck, 1981). After storage, typically about 50% of the slurry N is present as ammonium-N in the liquid fraction, and about 50% is present in organic forms in the particulate material. However, in some slurries, the proportion of the total N in the form of ammonium may be much less than 50%, due mainly to the loss of ammonia by volatilization (van Dijk and Sturm, 1983). The organically-bound N in the particulate material is not uniformly distributed: the concentration of N tends to increase with decreasing particle size and may be particularly high (> 4% N on a DM basis) in the fraction < 0.2 mm (Whitehead *et al.*, 1989a).

If the slurry is not aerated during storage, much of the ammonium-N remains in solution, though ammonia may volatilize slowly with a substantial loss of N over a period of several months. The volatilization of ammonia is curtailed if a surface crust forms on the slurry, or if the surface of the slurry is covered in some way (Sommer *et al.*, 1993). In a study of three cattle slurries stored without being covered for 6 months, losses ranged from 18 to 54% of the total N, and the addition of bentonite or biological additives to the slurry had little effect (Dewes *et al.*, 1990). If slurry is aerated during storage, there are increases in the decomposition of organic matter and in the amount of ammonia volatilized. If the slurry is digested anaerobically, a process that is normally associated with the production of biogas, there is also an increase in the rate of decomposition of organic matter but the volatilization of ammonia is curtailed. Digestion normally results in a narrowing of the C : N ratio; for example, with pig slurry, from 17 : 1 to 10.5 : 1 (Chaussod *et al.*, 1986). Examples of the change in composition (expressed in kg m^{-3}) resulting from normal storage, aeration and anaerobic digestion are shown in Table 4.11.

Usually none of the N in slurry occurs as nitrate, even if the slurry is aerated (Besnard, 1980; Velthof and Oenema, 1993; Whitehead and Raistrick, 1993a). This is probably because, at the normal pH of > 8, ammonium-N is

Table 4.10. Volume of slurry produced by dairy cows (m^3 slurry per cow per year) in relation to cow body weight and dry matter content of the slurry. (From van Dijk and Sturm, 1983.)

	Body weight (kg)		
% DM in slurry	500	600	700
10.5	14.6	17.5	20.4
7.5	20.4	24.4	28.5
6.0	25.5	30.6	35.8
5.0	30.6	36.8	43.0

Table 4.11. Changes in the composition of cattle slurry (kg m^{-3}) resulting from normal storage, aeration and anaerobic digestion. (From Vetter *et al.*, 1987.)

	Change (%) in composition		
	Normal storage	Aeration	Anaerobic digestion
Organic matter	−2.5	−20.7	−29.8
Total N	−3.2	−6.9	−6.4
Organic N	−13.8	+6.3	−26.1
NH_4-N	+7.9	−20.2	+13.3

more readily volatilized than nitrified. Nitrification is inhibited by high pH and by high concentrations of ammoniacal N (Engel and Alexander, 1960). However, nitrification has been found to occur in pig slurry that was both diluted and aerated (Loynachan *et al.*, 1976; Owens *et al.*, 1973) and because nitrification results in a lowering of pH, a pH of < 7 in stored slurry suggests that nitrification has occurred.

Nitrogen in Solid Manures

Manures of excreta plus straw, decomposing slowly under aerobic conditions, tend to lose N continuously over a period of months, due partly to ammonia volatilization and partly (unless the manure is protected from rain) to leaching. Reported losses have ranged from about 10% to 50% and, in some investigations, have been found to be inversely related to the C : N ratio of the manure (Kirchman, 1985). However, attempting to increase the C : N ratio of the manure by including additional bedding material, such as straw, is often not effective because more urine is absorbed. When manures are stored under anaerobic conditions (but without leaching), gaseous losses through ammonia volatilization and denitrification are low, and more N is retained in the manure (Kirchman, 1985).

Typical concentrations of N in manures consisting of excreta plus straw and composted for a few months are 0.4–0.8% on a fresh weight basis or 1.5–3.0% on a dry weight basis (Archer, 1988; Kirchman, 1985; Levi-Minzi *et al.*, 1986; van Dijk and Sturm, 1983), with almost all the N in organic forms. During storage for 3 months, the concentration of total N in the dry matter of six batches of farmyard manure increased, on average, from 2.1 to 2.7%, and the C : N ratio narrowed from 19.0 : 1 to 14.2 : 1 (Levi-Minzi *et al.*, 1986). A similar concentration of N (2.7%) and a C : N ratio of 14 : 1 were also reported in another investigation of composted cattle manure (Chaussod *et al.*, 1986).

In general, the use of solid manures on grassland is much less widespread than the use of slurry; solid manures are applied mainly to arable crops.

5 Amounts, Sources and Fractionation of Organic Nitrogen in Soils

Amounts of Organic N in Soils and Trends of Accumulation

Nitrogen is present in soil organic matter at a concentration of about 5% of dry weight, and differences in the content of total soil N therefore reflect differences in the content of organic matter. In most soils, more than 95% of the total N in soil is present in the soil organic matter, the remainder being in the inorganic forms, ammonium and nitrate. The actual amount of organic matter in a soil depends on a wide range of factors operating over many years. In general, the type of vegetation, the climate and the soil texture are major factors and, in agricultural soils, the past cropping history is important. In soils of temperate regions, the content of organic matter to a depth of 15 cm typically ranges from about 1.5% of dry weight (= 0.08% N) in sandy soils under arable cultivation to 8–10% (= 0.4–0.5% N) in clay soils under long-term grassland (Table 5.1). The content of organic matter normally declines with increasing depth (except in some peat soils) and is often negligible below 30–40 cm. Usually, total soil N amounts to between 2000 and 4000 kg ha^{-1} in

Table 5.1. Typical amounts of organic matter in the topsoil (% in dry weight; 0–15 cm) as influenced by soil texture and type of cropping. (From Archer, 1988.)

Soil texture	Arable	Ley/arable	Long-term grass
Sand	1.0–1.6	1.4–1.8	3–5
Light loam	1.4–2.5	1.8–3.0	4–6
Light silt	2.0–2.6	2.5–3.0	5–7
Medium loam / Medium silt	2.4–3.0	2.8–3.2	5–8
Clay	3.2–4.0	3.7–4.2	5–10

arable soils and between 5000 and 15,000 kg ha^{-1} in long-term grassland soils. The larger amounts of organic matter in grassland than in arable soils reflect a greater input of organic matter in dead plant material and animal excreta, combined with a slower rate of decomposition in the grassland soils. The relatively slow rate of decomposition is due to a lack of cultivation and hence less aeration in grassland than in arable soils. The larger amounts of organic matter in clay soils than in sandy soils are also due to a slower rate of decomposition in the clay soils. This results partly from the relatively poor aeration of clay soils and partly from the formation of complexes between clay and organic matter, but the relative importance of these two factors is uncertain.

Usually, the organic matter in soil has accumulated over many centuries and this accumulation indicates that the average annual input of organic material has exceeded the amount mineralized. Similarly, the fact that N has accumulated indicates that the average annual input of N to the soil has been greater than the annual output through crop removal and losses from the soil. Nevertheless, a small proportion of the total soil organic matter is mineralized each year, and a small proportion of the organic N is converted into inorganic forms which are available for uptake by plants and are susceptible to loss. In most grassland soils, as in undisturbed soils generally, the current annual rate of addition of organic N exceeds the rate of mineralization but, in many arable soils, the current rate of addition is less than the rate of mineralization.

When grass is sown on land that has been cultivated for many years, the contents of organic matter and N generally increase from year to year, so long as the grass remains unploughed. The increases are asymptotic, but eventually an equilibrium is reached when the annual additions of organic matter and N are equalled by the amounts that are mineralized. The time taken to reach equilibrium varies considerably with soil type, climate and sward management but, in temperate climatic conditions, is likely to be between 50 and 200 years. The situation at equilibrium can be expressed as $S = A/r$, where S = the total amount of organic matter (or N) present, A = the annual addition and r = the fraction of the total mineralized each year. Before equilibrium is reached, S will be less than A/r and will therefore increase, but at a diminishing rate, as the equilibrium position is approached. The content of organic matter in the soil is often particularly high in native uncultivated grassland because there have been no periods when the soil was cultivated and therefore subject to excessive aeration. Also, the susceptibility of soil to surface erosion (which tends to remove soil that is relatively high in N content) is much less with grassland than with arable cropping. However, when long-term grassland is ploughed, the rate of mineralization increases and there is a decline in the content of soil organic matter.

The changes that follow a change from arable cultivation to grassland or vice versa, although substantial, occur slowly in relation to the total amounts of organic matter and N, and are therefore difficult to measure. In order to assess the changes accurately, long-term investigations (ideally > 10 years) with a large number of replicate soil samples (ideally > 30 per plot, depending on plot size) are needed to overcome the variation that is inherent in soil sampling and analysis. When changes are assessed from the analysis of soil samples, it is also important to take the bulk density of the soil into account, since a soil under prolonged cultivation may well have a density 30% greater than that of a similar soil under long-term grassland. This problem can be overcome by expressing the results of analysis on the basis of sampled area rather than unit soil weight.

Data obtained at Rothamsted (UK) from plots on a silty clay loam, some sown to grass following long-term arable cultivation and some known to have been under grass for up to 300 years, indicated that more than 100 years were needed for a long-term arable soil to attain an equilibrium N content (Johnston, 1991). The average rate of increase was about 56 kg N ha^{-1} $year^{-1}$ during the first 40 years (Fig. 5.1). In another experiment in the UK, the increase in soil N under a grazed grass–clover sward sown on a previously arable soil was about 75 kg N ha^{-1} $year^{-1}$ during the first 10 years (Tyson *et al.*, 1990). Liming

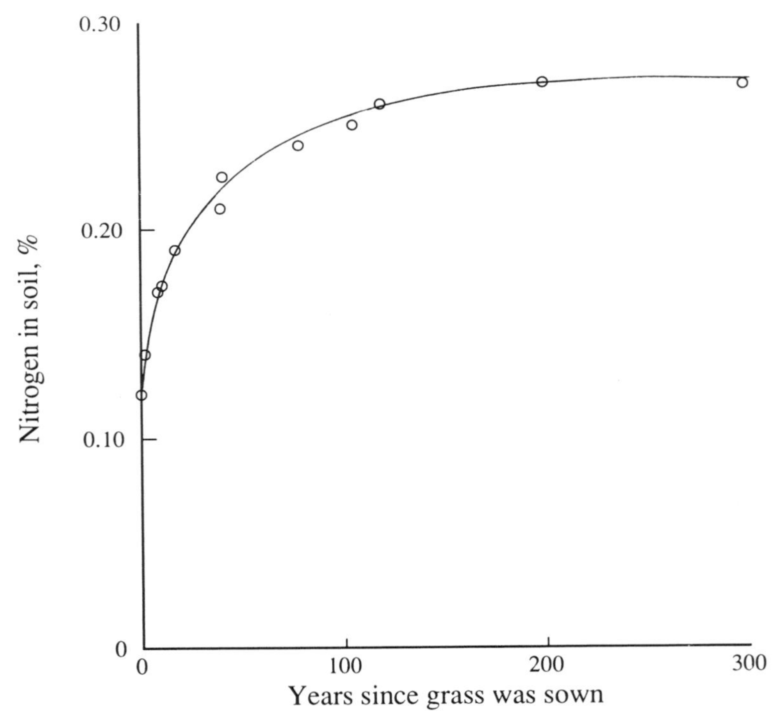

Fig. 5.1. The increase in soil N concentration (0–23 cm) after sowing grass on plots of a silty clay loam at Rothamsted, UK, previously under long-term arable cultivation. (Redrawn from Johnston, 1991.)

at this time to bring the pH from 5.6 to 6.7, followed by two further applications of lime at intervals of 10 and 5 years, apparently increased mineralization so that there was little further change in soil N during the next 15 years, after which an upward trend was resumed. In a 3-year investigation, soil N increased by about 70–180 kg N ha^{-1} N $year^{-1}$ under various grass and grass–clover swards, with the greatest increases being under grazed ryegrass with white clover swards (Clement and Williams, 1967). Similar results have been reported from other countries. In the Netherlands, soil N accumulated at rates of 70–100 kg N ha^{-1} $year^{-1}$ for several years in newly-sown grazed grassland, and the rate of fertilizer N had little or no effect (Hoogerkamp, 1973). In New Zealand, the average increase in soil N under grazed grass–clover swards was estimated to be 112 kg N ha^{-1} $year^{-1}$ (Walker, 1962), though larger increases occurred with a soil initially very low in organic matter (Sears *et al.*, 1965). In South Australia, the average increase in plots of native pasture improved by fertilizer P and grazed by sheep over 30–40 years was about 52 kg N ha^{-1} $year^{-1}$ (Russell, 1960), while the increase in pastures, based on subterranean clover and sown on soils low in organic matter, was 40–80 kg N ha^{-1} $year^{-1}$ (Simpson, 1987). The influence of climate on the N content of long-term grassland soils at equilibrium has been shown clearly in the USA, where the content along a transect from east to west was positively related to a humidity factor expressed as the ratio of rainfall to the saturation deficit of the air (Stevenson, 1986).

With agricultural soils, changes in management often prevent an equilibrium content of organic matter, or total N, being attained. Factors such as cultivation, the improvement of drainage and the application of lime, increase mineralization and therefore reduce the rate of accumulation. In some circumstances, the content of soil N in long-term grassland soils may decline without the soil being ploughed, as reported for grassland subject to increasingly intensive management in New Zealand (Ball and Field, 1987).

Decreases in Organic Soil N Following the Cultivation of Grassland

The ploughing of old grassland results in the mineralization of much of the total soil N and the consequent loss of N from the soil. Occasionally, as much as 70% of the original amount of total soil N may be lost during 20–30 years of subsequent cultivation, with most of the loss occurring during the first few years (Low, 1972). Detailed studies at three sites in the UK, taking into account changes in the bulk density of the soils, indicated that about 4000 kg N ha^{-1} was lost during 25 years, with half the decline occurring in the first 5.5 years and 90% within 18 years (Fig. 5.2). The decline over 25 years represented about 40% of the original soil N in the top 25 cm (Whitmore *et al.*, 1992). Similarly, studies in the USA showed that the decline in soil organic matter following

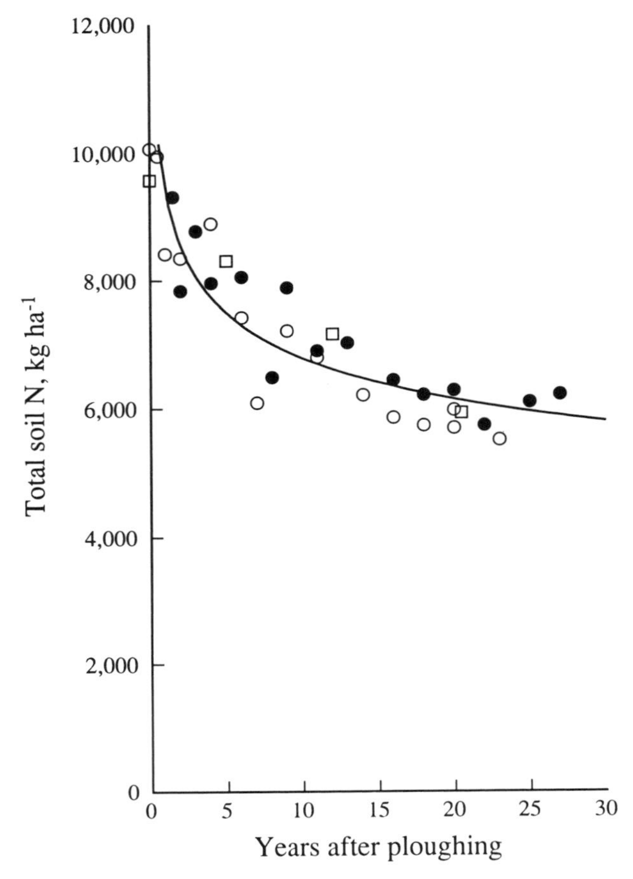

Fig. 5.2. Total soil N in three soils after the ploughing of old grassland and conversion to arable cropping (Blacktoft series, ○; Romney series, ●; Batcombe series, □). (Data from Whitmore *et al.*, 1992.)

the cultivation of long-term grassland soils occurred most rapidly during the first few years and approached a new equilibrium in 50–60 years (e.g. Hobbs and Thompson, 1971). In general, the rate of decrease is greater than the rate of increase that occurs if the soil is subsequently re-sown to grass (White *et al.*, 1976). In a comparison of 12 pairs of soils, one of each pair being under native prairie grassland and the other having been cultivated for at least 40 years, the concentration of N in the soil was between 31 and 56% lower in the cropped soil (Campbell and Souster, 1982). In another comparison, involving three pairs of soils, either cultivated or remaining under grass for 60–70 years, the amount of N (taking into account changes in bulk density) was 18–34% less in the cultivated soils, the actual loss of N being influenced by the extent to which legumes were grown during the period under cultivation (Tiessen *et al.*, 1982). Other studies have indicated that the loss of the potentially mineralizable fraction of the soil N (assessed by incubation) is greater than the loss of total N (Campbell and Souster, 1982; Smith and Young, 1975).

Sources of Organic N in Grassland Soils

The immediate sources of organic N in grassland soils are dead plant material, animal excreta and dead microbial biomass. However, the N in these various sources is derived either from biological fixation, or from deposition from the atmosphere or from fertilizers. The biological fixation of atmospheric N_2 and its conversion to plant N is described in Chapter 3, and the uptake and assimilation of inorganic N from fertilizers in Chapters 2 and 10. The supply of N from the atmosphere is described below.

The plant materials and animal excreta that form the immediate source of soil organic N vary widely in composition, and consequently in the extent to which their N accumulates in the soil. It is well established that organic materials with a high C : N ratio (> *c.* 30 : 1) are slow to release inorganic N and may cause some immobilization, at least temporarily, whereas materials with a low C : N ratio (< *c.* 20 : 1) usually release inorganic N rapidly through mineralization (see Chapter 6). Immobilization tends to promote the accumulation of soil N, in relation to the amount added in organic materials, while mineralization has the opposite effect. The balance between the immobilization and mineralization of N is also influenced by the time-period being considered, and material with a C : N ratio of between 20 : 1 and 30 : 1 may immobilize inorganic N initially but release a proportion of the immobilized N later in the decomposition process.

In some situations, the accumulation of C and N in grassland soils is limited by a deficiency of another nutrient, such as phosphorus (P) or sulphur (S), restricting the growth of clover and consequently the symbiotic fixation of N (Walker, 1965). The rate of accumulation of N in grassland soils in Australia has been increased by the application of P (Henzell *et al.*, 1966) and S (Williams and Donald, 1957).

The Atmosphere as a Source of Soil N

Nitrogen is transferred from the atmosphere to soil through the deposition of gases (mainly ammonia and nitrogen oxides) and aerosol particles (mainly ammonium and nitrate salts). The deposition may be either 'wet' (in rain or snow) or 'dry' (through sorption by vegetation or soil, or through the gravitational settling of aerosol particles). The ammoniacal N in the atmosphere is derived mainly from agricultural sources (see Chapter 8) with relatively small contributions from industry and fuel combustion. Nitrate in the atmosphere is derived mainly from the nitrogen oxides that are released to the atmosphere during the combustion of fuels, increasingly from those used in motor vehicles. Small amounts of nitrogen oxides are formed naturally in the atmosphere by lightning but this source of fixed N is thought to be insignificant in the UK (Royal Society, 1983).

On average in temperate regions, the atmospheric inputs of ammoniacal N and nitrate-N are probably similar (Fangmeier *et al.*, 1994). However, there is considerable regional variation, both within and between countries, in the deposition of the various forms of N, reflecting the distribution of major sources in relation to wind direction and rainfall. Recent evidence suggests that total wet and dry deposition amounts to about 30–35 kg N ha^{-1} $year^{-1}$ over much of southern and eastern England (Goulding, 1990). In northern and western parts of the UK, inputs of N from the atmosphere are less, possibly only about 15 kg N in areas remote from intensive agriculture and urban centres (Campbell *et al.*, 1990). Total wet plus dry deposition has been reported to amount to 40–80 kg N ha^{-1} $year^{-1}$ in parts of the Netherlands, these large amounts reflecting the intensive livestock production in that country (van Breemen and van Dijk, 1988). In other parts of western Europe, amounts may be similar to those in the UK with an average deposition of 30–40 kg N ha^{-1} $year^{-1}$ reported for parts of Germany (Fangmeier *et al.*, 1994). The average amounts of N deposited in the USA may well be rather less, with a reported range of about 5–30 kg N ha^{-1} $year^{-1}$ (National Research Council, 1978).

The inorganic forms of N deposited from the atmosphere are largely converted into organic soil N following their assimilation by either plants or soil microorganisms and, in the absence of cropping, will tend to accumulate in the soil. In areas of natural or semi-natural grassland, N deposited from the atmosphere is often the main external source of soil N.

Inputs of Soil N from Decomposing Leaf and Stem Material

Even when grass swards are harvested regularly by cutting or grazing, a considerable proportion of the herbage remains *in situ*, and eventually dies and contributes to the soil organic matter. The amount of herbage that dies is difficult to measure, partly because some senescing and dead material is consumed by earthworms and other soil fauna, and partly because the rate of decomposition varies widely with factors such as temperature and rainfall. However, estimates based on physiological measurements have indicated that, even with good sward management, the amount undergoing decomposition each year is similar to the amount harvested by cutting, or consumed by grazing animals, during a growing season (Parsons, 1988). With swards cut infrequently, and with lax grazing regimes, decomposition *in situ* is higher (Hunt, 1983; Parsons, 1988; Parsons *et al.*, 1983). In general, the proportion of the herbage decomposing *in situ* is probably about 20% greater under grazing than cutting (Hassink and Neeteson, 1991). Recent measurements in Northern Ireland indicated that rates of senescence in grazed swards during the growing season were usually between 15 and 45 kg DM ha^{-1} day^{-1} (Binnie and Chestnutt, 1994).

Because grass tillers generally live for no more than a year, it is possible with cut swards to estimate the minimum annual input to the soil organic matter from the amount of stubble remaining after a cut. Reported amounts of organic matter in the stubble of grass swards cut at a height of 5 cm at various times of year have ranged from about 1500 to 4000 kg ha^{-1} (Garwood, 1967b; Whitehead *et al.*, 1990). In addition to stubble, there is inevitably, as indicated above, some death and decomposition of leaves during the periods of regrowth, the amount depending partly on the interval between successive harvests and on weather conditions. In an Australian study, drought was found to be the main factor affecting the rate at which leaves died in a grass–clover sward in summer, whereas rainfall was the main factor affecting the rate of decay, though with a lag period of about 2 weeks (Bowman *et al.*, 1982). The death of herbage is increased by the presence of grazing animals, through the effects of trampling and the deposition of dung. In general, the senescence and death of herbage material of both grasses and clovers is greatest during late autumn and winter due to low temperatures and lack of light (Collins *et al.*, 1991; Widdowson *et al.*, 1966; Woledge *et al.*, 1990).

The concentration of N in stubble and dead herbage is usually within the range of 1.0–3.0% on a dry weight basis. With grasses, the concentration in stubble depends to some extent on N supply during the previous growth period and is generally 50–75% of that in the herbage above 5 cm (Whitehead *et al.*, 1990). The concentration of N in dead brown herbage is also increased by previous fertilizer application. For example, the concentration in dead ryegrass herbage was 1.1% N where no fertilizer N had been applied, and 2.5% N where fertilizer had been applied at 450 kg N ha^{-1} $year^{-1}$: the corresponding concentrations in the green herbage were 2.2% and 3.4% N (Whitehead, 1986). Similar data were reported in a subsequent study (Wilman *et al.*, 1994). With clovers, the concentration of N in dead material is higher than that in associated grass. For example, dead leaf + petiole material of white clover from a grass–clover sward receiving no fertilizer N contained 2.0–3.5% N, while dead herbage of the grass contained only 0.9–1.3% N (Whitehead, 1986). The C : N ratio of material containing 1.1% N would be about 44 : 1, whereas that of material containing 2.8% N would be about 17 : 1. In natural grasslands receiving no fertilizer N, there may be substantial differences between species in the N concentration and C : N ratio of the leaf litter. A study of five species grown under identical conditions in Minnesota, USA, showed N concentration in the litter to range from 0.3% to 1.1%, and C : N ratio to range from

Table 5.2. Estimated amounts of N from dead leaf and stubble transferred to the soil each year in swards of four different types.

	Amount of organic matter ($kg\ ha^{-1}$)	N in organic matter (%)	Approximate C : N ratio	Amount of N transferred ($kg\ ha^{-1}$)
Cut grass; 300 kg N ha^{-1} $year^{-1}$				
Dead leaf	500	1.9	25 : 1	10
Stubble	2500	1.9	25 : 1	48
Grazed grass, 300 kg N ha^{-1} $year^{-1}$				
Dead leaf	1000	2.0	24 : 1	20
Stubble	3000	2.3	21 : 1	69
Cut grass–clover; no fertilizer N				
Dead leaf	500	2.0	24 : 1	10
Stubble	3500	2.2	22 : 1	77
Grazed grass–clover; no fertilizer N				
Dead leaf	1000	2.0	24 : 1	20
Stubble	3500	2.2	22 : 1	77

44 : 1 to 122 : 1 (Wedin and Tilman, 1990). Herbage decomposing as a result of damage by grazing animals will tend to have a higher concentration of N, and a lower C : N ratio, than that decomposing as a result of senescence.

The actual amounts of N returned to the soil through the decomposition of leaf and stem material vary widely depending on climatic conditions and sward management. If it is assumed, for a grass sward receiving a moderate rate of fertilizer N (*c*. 250 kg N ha^{-1} $year^{-1}$), that the input of dead herbage material is 6000 kg DM ha^{-1} and that the average concentration of N is 1.7%, then the amount of N returned would be 100 kg N ha^{-1} $year^{-1}$. Estimates of the amounts of N transferred to the soil through the death and decay of herbage from different swards in UK conditions are given in Table 5.2. Other estimates for different types of grassland include 164 kg N ha^{-1} $year^{-1}$ for a grass–clover sward in a dairy farm system in New Zealand (Field and Ball, 1982) and 22 kg ha^{-1} $year^{-1}$ for a short-grass prairie in the USA (Woodmansee *et al*., 1981).

Inputs of Soil N from Roots

The transfer of N from plant material to the soil also occurs through the death and decomposition of roots, and the amount of N transferred is influenced by the rate of root production, the average length of life of the root material and its concentration of N. A newly-sown grass sward may produce up to 3000 kg root organic matter ha^{-1} in six months, increasing to 10,000 kg ha^{-1}, including both living and dead roots, after 3 or 4 years (Garwood, 1967a).

Under long-term grassland, the total amount of root material is often 10,000–20,000 kg ha^{-1} and in prairie grassland it may be as much as 25,000 kg ha^{-1} (Black and Wight, 1979). The rate of production of roots has been estimated to be about 4500 or 5600 kg DM ha^{-1} year^{-1} in established grassland in the Netherlands (Deinum, 1985; Goedwaagen and Schuurman, 1950) and 2800–6400 kg DM ha^{-1} year^{-1} in grazed natural grasslands in the USA and Canada (Sims and Coupland, 1979). Root production by timothy grown under young apple trees was estimated at 4100 kg year^{-1} (Atkinson, 1985). Fertilizer N has less effect on root production than on herbage production and, as noted in Chapter 2, the rate of fertilizer N producing maximum root growth is much lower than the rate producing maximum herbage growth. When white clover is grown alone, the weight of roots is much less than that produced by grass (Whitehead, 1983a) and, although under grass–clover swards, total root weights may be as high as those under grass alone (Garwood, 1967b), the weight of the clover component is much less than that of the grass (Young, 1958). In 4-year-old grass–clover swards, with white clover comprising 50% of the ground cover, the carefully separated roots of white clover (including nodules) represented only 6% of the total root weight and no more than 480 kg organic matter ha^{-1} (Young, 1958). With white clover relying entirely on N_2 fixation for its N supply, nodules may comprise 25–30% of the total root weight (Arnott, 1984).

There is a lack of information on the length of life of grass roots, due mainly to the difficulty of making measurements in the field. However, there is some evidence that, when swards are defoliated at intervals of 3–4 weeks, roots live, on average, for about 5–6 months (Garwood, 1967b; Troughton, 1981). When defoliation is less frequent, they may live for more than a year and the average length of life of grass roots in prairie grassland in USA has been estimated at between 2 and 5 years (Clark, 1977; Dahlman and Kucera, 1965; Sims and Singh, 1978). There is a greater turnover of root material when swards are intensively managed, at least partly because the death of roots is increased by defoliation (Eason and Newman, 1990; Evans, 1973; Oswalt *et al.*, 1959). Assessing the actual rate of turnover is complicated by the difference between the vascular stele and the cortex of roots: with mature roots, it is possible for the cortex of the older parts near to the crown to undergo decomposition even though the stele is continuing to transport water and nutrients absorbed by the younger portion of the root (Henry and Deacon, 1981). Under average conditions, grass roots probably live for about 1 year in typical productive lowland grassland and for 2–4 years in extensively managed or rough grassland. Clover roots in grassland probably have a maximum life of about 2 years (Harper *et al.*, 1991).

Although fertilizer N usually has little effect on the production of grass roots, it does increase the concentration of N in the roots. With little or no fertilizer N, root N concentration is often about 1.0% or less (Power, 1968; Power and Legg, 1984; Whitehead, 1970), and a range of 0.5% to 1.8% was reported for five grass species in Minnesota, USA (Wedin and Tilman, 1990). The concentration increases with increasing rate of fertilizer N, and tends to be greater in grazed than in cut swards (Table 5.3). Although the maximum concentration of N in grass roots is much less than the maximum concentration in herbage, the increase caused by fertilizer N will increase the rate of decomposition of the root material. The concentration of N is generally higher in clover roots than in grass roots, even when the grass is fertilized with high rates of N (Table 5.3), but the concentration in clover roots is reduced by defoliation and drought, probably due to the loss of root nodules (Whitehead, 1983a). The amounts of organic matter and N added to the soil through the decomposition

Table 5.3. Examples of concentrations of C and N (% of ash-free organic matter), and C : N ratios, in the roots of some grasses and legumes.

Species	Fertilizer N (kg ha^{-1} year^{-1})	C (%)	N (%)	C : N	Reference
Perennial ryegrass, cut	0	49.4	1.07	45.8	Whitehead, 1970
Perennial ryegrass, cut	200	48.4	1.22	39.6	Whitehead, 1970
Perennial ryegrass, cut	400	48.5	1.64	29.7	Whitehead, 1970
Perennial ryegrass, cut	650	48.1	1.92	25.0	Whitehead, 1970
Perennial ryegrass, grazed	*c.* 300	(48.4)*	2.21	21.9*	Whitehead, *et al.*, 1990
Perennial ryegrass, grazed	Nil	(48.4)*	1.50	32.3*	Garwood *et al.*, 1972
Rough fescue	Nil	48.5	1.53	31.7	Dormaar and Willms, 1993
Blue grama	Nil	49.0	1.17	41.9	Dormaar and Willms, 1993
White clover, cut	Nil	50.2	3.77	13.3	Whitehead, 1970
White clover, cut	Nil	(47.6)*	2.55	18.7*	Kuntze, 1964
White clover	Nil	47.6	3.42	13.9	Garwood *et al.*, 1972
Red clover, cut	Nil	48.4	2.79	17.5	Whitehead, 1970
Red clover, cut	Nil	47.0	2.11	22.3	Whitehead *et al.*, 1979
Lucerne, cut	Nil	47.4	2.49	20.3	Whitehead, 1970

*Estimated from other data.

Table 5.4. Estimated amounts of N from dead root material transferred to the soil each year in swards of four different types.

	Amount of organic matter (kg ha^{-1})	N in organic matter (%)	N transferred to soil (kg ha^{-1})
Cut grass; 300 kg N ha^{-1} year^{-1}	5000	1.6	80
Grazed grass; 300 kg N ha^{-1} year^{-1}	5000	2.2	110
Cut grass–clover; no fertilizer N	5000	1.1	55
Grazed grass–clover; no fertilizer N	5000	1.8	90

of roots vary with the composition and management of the sward (Table 5.4); and soil N has been reported to increase to a greater extent under grass–clover than under all-grass swards (Clement and Williams, 1967; Mullaly *et al.*, 1967; Simpson and Freney, 1967).

In addition to the input of soil N from the decomposition of roots, there may also be a small transfer of N from plant to soil through the exudation of amino acids from the roots, as shown to occur with wheat (Janzen, 1990).

Inputs of Soil N from the Excreta of Grazing Cattle

Large amounts of organic matter and N are transferred to the soil in excreta when swards are grazed. If grazing animals consume 10,000 kg dry organic matter ha^{-1} $year^{-1}$, with an average digestibility of 65%, the amount of organic matter returned in the dung is 3500 kg ha^{-1}. If the total amount of N consumed is 250 kg ha^{-1} and if 80% is excreted (see p. 71), then the return of N in dung and urine amounts to 200 kg N ha^{-1} (though some may be deposited in milking parlours, etc.). The actual amounts of N added to the soil in excreta in any situation will be influenced by factors such as stocking rate, the composition of the herbage and whether additional feed is provided. Usually more N is excreted in urine than in dung, and the proportion in the urine increases with increasing concentration of dietary N (see p. 71).

Assuming, for dairy cattle, a grazing intensity of 700 cow days ha^{-1} $year^{-1}$, a urination frequency of 10 times per day with 85% in the field being grazed, an average urination volume of 3 l and a concentration of 9 g N l^{-1} (see Chapter 4), then the return of N in urine would amount to 160 kg N ha^{-1} $year^{-1}$. With dairy cattle grazing high-yielding swards of herbage containing > 3% N, the amount of N returned in urine may be as high as 240 kg N ha^{-1} $year^{-1}$ (O'Connor, 1974; van der Meer, 1983). The ground area covered by a single urination is influenced by the volume of urine, and also by the infiltration rate of the soil and therefore by soil texture, porosity and moisture status. The area is greater with clay soils than with sandy soils, and is greater under wet than dry conditions. Grass growth is affected over a larger area than that actually covered by the urine, because some N can be taken up by grass roots growing laterally in the soil. In general with cattle, the area covered per urination is 0.2–0.5 m^2 (Haynes and Williams, 1993) and the area affected is about 0.5–0.7 m^2 (Lantinga *et al.*, 1987; Richards and Wolton, 1976b). Affected areas of 0.9–1.2 m^2 have been reported for cows grazing tall fescue in the USA (Lotero *et al.*, 1966). Assuming the urination volume, frequency and concentration to be as above, and the average area covered to be 0.4 m^2 with no overlapping, it can be calculated that urine would be deposited on 24% of the grazed area each year, supplying the equivalent of 670 kg N ha^{-1} in a single addition. If the concentration of N were 15 g l^{-1}, the rate of addition would be equivalent to

1110 kg N ha^{-1}. However, the N added in a urine patch is not distributed uniformly: the centre receives more, and the periphery less, per unit area than the average amount. Also, with areas that are grazed intensively, there may be appreciable overlapping of urine patches during the course of a grazing season (Afzal and Adams, 1992; Petersen *et al.*, 1956; Richards and Wolton, 1976b).

In general, most of the urea in urine is hydrolysed to ammonium within a few days (Holland and During, 1977; Stillwell and Woodmansee, 1981; Thomas *et al.*, 1988) and much of the ammonium is nitrified to nitrate within a few weeks. However, both reactions occur more slowly in cold conditions (see Chapter 6). Some of the inorganic N (ammonium and nitrate) is assimilated by plants and soil microorganisms, and thus contributes eventually to the soil organic N, but some of the ammonium is volatilized as ammonia, and some of the nitrate is leached and/or denitrified. A visible effect of urinary N on grass growth is often present for about 3 months (Richards and Wolton, 1976b). Although some urine N is quickly immobilized by the soil microorganisms (Holland and During, 1977; Whitehead and Bristow, 1990), much of the apparent immobilization detected by ^{15}N studies may be due partly to the turnover of the microbial biomass (see p. 120) (Ball *et al.*, 1979; Carran *et al.*, 1982) and may not necessarily result in a net gain in soil organic N.

With the N returned in dung, it can be calculated that, if the grazing intensity is equivalent to 700 cow days ha^{-1} year^{-1}, and if there is an average of 12 faecal excretions per cow per day with 85% in the field being grazed and each covering an area of 0.07 m^2 (Haynes and Williams, 1993; Lantinga *et al.*, 1987; Marsh and Campling, 1970; Wolton, 1979), then about 5% of the grazed area would be covered by dung each year. Where dairy cows graze intensively, as much as 8–10% of the area may be covered by dung (Bastiman and van Dijk, 1975). If 10,000 kg of herbage DM ha^{-1} containing 300 kg N is consumed, the return of N in dung is likely to amount to between 50 and 80 kg N ha^{-1} year^{-1}. Assuming that each deposition of dung amounts to 0.3 kg DM, that it covers 0.07 m^2, and that the concentration of N is 2.8%, then the return of N on the area actually covered would be 1200 kg ha^{-1}. Other estimates of the rate of N added to the area actually covered by dung include 750 kg N ha^{-1} (Holmes, 1968) and 1330 kg N ha^{-1} (O'Connor, 1974). In contrast to the N in urine, most of the N in dung is not readily mineralized to plant-available forms. The apparent recovery of dung N by ryegrass receiving fertilizer N at 250 kg ha^{-1} year^{-1} was 8.3%, but recovery was nil when the rate of fertilizer N was 400 kg N ha^{-1} (Deenen and Middelkoop, 1992). The C : N ratio of cattle (and sheep) dung is typically about 23 : 1 (Kirchmann, 1991) suggesting that mineralization of the organic N will not occur rapidly.

The N in dung affects the growth of grass in an area greater than that actually covered: the difference may be small in relatively dry areas such as North Carolina (Petersen *et al.*, 1956) but much greater (up to six times the area covered) in wetter areas such as southwest Scotland (MacLusky, 1960). The disappearance of dung from the soil surface depends partly on physical factors such as rain and frost, and partly on the activity of earthworms (Holter, 1983) or, in drier areas, of dung beetles. When the weather is dry after the deposition of dung, a hard crust forms (Weeda, 1967; Underhay and Dickinson, 1978) and this protects the dung pat from subsequent wetting and erosion by rainfall, and from the dispersing effect of birds (such as crows) seeking beetle grubs. In temperate climates, dung often decomposes less rapidly in summer, when a crust forms, than in winter (Weeda, 1967; MacDiarmid and Watkin, 1972b; Rowarth *et al.*, 1985), though it disappears most rapidly in warm moist conditions. In New Zealand, fresh sheep dung had completely degraded within 17 days in winter but remained visible for more than 100 days in summer (Rowarth *et al.*, 1985). However, with intensive grazing by dairy cows in the wetter climate of northwest England, the disappearance of dung pats required an average of about 60 days in spring, and about 40 days in summer, though there was considerable year-to-year variation (Bastiman and van Dijk, 1975).

Inputs of Soil N from the Excreta of Grazing Sheep

If sheep consume herbage similar, in terms of kg DM ha^{-1} $year^{-1}$, to the amount consumed by cattle, the rate of N deposited per unit area of a urine patch is lower, and the proportion of the total area covered is greater, than with cattle (Scott, 1977). With grassland intensively grazed by sheep (e.g. 14 ewes and 28 lambs per hectare grazed for 240 days) and assuming that the urination frequency of the ewes is 20 times per day, that the average urination volume is 150 ml, that the average concentration of N is 9 g l^{-1}, and that the average urine production from the lambs is half that from the ewes, then the return of N in urine would be equivalent to 180 kg N ha^{-1} $year^{-1}$. The area covered by a sheep urination is typically about 0.03–0.05 m^2 (Doak, 1952; Haynes and Williams, 1993; Morton and Baird, 1990) but the area affected is greater, with areas of 0.064 m^2 (Doak, 1952) and 0.09 m^2 (Thomas *et al.*, 1990) having been reported. With the return of 180 kg N ha^{-1} $year^{-1}$ and an area covered per urination of 0.04 m^2, 54% of the sward surface would receive urine in 1 year and at a rate equivalent to 330 kg N ha^{-1}.

The return of N in dung from sheep grazing three grass–clover swards, and a grass sward receiving 420 kg N ha^{-1} $year^{-1}$, during a grazing season of 22 weeks was estimated to range from 45 to 75 kg N ha^{-1} (Parsons *et al.*, 1991a). It is difficult to estimate the area covered, because sheep dung is often in the form of pellets which are easily dispersed. One report suggests an average area per patch of about 0.018 m^2 (Morton and Baird, 1990) but another that the average area affected per sheep per day was 0.05–0.07 m^2 (Frame, 1971). On the basis of the latter estimate, less than 5% of the sward surface would be affected, even with intensive grazing.

In many sheep grazing systems, and particularly in relatively hot and dry areas, much of the excreted N is deposited in 'camping' areas, to which the animals are attracted by a supply of water or by shade, and where they tend to congregate at night. This effect is widespread in New Zealand (Haynes and Williams, 1993) and in Australia where, in one instance, about half the total dung was deposited on 5% of the total grazed area (Hilder, 1966).

Influence on Soil N of Excreta Returned during Grazing

As the intensity of grazing is increased, there is increased consumption of the herbage available which results in a larger proportion of the herbage N being returned to the soil via animal excreta, and a smaller proportion being returned via death and decomposition. The excreta of grazing livestock are returned to the soil in numerous discrete patches which, as noted above, may amount per year to about 25–35% of the total area for cattle, and up to 60%, though often much less, for sheep.

The N in urine promotes the growth of grass when the supply of N is otherwise less than optimum, but it may reduce growth when the supply of N is excessive and/or when urine scorch occurs. An increase in grass growth, together with an increase in the concentration of N in dead material, will tend to increase the proportion of the urinary N that accumulates ultimately in the soil. Such increases are proportionately greatest in the absence of fertilizer N and have been demonstrated, for example, in prairie grassland (Jaramillo and Detling, 1992). Conversely, when simulated urine was applied to perennial ryegrass on a sandy soil receiving 400 kg fertilizer N ha^{-1} $year^{-1}$, reduced growth and urine scorch were observed, and the effect increased with increasing N concentration of the urine (Middelkoop and Deenen, 1990). When growth is depressed, there will be a reduced amount of plant material to contribute organic N to the soil, and the concentration of N in the plant will be high, thus increasing mineralization. These effects are probably involved when the total soil N content declines under intensive grassland management (see p. 85).

Most of the N in dung is in organic forms that are relatively resistant to decomposition, and often only about 25% is mineralized in the first year after deposition (Wolton, 1979). The remainder accumulates in the soil, at least temporarily. The rate at which dung is incorporated into the soil depends largely on the rate of physical disintegration and therefore on weather conditions (Haynes and Williams, 1993). In a study of the decomposition of cattle dung placed on grassland under natural rainfall conditions in northeast England, the concentration of N showed little change over a period of 85 days, indicating that mineralization of N occurred at the same rate as that of dry matter (Dickinson and Craig, 1990). However, when simulated rainfall was applied to double the normal rainfall, there was a preferential loss of N, particularly after the first 50 days of decomposition. At the end of 85 days, the dung exposed to natural rainfall retained 36% of its original N, dung receiving additional rainfall retained 23%, and dung protected from rainfall retained 67%.

Despite the large amounts of N returned in excreta, the impact on the accumulation of soil organic N is relatively small. For example, the increase in soil N under grass–clover swards over a 3-year period was about 160 kg N ha^{-1} $year^{-1}$ when the swards were grazed by sheep and 110 kg N ha^{-1} when the herbage was cut and removed (Clement and Williams, 1967). In other experiments, the return of excreta has been reported to have no significant effect on the accumulation of soil N, when compared with grazing in which the return of excreta was prevented (Metson and Hurst, 1953) or when compared with cutting (Watson and Lapins, 1964). However, in a 4-year study in the Netherlands, the accumulation of N in two soils (sampled to include roots and stubble) was about 150 kg N ha^{-1} $year^{-1}$ more under grazing than under cutting (Hassink and Neeteson, 1991). Excreta often have relatively little effect on soil N because the N, particularly that in urine, is susceptible to the volatilization of ammonia, to leaching and to denitrification, as well as to uptake by grass. Also, with grass–clover swards, excretal N tends to reduce N_2 fixation by clover. In the absence of both fertilizer N and clover, the return of excreta does promote grass growth, as shown for example by a 79% increase in production under grazing by dairy cows compared with cutting (Ledgard, 1991), and the increase in growth tends to increase soil N. An increase in the concentration of N in dead herbage material, from an average of 0.56% to 1.07% due to simulated urine applied more than one year previously, was reported for prairie grassland (Jaramillo and Detling, 1992).

In addition to the effects of excreta, the presence of grazing animals tends to increase the amount of herbage material undergoing decomposition in the sward and hence the accumulation of soil N. In a study in New Zealand, the amount of dead herbage residue was twice as high under set stocking as under rotational grazing, and the intensity of grazing was found to influence the content of soil N (Steele and Brock, 1985). Intensive rotational grazing to 500 kg residual dry matter ha^{-1} was found to result in a loss of 30 kg soil N ha^{-1} $year^{-1}$, whereas grazing to 1200 kg residual dry matter ha^{-1} was associated with a gain of 130 kg soil N ha^{-1} $year^{-1}$.

Inputs of Soil N from Animal Excreta Returned as Slurry

Slurry applied to grassland tends to increase the amount of organic N in the soil. Cattle slurry, at various rates supplying between 310 and 1250 kg N ha^{-1} $year^{-1}$ at intervals during each of four successive winters at two sites, increased soil N by between 90 and 570 kg N ha^{-1} year^{-1}, representing between 18 and 30% of the N applied in the slurry during the 4-year period (Gostick, 1981). Pig slurry applied during three winters at two other sites, at rates amounting to between 830 and 4700 kg N ha^{-1} year^{-1}, resulted in the accumulation of between 120 and 230 kg N in the soil, representing from 4 to 15% of the N applied in the slurry (Gostick, 1981). In another investigation, the application of cow slurry at a rate of 200 m^3 ha^{-1} (about 540 kg N ha^{-1}) year^{-1} to plots of perennial ryegrass for 17 years increased total soil N to a depth of 15 cm from 0.37% in the control plots to 0.45%, representing an annual increase of about 90 kg N ha^{-1} year^{-1} (17% of that applied); pig slurry had a similar effect (Christie and Beattie, 1989).

Slurry from housed poultry is sometimes applied to grassland. Much of the N in poultry excreta (40–70%) occurs as uric acid (Kirchmann and Witter, 1992) which is hydrolysed by microbial enzymes, via allantoin and allantoic acid, to urea and thence to ammoniacal N. Complete hydrolysis of the uric acid in poultry manure occurred during 10 days' incubation with a silt-loam soil at 25°C (Kirchmann, 1991) and the effects of uric acid are therefore likely to be similar to those of the urea in the urine of cattle and sheep. The accumulation of slurry N in the soil will occur mainly from the fraction that is insoluble in water and is not readily mineralized.

Impact of Fertilizer N, Liming and Drainage on the Accumulation of Soil N

The application of fertilizer N often has little effect on the rate of accumulation of soil N under grass and grass–clover swards (Clement and Williams, 1967; Hassink and Neeteson, 1991; Hoogerkamp, 1973; Wolton, 1955) though it has been found to promote accumulation on soils extremely low in organic matter (Sears *et al.*, 1965; Theron, 1965). The lack of effect of fertilizer N on most soils is consistent with the effects on plants described in Chapter 2. Thus, although fertilizer N generally increases the amount of herbage (though not of roots), and of livestock excreta if the sward is grazed, it also increases the concentration of N in these materials and therefore promotes mineralization rather than accumulation. In some situations, fertilizer N may also stimulate the soil microorganisms to mineralize organic matter already in the soil, thus tending to offset any increase in accumulation (see Chapter 6). A positive effect of fertilizer N on the accumulation of soil N is most likely on poorly-drained clay soils in which the mineralization of dead plant material is curtailed by poor aeration.

Changes in soil pH and in drainage status may also influence the equilibrium between accumulation and mineralization. For example, liming (Tyson *et al.*, 1990) and the installation of improved drainage (Scholefield *et al.*, 1988) both tend to increase the rate of mineralization and curtail the accumulation of soil N. They may also cause some redistribution of organic matter and N to greater depths within the soil profile, for example by encouraging a greater rooting depth and increasing the earthworm population.

Influence of Soil Fauna on the Incorporation and Decomposition of Organic Matter

The decomposition of plant and animal residues is brought about mainly by microbial enzymes but soil fauna, particularly earthworms, are often important in the early stages. Of the dead plant material and dung consumed by earthworms, some is converted into earthworm biomass and some is transferred, either before or after consumption, from the surface to below the surface of the soil. This transfer makes the material more accessible to the soil microorganisms, and the fraction that is consumed by the earthworms may also be partially digested. Most of the N in the plant material consumed is returned to the soil in casts with little chemical change, but some is metabolized and excreted in the urine, mainly as ammonia and urea. In addition, some of the N consumed is secreted by the worm as a mucoprotein, which acts as lubrication between the worm and the soil (Lee, 1985). Several studies have shown that the N content of earthworm casts is higher than that of the bulk of the surface soil, and that a higher proportion of the total N is in inorganic form (see p. 110). In a study of a permanent pasture in New Zealand, more than 600 kg DM ha^{-1} $year^{-1}$ of dead leaf litter was consumed by earthworms, and more than 70% of the N consumed was present in the worm casts (Syers *et al.*, 1979).

Grassland soils often contain large numbers of earthworms, usually many more than comparable arable soils, though reported numbers vary widely. For example, a survey of published data for grasslands indicated a range from 22,000 to more than 2,000,000 ha^{-1} (Lee, 1985). The usual range for grassland in the UK, Ireland, New Zealand and Australia is 250,000 to 750,000 ha^{-1} (Cotton and Curry, 1980; Edwards and Lofty, 1977; Lee, 1985; Tyson *et al.*, 1990) with more than 1,000,000 ha^{-1} having been reported in old grassland on a clay loam soil in the UK (Low, 1972) and in some clover-rich pastures in New Zealand (Lee, 1985). The usual range in the fresh weight of earthworms is 500–3000 kg ha^{-1} (Cotton and Curry, 1980; Edwards and Lofty, 1977; Lee, 1985). Dry weight as a proportion of fresh weight appears to vary but proportions between 13% and 30% have been reported (Lee, 1985; Sabine, 1983). An example of the difference between long-term grassland and continuous cereal plots in the numbers and weight of earthworms is shown in Table 5.5. The intensification of grassland management tends to increase the earthworm

Table 5.5. Numbers and fresh weight of earthworms in long-term grassland (28 years after sowing) and continuous arable plots (> 50 years) sampled in spring. (From Tyson *et al.*, 1990.)

	Numbers (ha^{-1})	Weight (kg ha^{-1})
Long-term grassland	769,000	390
Long-term arable plots	175,000	30

population by making more plant residues and dung available for decomposition (Anderson, 1988). Moderate applications of slurry also tend to increase the numbers and weight of earthworms, though large applications may cause a temporary decrease (Curry, 1976). The weight of earthworm biomass is increased by the drainage of poorly drained grassland soils, but is reduced when swards are ploughed and re-seeded (Scholefield *et al.*, 1988). In highly productive pastures, the weight of earthworms in the soil is often approximately equal to the weight of livestock carried on the surface (Haynes and Williams, 1993).

The concentration of N in earthworm tissue is about 10.5% on a dry weight basis (Edwards and Lofty, 1977; Lee, 1985; Sabine, 1983) and, with the dry weight of earthworms in grassland usually ranging from 50 to 500 kg ha^{-1} (Curry, 1976; Edwards and Lofty, 1977; Lee, 1985), the range in the amount of N in earthworm biomass is about 5–50 kg N ha^{-1}. The average length of life of earthworms in grassland in uncertain, and estimates of the rate of turnover of the earthworm population range from once to seven times per year (Anderson, 1988; Lee, 1985; Satchell, 1967). A turnover of twice per year would result in 10–100 kg N ha^{-1} of readily decomposable N becoming available each year, in addition to that in the casts. When the earthworm population is increasing, as in a newly established grassland, the population will tend to immobilize N. However, when a steady state population is achieved, the presence of earthworms will increase the rate of circulation of N and thus add to the amount of plant-available N in the soil at any time (Syers and Springett, 1984). The amount of N released annually into the soil, mainly as dead tissue protein and as excreted mucus, urea, uric acid and ammonia has been estimated to be between 18 and 92 kg ha^{-1} depending on the size and activity of the population (Syers and Springett, 1984).

In some parts of the world, dung beetles are important in the cycling of N in grazed grasslands. In a study in Japan, the mineralization of organic N in cow dung was found to proceed in two stages. The easily decomposable organic N was mineralized rapidly during the first five days, irrespective of dung beetle activity but subsequent mineralization was promoted by dung beetles (Yokoyama *et al.*, 1991).

Some soil microfauna (e.g. some nematodes and protozoa) feed on the fungi and bacteria and thus increase the rates of decomposition and N mineralization. They may also distribute nutrients more uniformly through the soil, thus counteracting the localization of nutrients caused by urine and dung patches. For example, it has been estimated that nematodes that feed on bacteria move about 1 cm day^{-1} and microarthropods about 5 cm day^{-1} (Anderson *et al.*, 1981).

Fractionation of the Soil Organic N

Soil organic matter includes stable humus material, plant and animal residues in various stages of decomposition and the microbial biomass. Macrofauna such as earthworms are usually excluded. However, soil organic matter determined by analysis often includes some living root material, simply because it is impossible to separate roots completely from the soil. In some grassland studies, all plant roots have been deliberately included in the soil organic matter whereas, in others, the roots have been separated as completely as possible.

In addition to the complication caused by living roots, there are difficulties in fractionating the actual soil organic matter. A major problem is that, in terms of its composition, soil organic matter is a continuum ranging from completely undecomposed fragments of plant material to colloidal humus, with at least part of the microbial population being distributed across this continuum. All schemes for fractionation therefore involve rather arbitrary boundaries. The scheme based on fulvic acid (material soluble in dilute alkali and not precipitated by dilute acid), humic acid (material soluble in dilute alkali and precipitated by dilute acid) and humin (material insoluble in dilute alkali) is widely used, though the three fractions are not well defined chemically.

In chemical terms, about 30–40% of the soil organic N is generally present in the form of amino acids that are released by hydrolysis with boiling 6M hydrochloric acid, and a further 8–10% is present in the form of amino sugars. The relative amounts of individual amino acids and amino sugars indicate that a large proportion of the soil N is derived from microbial material (Kowalenko, 1978). The proportion of the total soil N present as amino N is somewhat higher for grassland soils than for comparable soils under arable cultivation (Whitehead *et al.*, 1975). Some of the remaining N is released as ammoniacal N by acid hydrolysis but about half the total is chemically unidentified. Much of the unidentified N is thought to occur in complex humus polymers resulting from the reaction of amino compounds or ammonia, derived from plant and microbial proteins, with phenolic or quinone-type compounds derived from the degradation of lignins (Fig. 5.3). In addition, some humus polymers may be derived from the cell wall material of the soil microorganisms.

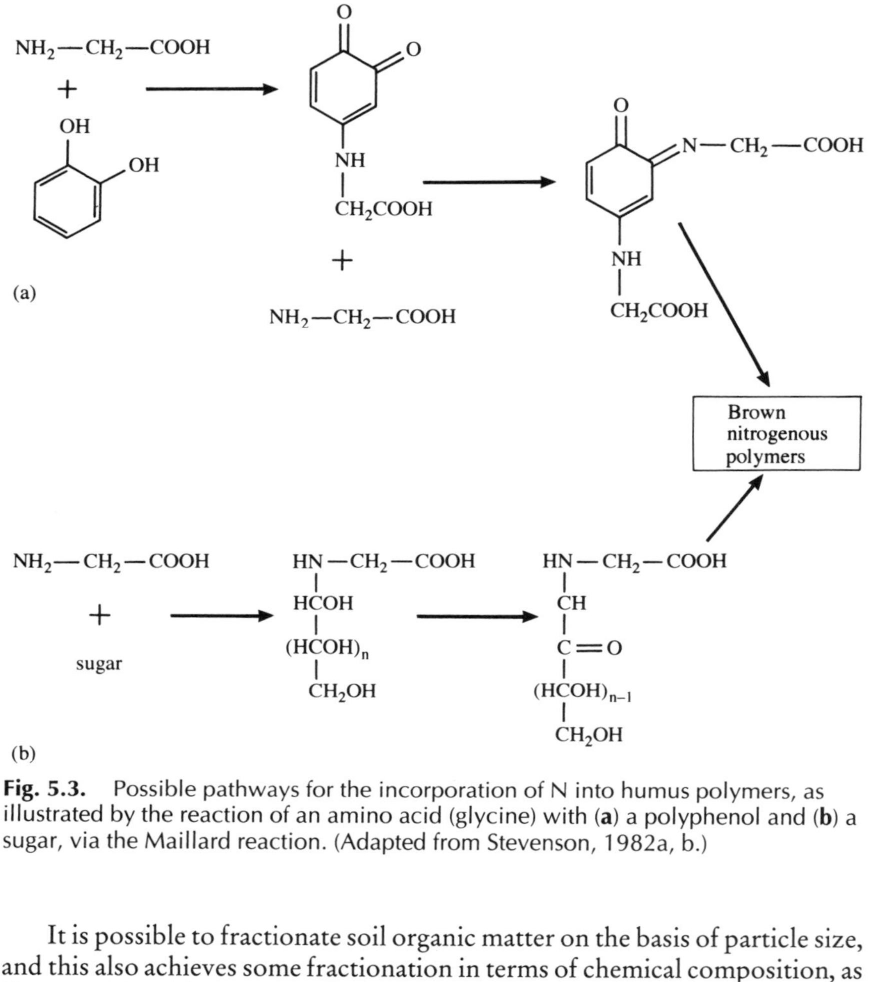

NH_2—CH_2—COOH + OH OH → (o-quinone with NH—CH_2COOH) + NH_2—CH_2—COOH → (quinone imine with N—CH_2—COOH and NH—CH_2COOH) → Brown nitrogenous polymers

(a)

NH_2—CH_2—COOH + sugar → HN—CH_2—COOH | HCOH | $(HCOH)_n$ | CH_2OH → HN—CH_2—COOH | CH | C=O | $(HCOH)_{n-1}$ | CH_2OH → Brown nitrogenous polymers

(b)

Fig. 5.3. Possible pathways for the incorporation of N into humus polymers, as illustrated by the reaction of an amino acid (glycine) with (**a**) a polyphenol and (**b**) a sugar, via the Maillard reaction. (Adapted from Stevenson, 1982a, b.)

It is possible to fractionate soil organic matter on the basis of particle size, and this also achieves some fractionation in terms of chemical composition, as illustrated by differences in C : N ratio. Thus with organic matter separated from soils under leys, the C : N ratio of the fraction > 0.2 mm was about 18 : 1, whereas the C : N ratio of the fraction < 0.002 mm was 7 : 1; and these differences were reflected in the extent of mineralization during incubation (Cameron and Posner, 1979). Various hypothetical fractionation schemes for soil organic matter, based on differences in the rate of turnover of the fractions, have also been proposed for modelling purposes but it is often difficult to relate these schemes to analytical data. One scheme involves five fractions of differing biological stability: (i) fresh residues from plants and animal excreta, (ii) residual material, e.g. lignin from previous additions, (iii) the soil microbial biomass, (iv) material sorbed to the soil colloids and therefore physically protected and (v) humified and therefore chemically resistant soil organic matter (Jenkinson and Rayner, 1977). Despite the impossibility of achieving a completely clear-cut and yet meaningful fractionation, two fractions that can be defined fairly precisely, and that can be measured, are the macroorganic matter fraction (corresponding approximately to (i) plus (ii) above) and the microbial biomass.

The Macroorganic Matter Fraction of Soil N

The macroorganic matter fraction in grassland soils is composed mainly of dead plant material in the early stages of decomposition, with a small contribution from dung. It is defined as the material separated by flotation with water and retained on a mesh, usually of 0.20 or 0.25 mm diameter (Barley, 1955; Warren and Whitehead, 1988). In some studies, plant roots have been deliberately included in the macroorganic matter (e.g. Adams, 1980; Garwood *et al.*, 1972) whereas in others, living roots have been separated as completely as possible before separation of the macroorganic matter fraction (e.g. Goodman, 1988: Warren and Whitehead, 1988; Whitehead *et al.*, 1990). Macroorganic matter generally constitutes a greater proportion of the total organic matter in grassland soils than in arable soils. The C : N ratio of the macroorganic matter under grass, when living roots are excluded, is generally between 16 : 1 and 27 : 1 with a mean of about 21 : 1 (Warren and Whitehead, 1988; Whitehead *et al.*, 1990). It is thus intermediate in N concentration (1.8–3.0%) and C : N ratio between living grass roots (see Table 5.3) and humified organic matter which normally has a C : N ratio of between 9 : 1 and 12 : 1. In UK grassland soils, the amount of N in the macroorganic matter, excluding living roots, is generally in the range of 100 to 400 kg N ha^{-1} and tends to increase with the age of sward (Whitehead *et al.*, 1990). The N in macroorganic matter is relatively labile compared with that in the humified organic matter.

The Microbial Biomass Fraction of Soil N

The microbial component of the soil includes bacteria, actinomycetes, fungi, algae, protozoa and microfauna. In many soils, the soil microbial biomass accounts for between 1% and 3% of the total organic matter, but the proportion is smaller in peaty soils (Sparling, 1985). However, the biomass makes a greater contribution to soil fertility than these proportions suggest. It has two important roles, first as the source of enzymes responsible for many biochemical processes in soils and, second, as a reservoir of potentially plant-available N, P and S. Usually, less than 25% of the biomass appears to be active at any one time, the main restriction being a lack of readily available organic material as a substrate for energy supply. However, the biomass in a particular microsite of the soil is activated when conditions are favourable. In general, there is much more microbial biomass in grassland than in comparable cultivated soils (Adams and Laughlin, 1981; Clark and Paul, 1970), due partly to the grassland soils containing more total organic matter (Brookes *et al*., 1985). The microbial biomass was found to be slightly greater in grazed than in mown plots of long-term grassland, but the rate of fertilizer N (up to 700 kg N ha^{-1} $year^{-1}$) had no consistent effect (Hassink, 1992a).

The average concentration of N in microbial cells is usually 5–10% on a dry matter basis (higher in bacteria than in fungi) and is higher than the concentration in the bulk of the soil organic matter. The biomass therefore contains a larger proportion of the soil organic N than it does of the total organic matter, often accounting for between 2 and 6% of the total soil N (Bristow and Jarvis, 1991; Brookes *et al*., 1985) though only 1–2% in ungrazed prairie grassland (Coupland and van Dyne, 1979). Assuming an intensively managed long-term grass–clover sward to have a total soil N of 12,000 kg ha^{-1}, and the biomass to account for 4% of the total N, the biomass N would amount to 480 kg N ha^{-1}.

If the supply of available C is not limiting, the soil microorganisms compete with plants for the inorganic N present in the soil and convert it into microbial tissue. This immobilized N (see p. 120) is unavailable to plants until there is a net decline in the microbial biomass when some microbial N is mineralized. Microbial N exists in two main forms, first as cytoplasmic protein which decomposes readily, and second as cell-wall components which are more resistant to decomposition. The cell walls of fungi contain chitin (a polymer of *N*-acetylglucosamine) in association with other polysaccharides and protein, while the cell walls of bacteria contain *N*-acetylglucosamine and *N*-acetylmuramic acid, in the form of chains that are linked to one another by amino acids (Parsons and Tinsley, 1975).

The soil microbial biomass is in a constant state of turnover and the cytoplasm of dead microbial cells is readily mineralized by the remaining microorganisms. The rate of turnover is increased by the activities of protozoa and nematodes which consume bacteria and fungi (Anderson *et al*., 1985; Clarholm, 1985). Field studies using ^{15}N have demonstrated that the microbial biomass can both immobilize and mineralize N and that some immobilized N may be re-mineralized rapidly (Jansson and Persson, 1982). Rapid changes in soil moisture and temperature, particularly freezing and thawing, may cause the death of a large proportion of the microbial biomass followed by a flush in the mineralization of N.

Agricultural practices which influence the amount of organic matter in soils also influence the amount of microbial biomass. Increases in microbial biomass occur as a result of the repeated application of slurry (Adams and Laughlin, 1981; Christie and Beattie, 1989) and decreases when grassland is converted to arable cultivation (Adams and Laughlin, 1981).

Differences in Contents of Organic Matter and N between Grassland and Arable Soils

A soil under long-term grassland differs in several respects from a similar soil that has been subject to many years of arable cultivation. A major difference is that the amount of organic matter (and hence of total N) is greater in the grassland soil, the difference being greatest in the top 5 cm of the soil. An example for adjacent plots established on a previously arable soil and then maintained under arable cultivation or under a grazed grass–clover sward for 20 years is shown in Fig. 5.4. Much of the difference is in roots and macro-organic matter. The higher content of organic matter in grassland is associated with a larger population of earthworms (see p. 99), more microbial biomass (see p. 104) and greater activity of various soil enzymes such as urease (see p. 153). Despite the high concentrations of total N and the large microbial biomass, the concentration of inorganic N is often lower in a grassland soil than in a comparable arable soil, due to the large capacity of grass to take up inorganic N.

Another difference between grassland and arable soils, at least with grazed grassland, is that the grassland soils show a much greater spatial variability in the concentrations of plant-available nutrients, including ammonium- and nitrate-N (Afzal and Adams, 1992) and K. This spatial variability results mainly from the return of excreta in numerous small patches. In the absence of fertilizer N (or with only small applications), the large amount of potentially plant-available N in urine results in increased grass growth and increased concentrations of N in both shoot and root material (Day and Detling, 1990; Thomas *et al.*, 1986; Whitehead and Bristow, 1990). With increasing rate of fertilizer N over the range of about 250–400 kg N ha^{-1} $year^{-1}$, the effects of urine on grass growth and N uptake decline and may become negligible (Deenen and Middelkoop, 1992). Initially after the deposition of urine, the concentration of ammonium in the soil solution rises rapidly due to the hydrolysis of urea (see p. 153) but nitrification (see p. 114) during the subsequent 1–3 weeks results in the conversion of much of the ammonium to nitrate (Haynes and Williams, 1992; Whitehead and Bristow, 1990). These transformations cause changes in soil pH, an initial rise being followed by a decline to below the pH of unaffected soil (Haynes and Williams, 1992).

In general, long-term grassland soils have a greater total porosity, and a greater air-filled porosity at field capacity, than do arable soils (Garwood *et al.*, 1977; Haynes and Swift, 1990). These differences are due partly to the grass roots enlarging pores by the physical pressure of growth and by their removing water and periodically drying the soil. Porosity is also promoted by earthworms, which not only form pores but also stabilize them by secreting mucus and transferring organic matter from the soil surface. The greater porosity of grassland soils results in their returning to field capacity more quickly after

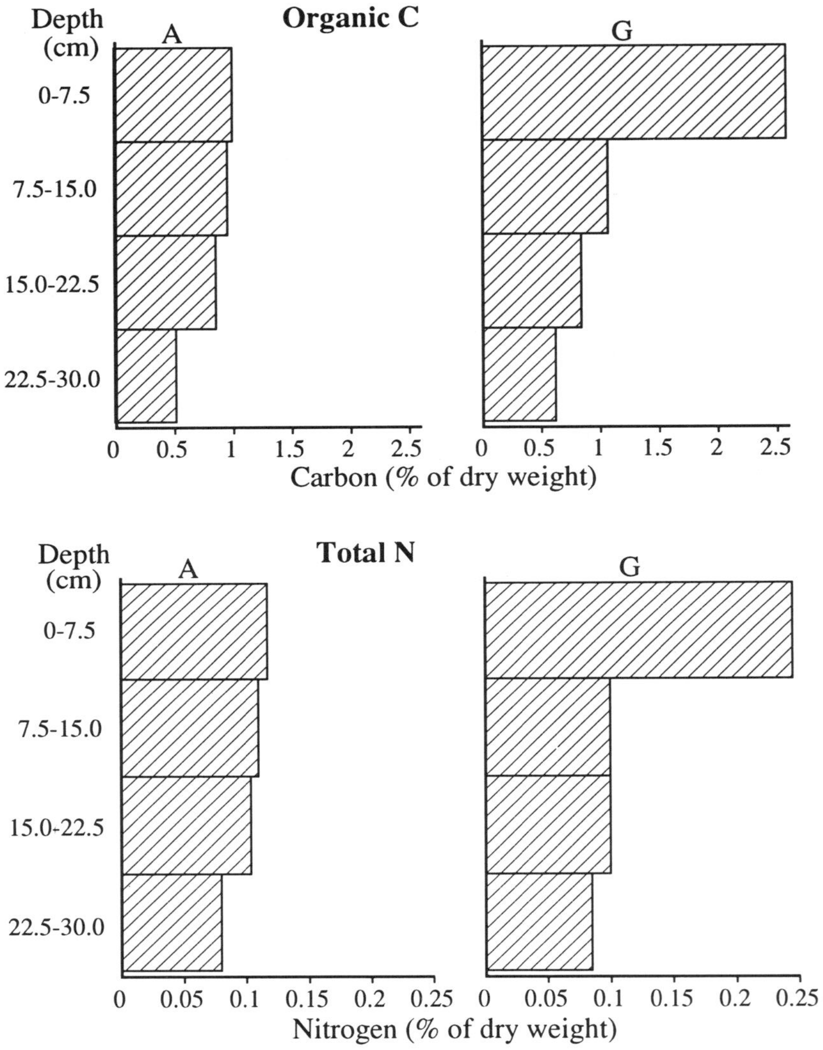

Fig. 5.4. Organic C and total N in soils from adjacent plots maintained for 20 years in (A) arable cultivation or (G) grazed grassland. (Data from Garwood *et al.*, 1977.)

heavy rainfall and consequently having less tendency to denitrify than do comparable arable soils (see p. 187). The accumulation of soil organic matter in grassland soils is also associated with an increase in aggregate stability (Clement, 1961; Haynes and Swift, 1990) and a decrease in bulk density (Watson, 1969).

The presence of a vegetation cover throughout the year, together with greater aggregate stability, ensures that grassland soils are relatively resistant to erosion by wind and water. And because the surface layer of the soil is at most risk from erosion and also has the highest concentration of N, improved resistance to erosion promotes the conservation of soil N. In temperate regions, the importance of grassland in the control of soil erosion is greatest in areas with a Mediterranean climate (Porqueddu *et al.*, 1994).

6 Mineralization, Immobilization and Availability of Nitrogen in Soils

Processes of Mineralization and Immobilization

The mineralization of N occurs when the organic forms in soil organic matter and in plant and animal residues are converted into inorganic or 'mineral' forms. Mineralization results almost entirely from microbial activity. It is essentially 'ammonification' (i.e. the conversion of organic N to ammonium N) but nitrification, the subsequent oxidation of ammonium to nitrate, may be considered as being part of the mineralization process. Ammonification is carried out by a wide range of heterotrophic microorganisms, most of which prefer aerobic conditions, and it therefore occurs most rapidly in well-aerated soils. Various protease and deaminase enzymes are produced and these hydrolyse specific substrates to produce ammonium-N. Some of the ammonium N is assimilated by the microorganisms, and some is released into the soil where it may be taken up by plants or may be nitrified. Nitrification is restricted to a smaller group of microorganisms and is carried out mainly, though not exclusively, by two groups of autotrophic bacteria for which aerobic conditions are essential.

Immobilization is the reverse of mineralization, i.e. the transformation of inorganic N into organic forms. One major route is through the assimilation of inorganic N by the soil microorganisms, and the term 'immobilization' is often used to refer solely to this process. However, the uptake by plants of inorganic N and the subsequent death and decay of plant tissue, also result in the conversion of inorganic to organic N and a tendency for organic N to accumulate in the soil. The distinction between mineralization/immobilization and the accumulation of soil organic N may be confusing, especially in grassland soils which often show net mineralization combined with an accumulation of N. However, this combination is possible provided that the total input of N (from fertilizer, biological fixation and/or deposition from the atmosphere) exceeds the removal of N in herbage or animal product. Although there may be a small contribution to immobilization from purely chemical reactions in the soil, in this chapter, the term immobilization refers to microbial immobilization unless the other processes are specifically mentioned.

Both mineralization and immobilization occur continuously in soils, and both processes are retarded by cold and by dry conditions. The balance between the two processes can vary widely from soil to soil, and from time to time in any one soil, depending mainly on the nature of the organic materials undergoing decomposition. The two processes are closely related through their dependence on the soil microbial population. An increase in the amount of inorganic N with time indicates that net mineralization is occurring, but a decrease in inorganic N does not necessarily indicate net immobilization. It may be due to plant uptake, leaching, denitrification and/or the volatilization of ammonia. Also, a constant amount of inorganic N does not necessarily imply a lack of microbial activity; it may be that vigorous mineralization is balanced by immobilization and/or by plant uptake and loss processes.

The decomposition of plant and animal residues in soil always results in some of the C and N being mineralized. During decomposition, some of the C is assimilated as small molecular compounds and immobilized by the microbial biomass, but some is inevitably mineralized to CO_2 due to microbial respiration. The net mineralization of C normally begins immediately after the addition of the organic material to the soil, and the CO_2 escapes to the atmosphere. However, the net mineralization of N often occurs more slowly, and there may be a lag period of months or even years before it is appreciable. The absence of net mineralization implies that all the inorganic N released from plant and animal residues is immobilized immediately by the microbial population. In contrast to the CO_2, most of the N that is mineralized remains in the soil–plant system, partly because both ammonium and nitrate are readily soluble in water, and partly because both are susceptible to uptake by plants and microorganisms. Although it is possible for some inorganic N to be lost through leaching, ammonia volatilization and/or denitrification, such losses are usually negligible from the low concentrations of ammonium and nitrate that occur naturally during the decomposition of plant residues. However, high concentrations of inorganic N may occur temporarily in the soil as a result of the mineralization of N in livestock excreta and in slurries containing urine, and losses of N may then be appreciable.

Despite the differences between CO_2 and the inorganic N released by mineralization, there is a close relationship between the C and N in soil organic matter due to the needs of the soil microorganisms. During the initial phase of decomposition there is an increase in microbial biomass, and both C and N are required for the synthesis of microbial cell constituents. Only the inorganic N that is surplus to this requirement is released as net mineralized N and, if the concentration of N in the decomposing organic material is low, there will be no surplus. The balance between mineralization and immobilization is therefore influenced to a large extent by the C : N ratio of the material undergoing decomposition. However, whatever the initial C : N ratio (provided it is wider than about 10 : 1), it becomes narrower during the decomposition process, largely due to the inevitable loss of CO_2. The soil microbial biomass that is synthesized as a result of decomposition usually has an average C : N ratio of between 5 : 1 and 8 : 1. Bacteria have a ratio of between 3 : 1 and 5 : 1, and fungi (particularly phycomycetes which have cellulose in their cell walls) have a wider ratio of up to 15 : 1 (Paul and Clark, 1989). For both bacteria and fungi, C constitutes about 45% of dry weight. Although both ammonium and nitrate can be immobilized by soil microorganisms, ammonium appears to be assimilated more rapidly when both forms are present (Davidson *et al.*, 1990; Jannson and Persson, 1982). Immobilization is dependent on the growth of the microbial population and therefore on the supply of available C, as well as the temperature and moisture content of the soil (Paul and Clark, 1989). After the initial phase of decomposition, the amount of microbial biomass tends to decline and this results in an increase in the net mineralization of N. In general, the rate of turnover of N from dead microbial cells is about five times that of the humified soil organic N (Stevenson, 1986). Soil fauna such as nematodes and protozoa tend to increase the mineralization of soil N by feeding on bacteria and fungi, utilizing some of the C and N for their own biomass and some of the C for respiration, but excreting some of the N in soluble forms (Clarholm, 1985; Woods *et al.*, 1982).

In addition to the microbial immobilization of N, there is the possibility that some immobilization may occur through the reaction of amino acids and/or ammonia with phenolic and quinone-type compounds derived from the degradation of lignins (see p. 101). Such reactions are thought to contribute to the formation of the more stable humified fraction of the soil organic matter.

Mineralization of N in Soils

Recently-added plant residues and animal excreta decompose quickly in comparison with humified soil organic matter. Also, despite often having a wide C : N ratio, they mineralize more N than their proportion of the total soil organic matter would suggest. In particular, the N in livestock urine, which has a low C : N ratio, mineralizes rapidly. As indicated in Chapter 5, some of the material added to the soil in plant residues and dung is consumed by earthworms or other soil fauna, and this process tends to accelerate the net mineralization of N. In comparison with the plant residues and dung being consumed, worm casts have a lower C : N ratio and a higher proportion of their total N as ammonium and nitrate (Parkin and Berry, 1994; Syers *et al.*,

1979). They also contain larger numbers of microorganisms (Knight *et al.*, 1989). The lower C : N ratio in worm casts than in litter is due mainly to the loss of C during respiration, and the relatively high proportion of inorganic N is due to the excretion of ammonium, urea and/or free amino acids (Edwards and Lofty, 1977). In an incubation study of leaf litter from grass and clover, added to soil at a rate equivalent to the maximum observed in New Zealand pastures, the presence of ten earthworms per kg soil increased the amount of inorganic N present after 11 weeks by about 50%; and this effect was reflected in the subsequent uptake of N by ryegrass grown in the soil after removal of the earthworms (Rus Jerez *et al.*, 1988). In a field experiment, in an area of New Zealand where the absence of earthworms had resulted in the development of a mat of peaty organic matter at the soil surface, the growth of grass was increased by more than 70% when earthworms were introduced, apparently due to their promoting the mineralization of nutrients present in the organic matter (Stockdill, 1982). Although this effect declined in subsequent years, there was a continued increase in grass yield attributable to the earthworms of about 25–30% (Syers and Springett, 1983), probably due partly to the improvement of soil physical conditions. In another investigation, on newly reclaimed Dutch polders, a 10% yield increase was observed with the introduction of earthworms (Hoogerkamp *et al.*, 1983).

Although earthworms and other soil fauna promote mineralization, the process is due mainly to microbial enzymes. When the decomposing material has a C : N ratio of about 25 : 1 to 30 : 1, the microbial demand for N during the first few months of decomposition is approximately equal to the amount released and there is little change in soil inorganic N. In some instances, there may be a small amount of net immobilization during the first few days or weeks, with mineralization occurring slowly after that time. When the C : N ratio is greater than 30 : l (equivalent to a concentration of N of < about 1.6% on an organic matter basis), net immobilization is likely to be more prolonged and to continue for several weeks or more. On the other hand, when the C : N ratio is less than 25 : l (equivalent to a concentration of N of > about 2.0%), net mineralization is likely to occur within a period of a few weeks. The lower the C : N ratio, the more rapid is the mineralization. Studies of the mineralization or immobilization of N during the incubation of grass roots in soil have shown results consistent with this general pattern (Dormaar and Willms, 1993; Power, 1968). Although the C : N ratio has a major influence on the balance between immobilization and mineralization, other aspects of the composition of the decomposing material are also significant. For example, highly lignified materials decompose slowly and release relatively little C for microbial growth, and they therefore tend to immobilize less N than would be expected on the basis of their C : N ratio (Fox *et al.*, 1990; Herman *et al.*, 1977). The ratio of readily available C to readily available N appears to be more important than the ratio of total C to total N, though the 'readily available' fractions are not easily determined by analysis.

The plant and animal residues that are present in grassland soils vary widely in their capacity to mineralize or immobilize N. When little or no fertilizer N has been applied, grass roots and leaf litter often have C : N ratios of between 40 : 1 and 60 : 1 (see Chapter 5) and their decomposition is therefore likely to result in some net immobilization. However, when fertilizer N has been applied at high rates (350–450 kg N ha^{-1} $year^{-1}$), the C : N ratio of grass roots is about 25–30 : l, and that of dead herbage is generally < 25 : 1, so that decomposition of the residues is likely to induce slow mineralization. The roots, stolons and herbage of white clover usually have C : N ratios in the range 13 : 1 to 20 : 1 (Table 5.3; Marstorp and Kirchman, 1991) and will therefore tend to mineralize N readily. In natural grasslands, differences between species in the C : N ratio of leaf litter and root material influence the amounts of N mineralized per year, with consequent effects on the competitive ability of the species (Wedin and Tilman, 1990).

The mineralization of N is slower from dung than from the corresponding plant material (Barrow 1961; Floate, 1970), a difference due partly to the more readily mineralizable fractions of both C and N being removed during the digestion process. Dung mineralizes most rapidly when it derived from highly digestible feed with a high concentration of N.

In the field, the mineralization of N from soil organic matter occurs partly from recently-added plant and animal residues, partly from more humified material and partly from the turnover of microbial biomass. The rate of mineralization is greater in sandy soils than in clay and loam soils. On non-sandy soils, the size of the microbial biomass was found to be related to the amount of N mineralized on incubation (Hassink, 1992a), and the difference between sandy soils and clay/loam soils appears to be related to differences in the activity and C : N ratio of the bacterial biomass (Hassink *et al.*, 1994). The overall rate of mineralization of soil N varies with season of the year, mainly in response to soil temperature and moisture status. When the moisture supply is adequate for the soil microorganisms, mineralization increases with increasing temperature over the range 5–30°C; and, when the temperature is above about 5°C, mineralization increases with soil moisture status between permanent wilting point and field capacity (Stanford and Epstein, 1974). At moisture contents higher than field capacity, there is a decline in the rate of mineralization because the growth of many microorganisms is restricted by poor aeration. In a comparison of grassland plots that differed in the depth between the soil surface and groundwater, annual mineralization was about 20% greater in 'dry' plots than in 'wet' plots (Berendse *et al.*, 1994). The peak rate of mineralization occurred in spring in the 'dry' plots and in summer in the 'wet' plots. When short wet and dry periods alternate, there is often an enhancement of mineralization, due partly to soil aggregates being disrupted during the drying phase (and making previously inaccessible organic matter more accessible to microbial enzymes) and partly to an increase in the turnover of the microbial population (Haynes,

1986b). When periods of freezing and thawing alternate, there is a similar effect (Campbell, 1978). The combined effects of temperature and soil moisture status are reflected in the seasonal differences in the rate of mineralization in the field. For example, in an investigation in Ireland, the mineralization of N in samples of grassland soil incubated at field temperature was greatest during the period from mid-April to the end of May, and there was a second but smaller peak from early August to the end of September (Herlihy, 1979). More mineralization occurred when the soil moisture content fluctuated than when it was maintained near field capacity. In a comparison (on both sandy and loam soils) of swards that had been either cut or grazed for at least two years previously, the rate of N mineralization in spring was 20–30% greater as a result of grazing (Hassink, 1992a).

As the mineralization of soil organic N is carried out by a wide range of microorganisms, it is not greatly influenced by natural soil pH unless conditions become strongly acid, when it occurs more slowly. At pH < 4.5, there is a decline in the population and activity of bacteria, sometimes accompanied by an increase in soil fungi. However, the liming of acid soils, even moderately acid soils, often causes an increase in the mineralization of N (Haynes, 1986b). For example, in New Zealand, liming a grass–clover sward on a soil of pH 5.6 at the rate of 7.25 t ha^{-1}, increased the amount of soil N taken up by the grass by 50 kg N ha^{-1} over 2 years (Edmeades *et al.*, 1986). An increase in the mineralization of soil N following the liming of grassland was also observed in a long-term investigation in southeast England (Tyson *et al.*, 1990). Incubation studies of 40 soils ranging in pH from 4.0 to 5.6 showed that liming to pH 6.7 approximately doubled the proportion of the soil N mineralized during 120 days (Nyborg and Hoyt, 1978). In contrast to the effect of liming, the contamination of soils by heavy metals (e.g. from sewage sludge) may reduce mineralization by causing a decline in the microbial population (Chang and Broadbent, 1982).

Measurement of N Mineralization in Soils

The capacity of soils to mineralize N is usually assessed by measuring the amount of NH_4-N + NO_3-N released during a period of incubation under uniform conditions. Published methods differ in incubation time, temperature, soil aeration and soil water status (see Keeney, 1982; Stanford, 1982) but ideally, in comparing a number of soils, the soil water tension and the oxygen concentration in the atmosphere around the soil should be uniform, and maintained constant during the incubation (Clement and Williams, 1962). Disturbance of the soil by sieving, drying and/or mixing may influence the rate of mineralization, and the effect is not always predictable. For example, sieving usually (but not always) increases the mineralization of N but to an extent dependent on soil type (Hassink, 1992b; Raison *et al.*, 1987). When the aim is to relate the measurement of net mineralization to other data from field experiments, it is possible to carry out a series of short-term incubations of soil cores in containers placed in the surface soil in the field (e.g. Hatch *et al.*, 1990b). An error may occur in incubation methods through the loss of NO_3-N due to denitrification (especially if conditions become anaerobic), but such a loss can be prevented by adding acetylene to inhibit nitrification and thus the formation of nitrate (Hatch *et al.*, 1990b; Warren, 1988). An alternative type of field incubation that causes less disturbance of the soil involves the insertion of tubes into the soil, rather than removing cores. If some, but not all, of the inserted tubes are covered to prevent leaching, and if there are tubes with and without plants, this method provides an estimate of plant uptake and leaching (though not of denitrification) in addition to mineralization (Raison *et al.*, 1987). Replication is important, particularly in grazed fields, because the balance between mineralization and immobilization may vary widely from point to point, even within distances of a few centimetres. At any one time, there may well be net mineralization at some micro-sites and net immobilization at others.

All incubation methods are time consuming and, if it were possible to relate mineralization to a specific fraction of soil N, a method of chemical analysis for this fraction would be valuable. However, because mineralization occurs from a wide range of components of the soil organic matter, and results from the action of a wide range of enzymes, it is unlikely that any single analysis would predict mineralization satisfactorily over a range of contrasting soils. Nevertheless, a number of chemical extractants have been examined, mainly in the context of predicting the supply of plant-available N in the field, as outlined on pp. 124–125.

Nitrification

Although both ammonium and nitrate ions are available for plant uptake, nitrification (the oxidation of ammonium to nitrate) is important because it influences the extent to which N is lost from the soil. Nitrate is much more mobile than ammonium, and is susceptible to losses through leaching and denitrification.

The main nitrifying organisms are two small groups of autotrophic bacteria. One group, which includes *Nitrosomonas*, oxidizes ammonium to nitrite:

$$NH_4^+ + 3[O] \rightarrow NO_2 + H_2O + 2H^+$$

The second group, which includes *Nitrobacter*, oxidizes nitrite to nitrate:

$$NO_2^- + [O] \rightarrow NO_3^-$$

Nitrite is usually oxidized rapidly and accumulates in the soil only in conditions that combine a high concentration of ammoniacal N with high pH. The biochemical pathway of the first stage of nitrification (the oxidation of ammonium to nitrite) has not been fully elucidated but both nitrous oxide (N_2O) and nitric oxide (NO) are produced as intermediates (Hutchinson and Davidson, 1993) and may volatilize from the soil (see p. 194).

Most nitrification in soils is thought to follow the pathway outlined above and, although some nitrification may be due to heterotrophic organisms, their contribution is thought to be negligible in most soils (Hutchinson and Davidson, 1993; Paul and Clark, 1989). In addition to these microbial pathways, there is also the possibility of the purely chemical oxidation of nitrite to nitrate (Haynes, 1986c; Schmidt, 1982).

The normal nitrification process occurs slowly at temperatures < 5°C, and the optimum temperature is about 30–35°C. Nitrification appears to be restricted to a greater extent than ammonification by low temperatures (Harrison *et al.*, 1994). It is inhibited by dry soil conditions, and the optimum soil moisture tension is normally between –0.1 and –1.5 MPa (Davidson *et al.*, 1990; Paul and Clark, 1989). However, nitrifying organisms can survive in inactive form when the soil moisture tension is < –9 MPa. Nitrification is curtailed when the pH is less than about 6.0, and becomes negligible below pH 4.5 (Paul and Clark, 1989). It is also inhibited by strongly alkaline conditions and, at pH > 8, the second stage of the process (nitrite to nitrate) is inhibited to a greater extent than the first stage, and so nitrite may accumulate. Low temperatures have a similar effect, allowing the accumulation of nitrite. Under waterlogged conditions, nitrification is inhibited by a lack of oxygen. When the water-table under grassland in the Netherlands was artificially raised, by an average of about 25 cm, the extent of nitrification (of the ammonium from mineralization) was reduced from 76% to 60% (Berendse *et al.*, 1994). The three factors, temperature, pH and moisture status probably account for much of the variation that occurs in the field (Schmidt, 1982; Stevens, 1988). It is likely that these factors, together with the supply of ammonium, acting over a period of years influence the population of nitrifying organisms and thus the short-term nitrification activity (Pang *et al.*, 1975).

Laboratory measurements of nitrification activity in a range of New Zealand grassland soils (mixed with sand and perfused with ammonium sulphate solution) showed a wide variation in the rate of nitrification, and differences in soil pH and organic matter explained only a small proportion of the variation (Steele *et al.*, 1980). In many of the soils, nitrification occurred slowly at the start of the perfusion and then increased exponentially until a steady rate was attained after about 4 days but, in other soils, there was a steady rate, either fast or slow, from the outset. Different patterns of nitrification were also observed in studies in the USA, and the differences between soils were related to soil pH and the numbers of nitrifying organisms (Morrill and Dawson, 1967). Studies on range grassland in California using ^{15}N as a tracer indicated that the rate of nitrification varied from 12% to 46% of the rate of gross mineralization during the growing season (Davidson *et al.*, 1990). Soils can be compared, in terms of their potential rate of nitrification at the time of sampling, by either incubating or perfusing them with an ammonium salt under standard conditions of temperature and moisture (Sarathchandra, 1978; Schmidt, 1982).

When a large amount of ammoniacal N is present in soil (e.g. following the addition of fertilizer, or livestock urine) there is often a lag period before the rate of nitrification responds, which may reflect the time taken for the population of nitrifying organisms to develop. For example, in studies with simulated urine in New Zealand, the hydrolysis of urea was complete in 3 days at all seasons but nitrification was not appreciable until after 7 days; then, at temperatures > 15°C, nitrification developed rapidly and was complete in 30 days while, at temperatures of 7.5–10°C, nitrification occurred slowly and required 60 days for completion (Holland and During, 1977). If the addition of ammonium is repeated, nitrification may occur without a lag period (Fleisher and Hagin, 1981). However, more information is needed on how the lag period in nitrification is influenced by soil characteristics and by the previous addition of ammoniacal N in fertilizers and manures.

There have been a number of reports suggesting that nitrification is inhibited in some grassland soils, particularly those under long-term or natural grassland, due to the release of organic compounds by the grasses. This view was based initially on the observation that the ratio of ammonium to nitrate was generally higher in such soils than in similar soils cultivated for arable crops. It was supported by some evidence that extracts from grass roots had an inhibitory effect on nitrification (Moore and Waid, 1971; Munro, 1966) though, in other investigations, no such effect was detected (e.g. Purchase, 1974; Ross and Cairns, 1980). A recent review of the evidence concluded that there was no justification for the view that nitrification is inhibited by organic compounds released from grasses (Bremner and McCarty, 1993). However, nitrification, especially the oxidation of nitrite, is inhibited by high concentrations of ammoniacal N in combination with high pH (Engel and Alexander, 1960; Schmidt, 1982), and by high osmotic pressure (Monaghan and Barraclough, 1992), and these factors may retard nitrification in urine patches. Laboratory studies indicated that the nitrification of urine N in most temperate grassland soils was unlikely to be restricted by high N concentration unless the concentration exceeded 16 g N l^{-1}, but that restriction due to high osmotic pressure could be severe in dry sandy soils (Monaghan and Barraclough, 1992). Small amounts of nitrite have been detected in urine patches during the first few weeks after application (Sherwood and Fanning, 1989) and in soil treated with cattle urine and incubated under laboratory conditions (Monaghan and Barraclough, 1992).

Appreciable concentrations of nitrite have also been reported in drainage and river waters in Northern Ireland, and specifically in drainage water from grazed and fertilized grassland. The results of detailed studies indicated that the nitrite was derived mainly from the nitrification of ammonium, though a contribution from the reduction of nitrate was also possible (Burns *et al.*, 1995).

In some situations, the nitrification of ammonium from fertilizers or slurries may result in substantial amounts of N being lost through nitrate leaching or denitrification. Such losses can be reduced if nitrification is restricted by incorporating a chemical nitrification inhibitor into the fertilizer or slurry. The most generally effective inhibitor currently available is nitrapyrin (2-chloro-6-trichloromethylpyridine), which was previously marketed as N-Serve (Dow Chemical Company). However, it has the disadvantage of being somewhat volatile and is therefore not suitable when the fertilizer is applied to the soil surface. Dicyandiamide (DCD) has been found to be useful for slurries (see p. 218) as it is water-soluble and mixes well with the slurry. It is however less tightly adsorbed than is NH_4^+ to the clay and organic matter in soils, and leaching may curtail its effectiveness. DCD contains 66% N and, as it is slowly hydrolysed to urea, it is effectively a slow-release fertilizer (Smith *et al.*, 1989). However, in most situations, the use of a nitrification inhibitor is not cost effective.

Nitrate and Ammonium in Grassland Soils

The concentrations of ammonium and nitrate in soil (excluding fixed ammonium – see p. 122) are normally assessed by analysing a soil extract prepared with a salt solution such as 1M or 2M KCl (Bremner, 1965). The extract includes both the water-soluble and the exchangeable fractions of the ammonium, and the total nitrate, which is readily soluble in water. In grassland that is fertilized regularly, the concentrations of ammonium and nitrate change rapidly during the growing season in response to fertilizer additions and subsequent plant uptake. In grassland that is grazed, the concentrations show large spatial variations due to the return of excreta (Afzal and Adams, 1992; Ball and Ryden, 1984; Williams and Clement, 1965) and there is often some accumulation of inorganic N towards the end of the season.

Studies in the Netherlands during 4 years indicated that the amount of inorganic N in the soil (0–60 cm) under long-term ryegrass swards immediately before the first application of fertilizer N in spring ranged from 18 to 84 kg ha^{-1} (Oenema *et al.*, 1989). Assessments made during the growing season in cut plots receiving up to 420 kg fertilizer N ha^{-1} $year^{-1}$, and with the soil sampled immediately before each application of fertilizer, showed the amount of mineral N to range from 20 to 75 kg ha^{-1}, with about half present in the top 20 cm of soil. On average, with this sampling regime, the difference between the unfertilized plots and the plots receiving 420 kg N ha^{-1} $year^{-1}$ was only 13 kg N ha^{-1} (Oenema *et al.*, 1989).

Table 6.1. Some examples of concentrations of ammonium and nitrate (mg N kg^{-1} in soil dry weight) in grassland soils.

Sward type	Fertilizer N (kg ha^{-1} year^{-1})	Sampling depth (cm)	NH_4-N (mg kg^{-1})	NO_3-N (mg kg^{-1})	Reference
Grazed ryegrass, 5 sites, UK	250	40	1–24	1–230	Jarvis and Barraclough, 1991
Grazed ryegrass, 5 sites, UK	450	40	1–48	2–140	Jarvis and Barraclough, 1991
Grazed grass–clover, SE England	Nil	10	8–60	3–13	Jarvis *et al.*, 1990
Grazed ryegrass, Wales	200	7.5	3	45	Cuttle and Bourne, 1992
Grazed pasture, Wales	100	7.5	10	75	Cuttle and Bourne, 1992
Long-term grass, Denmark	Nil	20	4.2–6.6	0.1–2.3	Christensen, 1983
Long-term grass, Denmark	262*	20	7.7–13.1	0.2–3.4	Christensen, 1983
Shortgrass prairie, Colorado, USA	Nil	10	0.4–2.7	0.8–4.9	Schimel and Parton, 1986
Grassland, California, USA	Nil	9	1.7–5.8	0.4–2.3	Davidson *et al.*, 1990

*As inorganic N in slurry.

When little or no fertilizer N is applied, the amounts of inorganic N in the soil generally remain low (< 15 mg N kg^{-1} soil) throughout the year though, in relatively warm and dry areas such as northern California, concentrations of inorganic N are often highest during the winter and early spring (15–20 mg N kg^{-1} soil) when moist conditions favour mineralization, and lowest during the summer (< 10 mg N kg^{-1} soil) after the period of maximum plant growth and uptake (Vaughn *et al.*, 1986).

The ratio of ammonium-N to nitrate-N is often greater in grassland than in arable soils and the concentration of ammonium is sometimes, though not always, greater than the concentration of nitrate (Table 6.1). There was more ammonium-N than nitrate-N in some grass and grass–clover swards, both cut and grazed, not receiving fertilizer N (Harrison *et al.*, 1994; Jarvis *et al.*, 1990; Marriott, 1988; Parsons *et al.*, 1991a; Thomas *et al.*, 1990) but not in a grazed grass–clover sward in Scotland (Younie and Watson, 1992). Sometimes ammonium has exceeded nitrate in grazed grass swards fertilized with N (Christensen, 1983; Parsons *et al.*, 1991a) and also in swards receiving slurry (Christensen, 1983). However, nitrate accounted for 70% of the inorganic N in soil under cut lucerne that received no fertilizer N, a proportion similar to that under a grass ley fertilized with calcium nitrate (Bergström, 1986). Soils under grazed grass swards fertilized with ammonium nitrate often, though not always, contain more nitrate-N than ammonium-N (Jarvis and Barraclough,

Table 6.2. Amounts of ammonium and nitrate (kg N ha^{-1}) to a depth of 90 cm in the soil below cut and grazed ryegrass swards receiving fertilizer N at 420 kg ha^{-1} $year^{-1}$, and below the centres of urine patches (mean and range values for samples taken in November). (From Ball and Ryden, 1984.)

	NH_4-N (kg ha^{-1})	NO_3-N (kg ha^{-1})
Cut sward (110 cores at random)	8 (6–11)	38 (22–74)
Grazed sward (39 cores at random)	10 (4–57)	160 (47–510)
Urine patches (9 cores)	258 (26–1090)	920 (200–1710)

1991; Jarvis *et al.*, 1990; Macduff and White, 1984; Younie and Watson, 1992). The ratio of nitrate-N to ammonium-N tended to increase with increasing rate of ammonium nitrate, even when there was little or no difference in the total inorganic N (Jarvis and Barraclough, 1991). Fertilizer N applied to cut swards during the growing season usually increases soil inorganic N for only 2–3 weeks, unless large amounts of fertilizer are applied. However, an accumulation persisting to the end of the growing season occurred, with cut swards, only when the rate of application exceeded 400 kg N ha^{-1} $year^{-1}$ (Prins, 1980).

The amounts of nitrate and ammonium in grassland soils are increased by the presence of grazing animals, particularly at the end of a grazing season and particularly in urine patches, as shown by the data in Table 6.2. Initially the effect of urine is to increase ammonium-N, but this effect declines as the ammonium is progressively nitrified to nitrate. Concentrations of ammonium-N in the surface 2.5 cm of a urine patch varied from 500 to 1000 mg N kg^{-1} (Vallis *et al.*, 1982). The concentration of nitrate after a period of several weeks depends not only on the concentration of N in the urine, and the rate of nitrification, but also on the factors (e.g. plant uptake, denitrification and leaching) that remove nitrate from the soil (Ball and Ryden, 1984; Stillwell and Woodmansee, 1981). Concentrations of inorganic N in the soil are also increased by dung, though the effect is much smaller than with urine (Afzal and Adams, 1992; Ryden, 1986; Spatz *et al.*, 1992). The application of slurry has a substantial effect, with concentrations of both ammonium and nitrate being increased for several months after the application (Eggington and Smith, 1986; Opperman *et al.*, 1989).

Mineralization/Immobilization Turnover Involving Fertilizer N

Field experiments that include plots with and without fertilizer N enable the 'apparent' recovery of the fertilizer N to be calculated from the amounts of N in harvested herbage. The retention of N in stubble and roots is generally ignored in calculations of recovery; with some justification, as the effect of fertilizer N on the amounts of N in these fractions is often relatively small. For grassland, the 'apparent' recovery is defined as the difference between the fertilized and unfertilized plots in the amount of N harvested in the herbage (usually per year), expressed as a percentage of the amount of N applied in the fertilizer. On this basis, the 'apparent' recovery of fertilizer N by grass is usually in the range 50–80% (see Chapter 10). However, the 'actual' recovery of the fertilizer N itself can be assessed if the fertilizer is labelled with ^{15}N, and usually the actual recovery in the herbage is lower than the apparent recovery (e.g. Barraclough *et al.*, 1985). Studies with ^{15}N-labelled fertilizer have also shown that more soil N is taken up in the presence of fertilizer N than in its absence, and that some labelled fertilizer N is retained in organic form in the soil at the end of the growing season (Dowdell and Webster, 1980; Hauck and Bremner, 1976; Jansson and Persson, 1982). These findings result largely, if not entirely, from a turnover between inorganic and organic forms of N in the soil, a turnover that is due to microbial activity (Jansson and Persson, 1982; Jenkinson *et al.*, 1985). When there is an active microbial population, the addition of labelled inorganic N is inevitably followed by some immobilization of the labelled N, and some mineralization of non-labelled N. As a result, there is more unlabelled inorganic N in the fertilized soil than in a corresponding unfertilized soil, even if the fertilizer N has no effect on net mineralization. The extent to which inorganic N is immobilized is related to the supply of readily available carbon (Okereke and Meints, 1985) and the relatively high proportion of the immobilized ^{15}N incorporated into amino acids and amino sugars (Campbell, 1978; Smith *et al.*, 1978) is consistent with the view that much of the N is incorporated initially into the soil microbial biomass. Microbial immobilization is greater with ammonium than with nitrate (Barraclough *et al.*, 1985: Jansson and Persson, 1982).

Some of the fertilizer N immobilized by the soil organic matter is subsequently remineralized though, after the first growing season, this occurs slowly. For example, during a 3-year lysimeter study in which ^{15}N-labelled calcium nitrate (at a rate of 400 kg N ha^{-1} $year^{-1}$) was applied to a ryegrass sward on a sandy loam soil, 51% was recovered in the first year, 6% in the second year, and 2% in the third (Dowdell *et al.*, 1980). Some of the N apparently immobilized in this type of experiment with grassland may be present in the stubble and roots. This effect was shown in an investigation with crested wheatgrass that received fertilizer N for 5 years, and in which appreciable amounts of fertilizer N were still present in root material 5 years after fertilization ceased (Power, 1983).

Although there is always a turnover between inorganic and organic N in soils when fertilizer N is added, it is also possible that fertilizer N sometimes increases the net mineralization of the soil organic N (Haynes, 1986b), an effect known as 'priming'. If and when a small 'priming' effect occurs, it is likely to reflect an increase in microbial activity due to a change in soil pH or nutrient status.

The extent of mineralization–immobilization turnover can be estimated from measurements of both the net and the gross rates of mineralization and immobilization. An effective procedure for measuring these processes in field soils involves the use of paired microplots treated with ammonium nitrate, one of each pair receiving labelled ammonium and unlabelled nitrate, and the other receiving labelled nitrate and unlabelled ammonium (Barraclough, 1988, 1991). Analysis for total N and ^{15}N in ammonium and nitrate in the soil, and for total N and ^{15}N in the crop, at the beginning and end of the experimental period then enables the net and gross rates of mineralization and immobilization to be calculated.

Mineralization/Immobilization Involving Organic Manures

The application of organic manure or slurry usually involves a substantial input of carbonaceous material, and this may result in the immobilization, at least temporarily, of some inorganic N already present in the manure or slurry, or in the soil. Such immobilization of inorganic N by cattle and pig slurries has been reported in pot experiments (Flowers and Arnold, 1983; Whitehead *et al.*, 1989a). However, the balance between immobilization and mineralization changes with time, and increasingly favours mineralization. The critical C : N ratio for the temporary immobilization of N by organic manures may be as low as 15 : 1, though it is influenced by the proportion of the total N present in ammoniacal form. If the C : N ratio is $> 15 : 1$, and if ammoniacal N is $< 20\%$ of total N, immobilization may be appreciable (Beauchamp and Paul, 1989). The balance between mineralization and immobilization is also influenced by the composition of the soil as well as by the composition of the manure. Mineralization is increased by a low C : N ratio in the soil, as well as by a low C : N ratio in the manure (Barbarika *et al.*, 1985). Cattle dung with a C : N ratio of about 19 : 1 showed slight immobilization of N during 10 weeks incubation with soil at 25°C, whereas pig dung with a C : N ratio of about 16 : 1 showed slight mineralization (Kirchmann, 1991).

In general, following the application of manure or slurry, the amount of N retained in organic form in the soil after a period of a year is directly related to the amount of C retained. The C retained is often about one-third of the C applied (though it may be as much as half when manures are applied at high rates) and the ratio between the C and N retained is usually between 10 : 1 and 15 : 1 (Jenkinson, 1982). Thus with an application of manure or slurry equivalent to 9 t dry organic matter (and assuming 50% C in the organic matter), about 1500 kg of C and 100–150 kg N will be retained in the soil organic matter after one year. Further decomposition occurs during subsequent years, with a proportion of the N remaining at the end of each year being mineralized during the succeeding year (see p. 215).

Non-exchangeable Fixation of Ammonium by Clays

Many soils have some capacity to retain NH_4^+ (and K^+) cations in a form that is not subject to normal cation exchange, and in which replacement of the ions occurs slowly. The capacity to fix NH_4^+ and K^+ ions is due to the presence of clay minerals that have a three-layer lattice structure (e.g. vermiculite, illite, montmorillonite). The exposed lattice surfaces of clays of this type contain hexagonal spaces with a diameter of about 2.8 nm, which is similar to the diameters of K^+ and NH_4^+ ions (Nommik and Vahtras, 1982). When these ions are present in sufficient numbers, the lattice layers are bound together, thus inhibiting subsequent hydration and expansion. Soils vary in their capacity to fix NH_4^+ from a few kg per ha to several hundred kg per ha, depending on factors such as their contents of the various clay minerals, their K status (K^+ reduces NH_4^+ fixation) and whether they are subject to marked wetting and drying cycles. Fixed ammonium in many topsoils represents between 3% and 10% of the total soil N, and for any one soil the proportion remains fairly constant. In subsoils, the proportion of the total N present as fixed ammonium is often higher than in topsoils and may exceed 50%.

The presence of a high concentration of ammonium, combined with drying conditions, tends to increase the amount of fixed ammonium and, although this ammonium is much less available to plants and microorganisms than is the exchangeable ammonium, it may become available slowly. And recently-fixed ammonium may be released quite rapidly during the nitrification of exchangeable ammonium (Green *et al.*, 1994). In some situations, the fixation of ammonium may have the advantage of providing a continuous but limited supply of available N over a long period (Nommik and Vahtras, 1982).

Assessment of Available Soil Nitrogen from Grass Uptake in the Field

In the field, the amount of N mineralized depends on a range of factors including soil type, weather conditions and the previous management of the soil. An important factor for grassland is whether or not the soil has been ploughed in recent months or years. With soils under grass, the plant-available N can be assessed retrospectively from the amount of N contained in the herbage harvested at intervals during the growing season, from unfertilized plots containing no clover. Grass takes up virtually all the inorganic N from mineralization through its well developed and extensive root system. Although the uptake during a growing season includes inorganic N present at the beginning of the season, as well as N mineralized during the season, in temperate regions with a substantial winter rainfall, there is usually little

Table 6.3. Examples of amounts of available soil N (kg N ha^{-1} $year^{-1}$; mean and range values) assessed from the amount in harvested herbage (without any adjustment for the amounts in stubble and roots, or for the deposition of N from the atmosphere) from grass swards receiving no fertilizer N in the years of assessment and grown in fields differing in previous management.

Sward type and previous management	No. of sites	Year of assessment 1	2	3	4	Mean	Source of data*
Continuous long-term grassland (most with low N inputs previously)	16	95 (46–149)	143 (37–255)	111 (70–161)	–	116	1
'Leys' with sward age 5–15 years (most with moderate to high N inputs previously)	8	131 (68–204)	104 (19–183)	83 (12–169)	–	106	2
Swards recently sown from:							
long-term arable cultivation	13	29 (15–55)	24 (8–71)	29 (6–90)	32 (7–56)	29	3
ley/arable rotation	21	66 (10–140)	54 (18–139)	62 (10–131)	69 (18–169)	63	3
'leys' aged 5-12 years (most re- sown from previous grass)	8	101 (38–180)	88 (16–131)	75 (6–165)	–	88	2
long-term grassland	16	112 (32–219)	88 (51–129)	92 (41–160)	–	97	1

N.B. Data for site/years of the investigations known to have been influenced appreciably by the presence of clover have been excluded, but a small contribution of clover to the above data is possible.
*1, Hopkins *et al.* (1990) and A. Hopkins (personal communication); 2, Hopkins *et al.* (1995) and A. Hopkins (personal communication); 3, Morrison *et al.* (1980) and unpublished.

carry-over of inorganic N from one season to the next. Most of the ammonium present in the autumn is nitrified to nitrate and, in general, the nitrate is either leached or denitrified during the winter. However, in unusually dry winters, a substantial amount of inorganic N may remain in the root zone and contribute to uptake in the following spring.

In assessing plant-available N from the amount of N in the herbage harvested in a succession of cuts during the season, it is often assumed that the amount of N in roots and stubble is the same at the end as at the beginning of the season and that net transfer of root/stubble N to soil organic N has been negligible. The second assumption, in particular, is likely to result in an under-estimation of uptake. Alternatively, it may be assumed that the amount of N transferred from roots and stubble is equal to the atmospheric input of N which is then ignored. Measurements of this type were made as part of several multi-centre experiments carried out in the UK during the period 1970–1986, with each experiment continuing for 3 or 4 years. Data for 280 site/years show a range in soil N supply of 6–255 kg N ha^{-1} $year^{-1}$ (Table 6.3), though in some instances the apparent soil N supply has undoubtedly been augmented by the presence of clover. The data, categorized on the basis of the recent cropping history of the field, show a considerable overlap in apparent soil N supply between fields previously under long-term arable cultivation, fields previously under ley/arable rotations and fields previously under long leys or old grassland. Also, the categories are not strictly comparable as the data were obtained at different locations and in different years. However, on average, the smallest amounts of soil N are provided by fields previously under arable cultivation, and the amounts tend to increase with increasing proportions of grass in the previous cropping history (Table 6.3). Amounts of plant-available N greater than 100 kg ha^{-1} $year^{-1}$ are generally associated with long-term grassland, either continuing as such or after ploughing and re-seeding. The variation in the amount of available soil N among different sites is due partly to differences in total soil N, a factor often related to the length of time under grass, and partly to differences in the proportion of the soil N mineralized. As an example of variation that may occur in the extent of mineralization, the percentage of the soil N taken up by ryegrass over six harvests (including the N in the roots and stubble which were recovered at the final harvest) from 21 grassland soils, that had received no fertilizer N in the year before sampling, varied from 1.5 to 4.0% (Whitehead, 1984).

Assessment of Available Soil N by Laboratory Methods

No simple method of chemical analysis will provide a reliable assessment of plant-available N in grassland soils, partly because no single extractant will predict potential mineralization by the soil microorganisms and partly because the actual amount of N mineralized and taken up is influenced by weather during the growing season. Rainfall is a particularly important factor, and periods of drought that result in a dry upper layer of soil can greatly restrict the amount of N actually available (Garwood and Tyson, 1973; Garwood and Williams, 1967). An additional problem is the difficulty of obtaining a representative sample of soil, particularly from areas that have been grazed and are therefore highly heterogeneous. Even if emphasis is placed on assessing the potentially available N, the assessment should include both the inorganic N at the time of sampling and the mineralizable fraction. The latter can be assessed from the increase in NH_4^+-N + NO_3^--N during a period of incubation under uniform conditions (see p. 113) but incubation methods, though useful in research investigations, are too time-consuming for routine use.

Despite the difficulties outlined above, the potential benefits of assessing plant-available N by rapid chemical analysis have encouraged studies of various extraction methods. Examples include the measurement of NH_4^+-N + NO_3^--N released from the soil by autoclaving with 0.01M $CaCl_2$ (Stanford, 1977), or by heating at 80°C with a solution of 2M KCl (Øien and Selmer-Olsen, 1980), and determination of the NH_4^+ produced by steam distillation of the soil with a buffered solution of pH 11.2 (Gianello and Bremner, 1988). With a range of soils from different locations, the results of such analyses can be converted to predict the plant-available N by applying an adjustment factor to take into account site-to-site differences in average weather conditions (Stanford, 1977; Whitehead, 1983b). It is, of course, impossible to predict year-to-year variation in weather at an individual site. While some of these extraction methods have produced good correlations with plant uptake with limited numbers of soils in restricted areas, none of the methods has proved sufficiently reliable to be adopted widely.

Impact of Ploughing of Grassland on Mineralization

When grass swards are ploughed, either for re-sowing to grass or for an arable crop, there is a large increase in the amount of N mineralized and the increase may persist, though to a diminishing extent, for several years. The size and persistence of the increase depend to a large extent on the age of the sward. For example, in a study on a sandy loam soil, the amount of nitrate-N in the soil (to 90 cm depth) in the October after ploughing in July ranged from 93 kg ha^{-1} after a 1-year grass–clover ley to 230 kg ha^{-1} after a 6-year ley, though the difference between sward ages of 3 and 6 years was rather small and inconsistent (Johnston *et al.*, 1994). With long-term grassland swards that have been unploughed for many years, the amount of N mineralized may be as much as 400 kg N ha^{-1} in the first year after ploughing. Some of the mineralized N is normally taken up by the subsequent crop, reducing its need for fertilizer N (Clement and Back, 1969; Johnston *et al.*, 1994) but some is leached in the form of nitrate (see Chapter 7). When some very old grassland at Rothamsted was ploughed, the amount of N mineralized was estimated at about 4000 kg ha^{-1} over 20 years (Whitmore *et al.*, 1992).

While some of the increase in mineralization after ploughing occurs from humified organic matter as a result of increased aeration, a substantial proportion is derived from the stubble and roots of the grass, plus the leaf litter and soil macroorganic matter. The actual amount of N mineralized is therefore influenced by the amounts of these fractions and their concentrations of N. In a ryegrass sward that had been established for 8 years, and sampled at intervals over a period of about 1 year, total N in the stubble, roots, leaf litter and soil macroorganic matter amounted on average to about 500 kg N ha^{-1}: with a similar 15-year-old sward, the N in these fractions amounted to 600 kg ha^{-1} (Whitehead *et al.*, 1990). As noted in Chapter 5, the concentrations of N in stubble, roots and leaf litter increase with increasing rate of fertilizer N, and the higher the concentration of N, the greater the proportion of the N mineralized during the first year after ploughing. This effect is illustrated by a laboratory incubation study in which the amount of N mineralized from a range of grassland soils (with stubble and roots incorporated) was found to increase with increasing rate of fertilizer N applied during the previous 3 years, and to be greater when the sward was grazed rather than cut (Table 6.4). In another investigation, the amount of N mineralized on incubation was between 10 and 90% greater in soils from plots that had been grazed rather than cut during the preceding three years (Hassink *et al.*, 1990). In a recently reported field experiment, increasing the rate of fertilizer N applied to plots of grazed grass over a 4-year period, was found to increase the amount of soil inorganic N after ploughing, and the uptake of N by a subsequent wheat crop, apparently due to greater mineralization from the grass residues (Webb and Sylvester-Bradley, 1994). Increasing the depth of ploughing tends to decrease the rate of mineralization of N in the plant residues (Richter *et al.*, 1989).

It is possible to predict the amounts of N likely to be mineralized, when swards of different ages and types of management are ploughed, from typical values for the amounts of stubble, leaf litter, roots and soil macroorganic

Table 6.4. The influence of the rate of ammonium nitrate fertilizer, applied to cut and grazed grass–clover swards, on the amount of ammonium plus nitrate (mg N kg^{-1}) in soils (0–15 cm) sampled immediately before ploughing and then incubated for 3 weeks. (From Williams and Clement, 1965.)

	Fertilizer N (kg ha^{-1} $year^{-1}$)			
Management	0	80	160	320
Cut	39	40	35	40
Grazed frequently	40	55	53	70
Grazed infrequently	54	53	58	63

matter, and the concentrations of N in these fractions. If it is assumed that there is a linear relationship between the concentration of N in the plant residue and the amount of N mineralized per unit dry weight within 1 year (Jenkinson, 1982), and also that one-third of the organic matter remains in the soil after 1 year and that the C : N ratio of the residual material is 15 : 1, then there is a curvilinear relationship between N concentration and the proportion mineralized (Fig. 6.1). This relationship can be used, together with information on the amounts of the various organic fractions, to predict the amounts of N likely to be mineralized on ploughing some typical grass swards (Table 6.5). In general, the amount of N mineralized will increase (i) with increasing age of the sward, (ii) with increases in the rate of fertilizer N applied during the year(s) before ploughing, and (iii) with grazing as compared with cutting management in the year(s) before ploughing. The presence of clover in the sward is likely to have a relatively small effect because, although the stolons and roots of clover have a higher concentration of N than do the stem bases and roots of grasses, their weight is usually much less than the weight of grass stubble and roots.

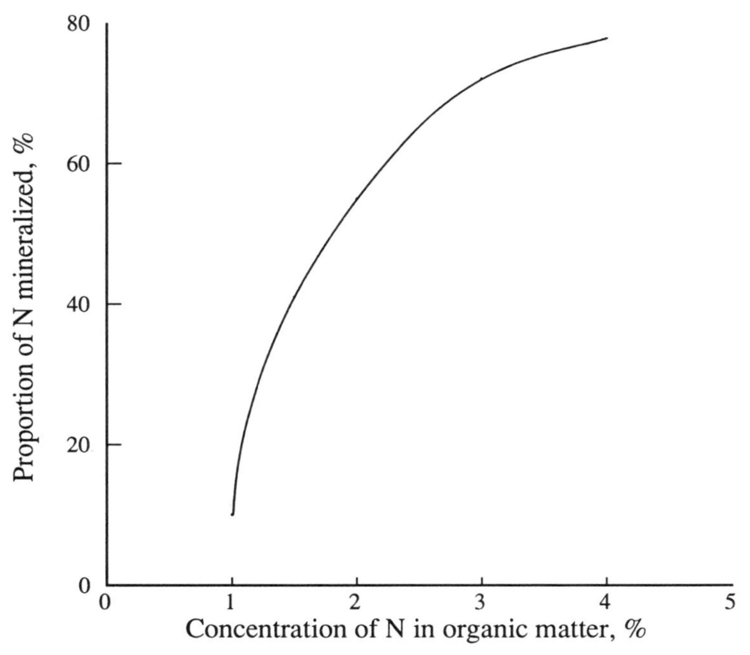

Fig. 6.1. Proportion of the N in plant material mineralized during 1 year's decomposition, in relation to N concentration, assuming (1) that organic matter in plant material = 90%, (2) that one-third of the plant material remains in the soil after 1 year and (3) that the residual organic material has a C : N ratio of 15 : 1. (Adapted from Jenkinson, 1982.)

Table 6.5. Amounts of N estimated as likely to be mineralized from several types of grass sward during the first year after ploughing. (Based on Whitehead *et al.*, 1990.)

Age of sward	Cut or grazed	Fertilizer N (kg N ha^{-1} $year^{-1}$)	Estimated amount of N mineralized (kg N ha^{-1})
1	Cut	50	45
1	Cut	300	75
3	Cut	300	120
3	Grazed	300	200
8	Grazed	300	300
15	Grazed	300	360

As indicated above, enhanced mineralization continues for more than a year after ploughing but not at a uniform rate. The time of year at which ploughing is carried out influences the seasonal distribution of mineralization and hence the relative amounts of the inorganic N taken up by a subsequent crop and lost through leaching and denitrification (Francis *et al.*, 1992). Although the effect of the readily decomposable organic materials on N mineralization normally persists for only a few years, an increase in the mineralization of soil N due to the ploughing of old grassland, in comparison with continuous arable cultivation, may be detected for more than 27 years (Richter *et al.*, 1989). Further data on the impact of ploughing on mineralization and nitrate leaching are included in Chapter 7.

The boost to the mineralization of N induced by ploughing is utilized in the practice of ley farming, in which periods of a few years under grass are alternated with a few years under arable cropping. Benefits for the subsequent arable crops of a period under grass include an increased supply of available N, and improvements in soil structure and water holding capacity. The relative importance of N supply and soil structure depends on soil type and fertilizer practice but, with a sandy clay loam soil, the increase in soil N supply resulting from a 3-year ley had a much greater effect on a subsequent cereal crop (not receiving fertilizer N) than did the increases in the stability of soil aggregates and water holding capacity (Clement, 1961).

Although the conversion of grassland to arable cropping increases the mineralization of soil organic matter and N, it greatly reduces the population of earthworms. Three years' cultivation of previously long-term grassland reduced the earthworm population by about half, and fields on the same soil series cultivated for 25 years had only about 14% of the earthworm population of the long-term grassland (Low, 1972). The reduction in the earthworm population will tend to reduce the mineralization of soil N during the later years of a period of arable cropping.

7 Leaching of Nitrogen from Soils

The Process of Leaching in Relation to N

Soluble soil constituents such as nitrate are leached when a soil, already at field capacity, receives more water from rainfall or irrigation than is needed for evapotranspiration. When a soil is at field capacity, no more water can be retained and the surplus therefore moves downwards. In most temperate regions, leaching occurs most frequently during the winter. Of the various forms of soil N, only nitrate is leached to a major extent, though some ammonium may be leached when concentrations are particularly high (e.g. in urine patches) and some urea may be leached under cool conditions ($< c.$ 5°C) when its conversion to ammonium is slow. It is unusual for other organic forms of N to be leached below the root zone, though organic N from slurry is sometimes leached through soils that have vertical cracks or large-diameter pores (macropores). In addition, small amounts of nitrous oxide produced by nitrification/denitrification (see p. 182) may dissolve in the soil water and be subject to leaching.

The nitrate in soils may be derived either from direct application as fertilizer or from the nitrification of ammonium; and ammonium may result from fertilizer application, from the hydrolysis of urea, or from the mineralization of soil organic matter. The nitrate ion is readily susceptible to leaching because it is negatively charged: it is therefore not adsorbed by the clay and organic colloids which are also negatively charged. In contrast, the ammonium ion is positively charged and tends to be retained on the clay and organic colloids of the soil by cation exchange.

The leaching of nitrate is important for two reasons, first because it represents a loss of plant-available N from the soil and, second, because the leached nitrate enters streams and groundwater where its presence in more than trace amounts is undesirable. In the UK, about 30% of the total domestic water supply is pumped from groundwater in chalk, limestone and sandstone strata, and such groundwater contains nitrate leached from the soils above. In recent years the concentration of nitrate in some sources of groundwater has exceeded 11.3 mg NO_3-N l^{-1}, the maximum concentration permitted (under EU regulations) in the water supplied for human consumption; and it is likely that most of this nitrate has been leached from agricultural soils (Heathwaite *et al.*, 1993; Royal Society, 1983). Increases in the concentration of nitrate in groundwaters have also been observed in other countries in recent decades (e.g. in Denmark and the USA) and, as in the UK, these increases have been attributed to the intensification of agriculture (Hallberg, 1989; Overgaard, 1984).

The process of leaching can be envisaged most simply when water is applied to a block of moist sand, with the sand particles having a uniform particle size and from which the water can drain freely. Water applied evenly to the surface moves steadily downwards and, in doing so, displaces an equal volume of water, plus any soluble material, from the layer below. If half the volume of the moist sand is occupied by water, an application of 10 mm of water displaces the existing water 20 mm downwards. However, in soils, the movement of water is much more complex. This is partly because soils vary in texture (i.e. in the proportions of sand-, silt- and clay-size particles) and partly because, whatever their texture, soils have a physical structure in which many of the particles form aggregates of various sizes. Soil texture has a considerable effect (though modified by structure) on the depth to which the existing soil water is displaced by 10 mm of additional water. The depth is much greater with sandy than with clay soils (Table 7.1). In well-structured soils, water moves most readily between the soil aggregates, which have effective diameters ranging from less than 0.002 mm to more than 2 mm and are separated from one another by pores that vary widely in size and shape. However, the aggregates themselves contain numerous micropores through which water moves slowly. Soils differ widely both in total pore space and in the distribution of pore sizes.

When water is applied to the surface of a well-structured soil, it does not move uniformly downwards but moves rapidly through large pores (macro-

Table 7.1. The influence of soil texture on the depth to which the existing water is displaced when 10 cm of additional water is added to soils. (From Burns, 1979.)

Soil texture	Water content at field capacity (%)	Mean displacement (cm)
Sandy	15	45
Medium	27	30
Clay	39	20

pores) with an effective diameter of more than about 0.05 mm, if these are continuous in the vertical direction. Movement occurs slowly, if at all, through the micropores. The existing soil water can therefore be categorized broadly into mobile and non-mobile phases (in macropores and micropores respectively), with the mobile phase being displaced much more readily by additional water. Any nitrate in the mobile phase is highly susceptible to leaching whereas nitrate in the non-mobile phase is much less susceptible. However, nitrate does tend to diffuse out of the non-mobile phase if there is a higher concentration there than in the mobile phase. On the other hand, if the concentration of nitrate in the mobile phase is high (e.g. after the application of fertilizer N) nitrate will diffuse from the mobile into the non-mobile phase. The size of the mobile phase, and the size and shape of the pores making up this phase, have a large influence on the extent to which leaching occurs after rainfall. Thus, when the soil contains large and continuous macropores or cracks, there is a rapid downward movement of water, known as 'by-pass' flow and, depending on the distribution of nitrate between the mobile and non-mobile phases, this may either increase or decrease the amount leached after a period of rainfall. If there is a high initial concentration of nitrate in the surface soil, its movement down the soil profile following the addition of water is inevitably accompanied by some vertical dispersion of the nitrate. In general, the greater the average depth of leaching, the greater the vertical distance over which the nitrate is dispersed (Addiscott *et al*., 1991).

However, although 'by-pass' flow may influence the short-term leaching of nitrate, the amount leached during a winter season as a whole is usually determined mainly by two factors: (i) the total quantity of water passing through the soil profile, which represents the excess of rainfall over evapotranspiration, and (ii) the concentration of nitrate in the soil at the onset of leaching. In temperate regions, a soil water deficit normally develops during the spring and summer, and the soil returns to field capacity at some time during the autumn or winter. Usually, grass growth has almost ceased by this time, and the uptake of N is therefore negligible. The leaching of nitrate is then caused by the excess winter rainfall (i.e. that falling in the period between the return to field capacity and the development of a water deficit in the following spring, minus the loss by evapotranspiration during the winter). However, there are exceptions to this general seasonal pattern. One exception occurs in soils with well-developed macropores, as these may allow some leaching of nitrate to occur in the autumn before field capacity has been attained. A second exception to the normal seasonal pattern occurs in dry winters (and in normal winters in dry areas) when some nitrate may remain throughout the winter in the non-mobile phase of the soil water.

The amount of excess winter rainfall necessary to leach almost all nitrate through the profile is greater with clay than with sandy soils, as relatively more nitrate is present in the non-mobile phase of clay soils. With a clay soil, it was found that cumulative drainage of 200 mm would leach only about 50% of the nitrate to below 55 cm, and that about 400 mm was necessary to leach 80% of the nitrate to below this depth (Scholefield *et al.*, 1993). The amount of excess winter rainfall varies widely from one area to another. For example, in 67 agroclimatic areas of England and Wales, it varies from an average of 125 mm in the area immediately to the north of the Thames estuary to an average of 1300 mm in the mountainous area of north Wales (Smith, 1984). Also, in any one area, there is considerable variation from year to year. Areas that differ in excess winter rainfall also differ in the amount of nitrate-N (expressed in kg NO_3^--N ha^{-1}) that can be leached before a given concentration in the groundwater (e.g. 11.3 mg NO_3^--N l^{-1}) is attained.

When grassland is mown or lightly grazed, the amount of nitrate leached is normally less than 20 kg N ha^{-1} $year^{-1}$, even if fertilizer N is applied. There is less leaching from mown or lightly grazed grassland than from areas under arable cultivation receiving similar rates of fertilizer N. However, when grassland is fertilized regularly and grazed intensively, more than 100 kg N ha^{-1} $year^{-1}$ may be leached as nitrate, much of it being derived from the N in urine. Similar amounts may be leached from productive grass–clover swards that are grazed intensively. In general with grassland, substantial leaching indicates that the total supply of N during the season has exceeded the requirements for growth. The supply is usually dominated by fertilizer application and the return of N in urine, but it also includes the N mineralized from organic sources as well as inputs from the atmosphere.

The main reason why relatively little nitrate is leached from mown grassland is that grass has a large capacity to take up nitrate, and this capacity is maintained for much of the year. However, with grazed grassland, the concentration of nitrate that develops in the soil of a urine patch (see p. 93) greatly exceeds the uptake capacity of the grass. Nitrate derived from urine, together with any surplus fertilizer nitrate, tends to accumulate until the autumn (Garwood and Ryden, 1986; Sherwood and Fanning, 1989) and is then susceptible to leaching and denitrification. The mineralization of N from plant residues and soil organic matter is particularly important as a source of nitrate when long-term grassland is ploughed, as the amount mineralized often exceeds the amount needed by the re-seeded grass or subsequent arable crop.

Methods for the Assessment of Leaching

Methods for assessing the amounts of N leached from soils are of three main types: (i) the use of lysimeters, (ii) the sampling of soil or soil solution to depths of 1 m or more in the field, and (iii) the analysis of water from field drains (Addiscott *et al.*, 1991; Goulding and Webster, 1992).

Lysimeters are columns of soil, usually in cylindrical containers and usually 1 m or more in depth, at the base of which there is a system for collecting the water that has passed through the soil. They have the advantages that their soil volume is easily defined, and that the inputs and outputs of water and N (other than gaseous forms of N) can be measured accurately. In some investigations, lysimeters have contained intact monoliths of soil while, in others, the soil has been re-packed into the lysimeters. Ideally, lysimeters should be installed in the field so that the soil surface is at ground level, and soil temperatures within each lysimeter are close to those occurring naturally. Despite the advantages of lysimeters for assessing the leaching of nitrate, they have some disadvantages. First, cracks may develop at the inside edge of the lysimeter, particularly with clay soils, allowing water to by-pass the bulk of the soil: for this reason, lysimeters are most effective with freely-draining soils (Goulding and Webster, 1992). Second, with re-packed lysimeters, the disturbance to the soil may increase the mineralization of soil N and hence the amount of nitrate leached. Third, in lysimeters which have no suction applied to the base, the amount of water draining through the soil is less than in the field, and consequently conditions are more anaerobic, and likely to increase denitrification and reduce nitrate leaching (Addiscott *et al.*, 1991). These various disadvantages are avoided or minimized in large monolith lysimeters designed so that suction can be applied at the base to match the surface tension lost by breaking the vertical pores in the soil (Belford, 1979). Such monolith lysimeters generally provide the most effective method for assessing nitrate leaching from many soil types, but they are expensive and time-consuming to install.

The second type of method involves measuring the concentration of nitrate-N at various depths in the soil at intervals during the winter, together with the calculation of the amount of water moving downwards through the soil from a water balance. The nitrate concentrations can be measured either by extracting samples of soil with a suitable solution (e.g. 2M KCl) and analysing the extracts for nitrate (Bremner, 1965) or by analysing samples of soil solution obtained using porous ceramic cups (Addiscott *et al.*, 1991; Hansen and Harris, 1975). The ceramic cups, fitted with plastic or metal tubes, are inserted at various depths in the soil, and samples of soil solution are obtained by applying suction to the tubes. It is important to install the ceramic cups at an angle of 45° or less to the soil surface, and to back-fill the hole with soil or silica flour, in order to minimize the possibility of water by-passing the cups (Addiscott *et al.*, 1991). Whether the measurement of nitrate-N is based on the extraction of soil samples, or on obtaining samples of soil solution, considerable replication is necessary to obtain a reliable estimate of the amount leached (Cameron and Scotter, 1988). Ceramic cups are most effective on sandy and other unstructured soils in which water moves fairly uniformly down the profile but they may underestimate the amount of leaching in structured clay soils, as they tend to be by-passed by water and nitrate moving through macropores (Goulding and Webster, 1992).

The third method of assessing nitrate leaching, based on the analysis of water from tile drains under an individual field, is applicable only in certain circumstances. It is most effective with well-structured clay soils for which the other two types of method are less suitable. It also has the advantage for grazed grassland of allowing for the heterogeneity caused by livestock. The amount of nitrate leached is estimated by combining data on nitrate concentrations with measurements of the water flow through the drains. However, it is unlikely that all the percolating water will be intercepted by the drains, and the proportion that is intercepted is difficult to assess. Another problem with the method is the need to relate the frequency of sampling for nitrate to the rate of flow of the drainage water (Cameron and Scotter, 1988). In some investigations based on this method, use has been made of drains already present but, in others, drainage systems have been installed specifically to study the effects of management on leaching and surface runoff (Goulding and Webster, 1992). A similar method, not requiring drains to be installed, can be used for small catchment areas with underlying clay, in which all drainage water leaves the catchment by a single ditch or stream. Again, by monitoring both nitrate concentration and flow rate, it is possible to estimate the amount of nitrate leached.

Amounts of N Leached from Mown Grassland

Lysimeter studies with cut grass swards have shown that the leaching of nitrate is negligible (< 5 kg nitrate-N ha^{-1} year^{-1}) when no fertilizer N is applied (Low, 1973; Webster and Dowdell, 1984a), a conclusion confirmed by the low concentrations of nitrate (< 3 mg nitrate-N l^{-1}) in groundwater below unfertilized grassland (Foster *et al.*, 1982). In general, the amount of nitrate leached from mown grassland is small provided that the rate of fertilizer N is no more than the optimum for grass growth at the particular location. Several investigations have shown less than 20 kg nitrate-N ha^{-1} year^{-1} to be leached with rates of fertilizer N of less than about 250 kg N ha^{-1} year^{-1} (Barraclough *et al.*, 1984; Bergström, 1987: Garwood and Ryden, 1986; Triboi, 1987; Webster and Dowdell, 1984a), though larger amounts may be leached when weather conditions are exceptional (see below). Studies in the Netherlands have shown that, with favourable soil and weather conditions, it is often possible to apply fertilizer to mown grass at rates up to 400 kg N ha^{-1} year^{-1} without appreciable loss by leaching (Prins *et al.*, 1988a). However, when fertilizer is applied at rates greater than about 400 kg N ha^{-1} year^{-1}, substantial amounts of nitrate may be leached from well-drained soils (Table 7.2). The amount of nitrate leached at a particular rate of fertilizer application is influenced by soil and weather factors and, with mown ryegrass on a soil overlying clay in the UK, there appeared to be little leaching even when fertilizer was applied at a rate

Table 7.2. Amounts of nitrate-N (kg ha^{-1} year^{-1}) leached from perennial ryegrass swards receiving two rates of fertilizer N, in lysimeters of a freely-drained loam soil during the period 1970–1982, with separate values for the winter of 1976–77 which followed a prolonged drought. (From Garwood and Ryden, 1986.)

	Fertilizer N (kg ha^{-1} year^{-1})	
	250	500
Nitrate-N, excluding 1976–77	15	145
Nitrate-N, during 1976–77	184	590

Table 7.3. Nitrate-N leached from plots of perennial ryegrass on a sandy loam soil overlying clay, at three rates of fertilizer N (mean values for 3 years). (From Barraclough *et al.*, 1983.)

	Fertilizer N (kg N ha^{-1} year^{-1})		
	250	500	900
Nitrate-N leached (kg ha^{-1} year^{-1})	4	27	151
Mean nitrate-N in leachate (mg N l^{-1})	2	13	86

of 500 kg N ha^{-1} year^{-1} (Table 7.3). Of the leaching that did occur, most was associated with heavy rain following the application of fertilizer N in early spring (Barraclough *et al.*, 1983).

When grassland is irrigated, particularly on sandy soils, it is necessary to control the amount of irrigation water to avoid the leaching of fertilizer N during the growing season. The irrigation of grassland is a widespread practice in parts of the USA, and appreciable nitrate leaching from irrigated cocksfoot has been reported, for example, from Nebraska (Watts *et al.*, 1991). However, although irrigation may increase the amount of nitrate leached, it also increases the volume of drainage water and often does not increase the concentration of nitrate in the leachate (Burden, 1982). In some situations, summer irrigation that is not excessive may well reduce subsequent leaching, by improving the uptake of N during the growing season (see p. 142).

Substantial amounts of nitrate may be leached from fields where a legume is grown alone, particularly after the first year from sowing. In a lysimeter study, white clover grown alone without fertilizer N resulted in an average concentration of nitrate-N in the drainage water of about 25 mg nitrate-N l^{-1} whereas the concentration under meadow fescue was less than 1 mg nitrate-N l^{-1} (Low, 1973). When, after 4 years, the clover was removed by pulling the plants from the soil, the nitrate leached during the following winter was equivalent to 130 kg N ha^{-1} (Low and Armitage, 1970). A marked increase in the leaching of nitrate from clover occurred after the death of herbage during the winter in the USA (Karraker *et al.*, 1950) but there was little leaching of nitrate from white clover grown for 1 year in a ley–arable rotation in New Zealand (Adams and Pattison, 1985). Considerable amounts of nitrate were leached from lucerne, both from the growing crop and after its incorporation into the soil (Robbins and Carter, 1980).

Amounts of N Leached from Grazed Grassland

As noted in Chapter 5, the urine patches in swards grazed by cattle often receive N at rates equivalent to more than 500 kg N ha^{-1} in a single application. Although some of the urinary N is lost as ammonia, some is taken up by the sward and some is immobilized in the soil organic matter, a substantial proportion is nitrified to nitrate. Some of this nitrate then remains in the soil until it is lost, during the autumn and winter, through a combination of leaching and denitrification. Evidence that substantial amounts of nitrate could be leached from intensively grazed grass–clover swards, even in the absence of fertilizer N, was first reported from New Zealand (Walker, 1962), and grazed grassland was subsequently identified as the major source of nitrate in groundwater in New Zealand (Burden, 1982; Steele *et al.*, 1984). Examples of the amounts of nitrate leached from grazed grass–clover swards, not fertilized with N, include 88 kg nitrate-N ha^{-1} $year^{-1}$ from paddocks grazed by beef cattle (Steele *et al.*, 1984) and 60–80 kg nitrate-N ha^{-1} $year^{-1}$ from pastures grazed by sheep (Field *et al.*, 1985). In both these studies, additional nitrate was leached when fertilizer N was applied.

In the UK, the impact of grazing animals on the leaching of nitrate was shown by measurements of the nitrate content of the soil, and the underlying chalk, from the plots of a field experiment in which four sward types were examined during the period 1976–1982. The four sward types were ryegrass receiving fertilizer N at 420 kg ha^{-1} $year^{-1}$ and ryegrass–clover receiving no fertilizer N, each divided into cut and grazed plots (Garwood and Ryden, 1986). The nitrate content of each successive soil horizon of 0.33 m, sampled to a depth of 4 m, was much greater under the grazed than under the cut swards (Fig. 7.1). The actual amounts of nitrate leached were calculated from the rate of percolation of water through the profile (on average, 0.8 m $year^{-1}$) combined with the nitrate concentration and bulk density of the various soil and chalk horizons, and these amounts ranged from 3 to 162 kg N ha^{-1} $year^{-1}$ (Table 7.4). Estimates from some of the same plots made in later years, and based on weekly measurements of nitrate in samples of soil solution, combined with drainage volumes measured in adjacent lysimeters, showed a similar leaching loss from the grazed sward receiving 420 kg N ha^{-1} $year^{-1}$ during the winter of 1987/88 but a lower loss during the warm and relatively dry winter of 1988/89 (Table 7.4). However, this investigation was carried out on a freely-drained soil overlying chalk, and the amounts of nitrate leached are therefore likely to have been greater than the amounts for less permeable soils. In a study on a clay-loam soil in Northern Ireland, leaching from individually drained plots, all grazed by beef cattle and receiving calcium ammonium nitrate at rates of 100, 200, 300, 400 and 500 kg N ha^{-1}, averaged 18, 25, 42, 72 and 64 kg nitrate-N ha^{-1} per winter respectively over 2 years (Watson *et al.*, 1992).

Studies in the Netherlands and Germany also indicated that the leaching of nitrate was greater from grazed than from cut grassland, often by about 30–60 kg N ha^{-1} $year^{-1}$ at rates of fertilizer between 250 and 450 kg N ha^{-1}

Table 7.4. Amounts of nitrate-N leached (kg N ha^{-1} $year^{-1}$) from cut and grazed plots of N-fertilized grass, and grass–clover receiving no fertilizer N, over the period 1976–1989. (Data from Garwood and Ryden, 1986; Ryden *et al.*, 1984; Macduff *et al.*, 1990.)

	1976–82	1987–88	1988–89
Grass; 420 kg N ha^{-1} $year^{-1}$, cut	29	n.d	5
Grass; 420 kg N ha^{-1} $year^{-1}$, grazed	162	199	20
Grass–clover; cut	3	n.d.	n.d.
Grass–clover; grazed	23	3	4

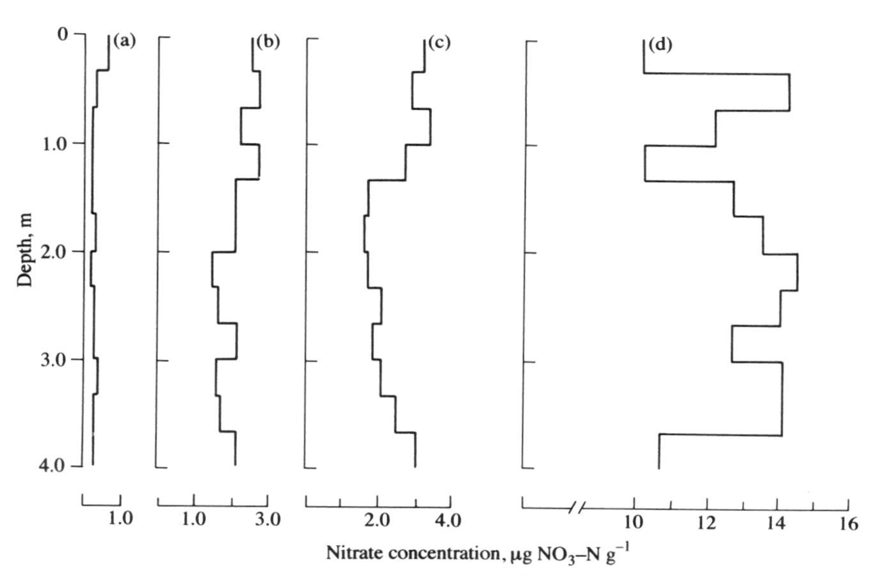

Fig. 7.1. Concentrations of nitrate (μg NO_3^--N g^{-1}) in soil and chalk sampled in winter from successive depths below grass–clover swards, both cut (**a**) and grazed (**b**), receiving no fertilizer N, and below all-grass swards, both cut (**c**) and grazed (**d**), receiving 420 kg N ha^{-1} $year^{-1}$ as ammonium nitrate. The grass–clover swards fixed about 270 kg N ha^{-1} $year^{-1}$ when cut and about 160 kg N ha^{-1} $year^{-1}$ when grazed. (Reproduced from Garwood and Ryden, 1986.)

Table 7.5. Amounts of nitrate-N leached (kg N ha^{-1} year^{-1}) from cut and grazed plots of long-term grassland receiving five rates of fertilizer N on a sandy soil, and from grazed plots of grassland on a clay soil, all in the Netherlands (mean values for duplicate plots and 3 years). (From Macduff *et al.*, 1990.)

	Fertilizer N (kg N ha^{-1} year^{-1})				
	0	250	400	550	700
Sandy soil, cut sward	10	25	30	73	78
Sandy soil, grazed sward	–	48	61	116	145
Clay soil, grazed sward	–	8	29	41	49

year^{-1} (Benke *et al.*, 1992; Steenvoorden *et al.*, 1986). The amounts of nitrate leached from a cut and grazed sward on a sandy soil, and from a grazed sward on a clay soil in the Netherlands, are compared in Table 7.5.

The potential effect of urine patches on nitrate leaching from grazed grassland is shown by the finding that soil samples, taken in November from previously marked urine patches in paddocks grazed by cattle, contained substantial amounts of nitrate at depths below 1 m, despite a soil water deficit of about 15 mm (Garwood and Ryden, 1986). Some of the nitrate may have resulted from the percolation of urine through macropores in the dry soil, followed by the hydrolysis of urea and nitrification of the resultant ammonium. This possibility is supported by the presence of ammonium with a peak concentration at a slightly shallower depth in the soil profile (about 1 m) than that of nitrate (Ryden *et al.*, 1984). The effect of urine on the contents of nitrate and ammonium in the soil was detectable up to 15–30 cm laterally from the centre of the patch. In an investigation with cores of undisturbed topsoil in New Zealand, urine moved rapidly to a depth of more than 15 cm in soils with an appreciable number of macropores, and less than half the urea-N was hydrolysed during this movement (Williams *et al.*, 1990). Urea that moves to the subsoil is hydrolysed more slowly than that remaining near the surface (Cuttle and Bourne, 1993). The time of year at which the urine is deposited has a large influence on the extent to which the N is leached. Thus, when cattle urine was applied at a rate of 3 l m^{-3} to grassland plots in Ireland on seven occasions between May and November, subsequent analysis of soil samples taken to a depth of 90 cm in November/December showed that less than 3% of the N from the urine applied between May and August remained in the soil profile, but that 30–50% of that applied between September and November remained and was therefore susceptible to leaching (Sherwood, 1986). In a similar investigation with artificial urine, only applications during or after September increased the amount of inorganic N present in the soil in late November (Cuttle and Bourne, 1993).

With fertilized grass swards, both cut and grazed, the amount of nitrate leached often increases gradually with increasing rate of fertilizer to a certain point, and then increases more sharply. At any one site, grazing is likely to reduce the rate of fertilizer N at which leaching becomes appreciable (Barraclough *et al.*, 1992). Thus, studies on grazed grassland in the Netherlands indicated that little nitrate leaching would occur at rates of fertilizer N of less than 200–300 kg N ha^{-1} but that leaching would increase substantially at higher rates (Kolenbrander, 1981). About 75% of the total loss was considered to be associated with the N returned in excreta. A similar conclusion, that little leaching occurred when the rate of fertilizer N was less than about 200 kg N ha^{-1} $year^{-1}$, but that leaching increased substantially above this rate, was reached from studies in Germany (Walther, 1989). The increase was greater with sandy soils than with loams.

When field drainage is installed, the amount of nitrate leached generally increases as shown with grazed plots of long-term grassland on a clay loam soil in Devon, UK (Scholefield *et al.*, 1993). The average amount of N leached per year during a 7-year period was almost three times greater from the plots with mole drains installed at a depth of 55 cm than from the undrained plots (Fig. 7.2), an effect probably due partly to increased mineralization, partly to enhanced transport through the soil and partly to reduced denitrification. Mineralization would have increased, and denitrification would have declined, as a result of improved aeration in the soil.

As noted above, results from New Zealand indicate that substantial amounts of nitrate may be leached from grazed grass–clover swards not fertilized with N. In a 3-year comparison in southwest England of grazed grass–clover with grazed grass paddocks, the latter receiving 200 kg fertilizer N ha^{-1} $year^{-1}$, the average amounts of N leached per winter were 16 kg and 79 kg N ha^{-1}: the annual liveweight gain of the grazing beef cattle on the grass–clover was 87% of that on the fertilized grass, and a simulation model indicated that with equal liveweight gains there would be little if any difference in leaching (Scholefield and Tyson, 1992). In a study in Wales, the amount of nitrate-N leached from a grazed grass–clover sward declined, over 3 years, from about 35 kg N ha^{-1} per winter to 6 kg N ha^{-1}, as the proportion of clover in the herbage harvested during the previous summer declined from 35% to 4% (Cuttle *et al.*, 1992). With swards that are used mainly for grazing, it is possible to reduce the amount of nitrate leached during the winter by introducing a cut for silage, preferably late in the growing season (Garwood and Ryden, 1986).

Little leaching of nitrate occurs from lightly grazed rough grassland but when such grassland is 'improved' by cultivation, and/or drainage and seeding with more productive species, leaching tends to increase. The increase depends on the intensity of the cultivation: disc harrowing has a relatively large effect but, at least in high rainfall areas, the increase is unlikely to produce an average concentration of nitrate in the drainage water of > 11.3 mg N l^{-1} (Roberts *et al.*, 1989). However, the disc harrowing and re-seeding of an area of rough grassland in Wales resulted in a leaching loss of about 97 kg N ha^{-1}

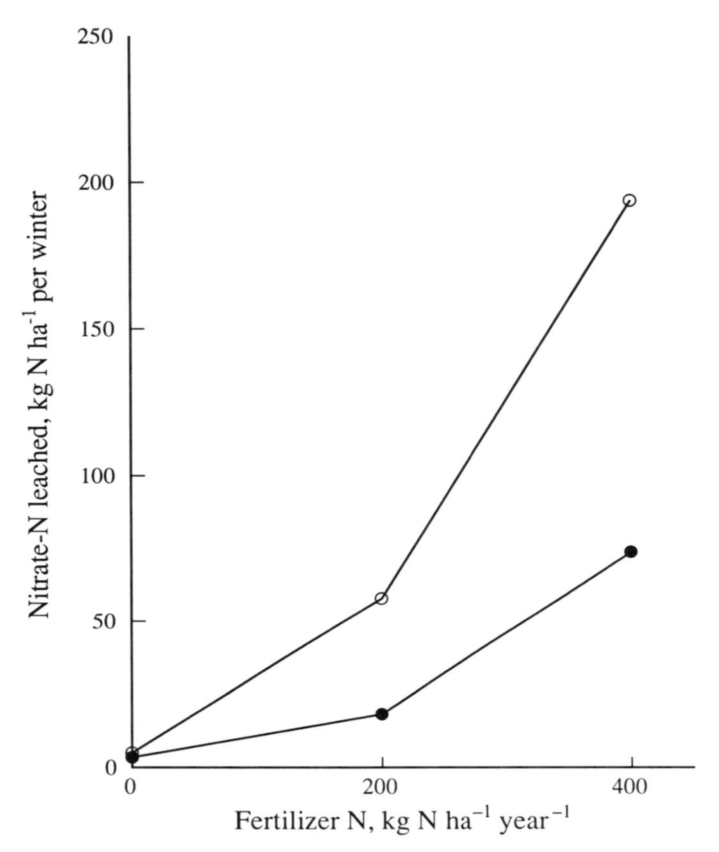

Fig. 7.2. Amounts of nitrate-N leached from plots of grazed long-term grassland on a clay loam soil: plots with mole drains at a depth of 55 cm (○); plots with no drainage system (●); means of 7 years. (Redrawn from Scholefield *et al.*, 1993.)

during a 3-year period, compared with about 30 kg N ha^{-1} for undisturbed grassland (Roberts *et al.*, 1989).

Influence of Soil Texture, Structure and Drainage on Leaching from Grassland

The actual amount of nitrate leached at a particular site is influenced by the water holding capacity and porosity of the soil, and therefore by its texture, structure and drainage status. The water holding capacity, which may be two to three times greater in loam and clay soils than in sandy soils, has two effects. One effect is on the amount of water required to re-wet the root zone after a dry period, and hence on the proportion of the annual rainfall that contributes to leaching. The second effect is on the amount of water required to displace the soil solution to a given depth, e.g to below the root zone (see Table 7.1). With a uniform input of water, the downward movement is two to three times greater in a sandy soil than in a clay soil. The pore distribution has a large effect on the ratio of mobile to non-mobile water and hence on the extent to which the water and nitrate in micropores can be by-passed. For example, the formation of macropores by earthworms has been reported to increase the rate of water movement through soils and the amount of N leached (Knight *et al.*, 1989).

The influence of soil type and irrigation on nitrate leaching from cut grass swards is illustrated by data from a lysimeter study that involved monoliths of four contrasting soils, ranging from a free-draining sand to a silty clay with poor natural drainage (Garwood and Ryden, 1986). All the lysimeters received fertilizer N at 420 kg ha^{-1} year^{-1} and, for each soil, half were irrigated to maintain the soil water deficit in summer to within 25 mm of field capacity while half received natural rainfall only. The leaching of nitrate was greatest from the sandy soil and least from the deep loam soil (Table 7.6) though, even when the sandy soil was irrigated, there was little downward movement of N during the growing season (Garwood and Ryden, 1986). In another lysimeter study with monoliths of long-term grassland on three soil types, each fertilized with 400 kg N ha^{-1} as calcium nitrate, a rather different result was obtained. Leaching losses in the first year were equivalent to 18 kg N ha^{-1} from a sandy loam, 39 kg N ha^{-1} from a silt loam and 44 kg N ha^{-1} from a clay soil, amounts that reflected a greater uptake of fertilizer N from the sandy loam than from the other two soils (Dowdell *et al.*, 1980).

In general, the ratio of nitrate leaching to denitrification is greater if the soil is well-drained. The ratio was found to be increased when a drainage system was installed under grazed paddocks on a clay loam soil in southwest England (Scholefield *et al.*, 1988). Some of the paddocks comprised long-term

Table 7.6. Amounts of nitrate-N leached from perennial ryegrass swards grown in lysimeters on four contrasting soils, each receiving 420 kg N ha^{-1} year^{-1}, during the period 1980–1984. (From Garwood and Ryden, 1986.)

	Soil series and type			
	Newport (free-draining sand)	Frilsham (sandy-loam over chalk)	Radyr (deep loam)	Tedburn (silty clay surface-water gley)
Available water capacity (mm)	96	137	162	177
Nitrate-N leached (kg N ha^{-1} year^{-1}):				
without irrigation	87	43	34	58
with irrigation	64	20	14	27

Table 7.7. Influence of drainage, and of re-seeding, of old grassland on the estimated amounts of nitrate-N leached and denitrified (kg N ha^{-1} $year^{-1}$) at two rates of fertilizer N (means of 4 years). (From Scholefield *et al.*, 1988.)

	Nitrate-N leached (kg ha^{-1})	Denitrification (kg ha^{-1})	Proportion (%) of leaching + denitrification as leaching
Old grassland; undrained; 200 kg N ha^{-1}	20	86	19
Old grassland; drained; 200 kg N ha^{-1}	56	56	50
Old grassland, undrained; 400 kg N ha^{-1}	48	110	30
Old grassland; drained; 400 kg N ha^{-1}	187	80	70
Re-seeded; undrained; 400 kg N ha^{-1}	24	113	13
Re-seeded; drained; 400 kg N ha^{-1}	74	78	49

grassland, receiving either 200 or 400 kg N ha^{-1} $year^{-1}$, and some comprised recently sown perennial ryegrass receiving 400 kg N ha^{-1} $year^{-1}$. For each type of sward, the installation of a soil drainage system lowered the water table, reduced the period during which the soil was waterlogged in winter, reduced denitrification and increased nitrate leaching (Table 7.7). The effect was greatest with the long-term sward receiving 400 kg fertilizer N ha^{-1} $year^{-1}$. However, the reduction in denitrification was much less than the increase in leaching, indicating that the improvement in drainage also increased the amount of nitrate mineralized from the soil organic N.

Influence of Rainfall and Irrigation on Leaching from Grassland

Rainfall or irrigation applied during the growing season generally reduces the amount of N that is leached during the following winter, by increasing grass growth and therefore the uptake of N. This effect is illustrated by the lysimeter study involving four soils with and without irrigation during the summer (Table 7.6). Conversely, when uptake during the growing season is seriously restricted by drought, the amount of N leached subsequently is likely to be high (Jordan and Smith, 1985). An extreme example of this effect was shown by a long-term lysimeter study in which large amounts of nitrate were leached during the winter of 1976/77 following two dry summers and a dry winter in 1975/76 (Table 7.2). The return of warm moist conditions in autumn 1976 undoubtedly promoted the mineralization of soil organic N, adding to the nitrate that was leached during the winter of 1976/77. However, the actual amounts of nitrate leached in this study were greater than would have occurred in normal farming practice because fertilizer N was applied at the pre-determined rates despite the summer drought conditions.

The influence of N uptake on leaching was also shown in a lysimeter study in which various patterns of drought and irrigation were imposed on cut swards fertilized with calcium nitrate at 400 kg N ha^{-1} $year^{-1}$ (Webster and Dowdell, 1984a). Irrigation equivalent to 120% of the average rainfall caused no increase in the leaching of nitrate compared with average rainfall, but there was a substantial increase when drought was imposed for 2 weeks either before or after each cut, within an irrigation regime that supplied the equivalent of the average total annual rainfall. These results suggest that the timing of fertilizer applications, in relation to rainfall during the growing season, will influence the amount of N leached during the following winter. As noted above (p. 131), the amount of winter rainfall required for nitrate to be leached almost completely to below a certain depth (e.g. 55 cm) varies with soil type: more than 400 mm of excess winter rainfall may be necessary on clay soils (Scholefield *et al.*, 1993).

Leaching of Fertilizer N Assessed by ^{15}N

In some investigations ^{15}N-labelled fertilizer has been used to assess the extent of nitrate leaching from the fertilizer itself. As would be expected, the proportion of the nitrate leached that is derived directly from the fertilizer increases with increasing rate of fertilizer application. For example, with perennial ryegrass grown on a sandy loam soil in lysimeters, the proportion was 9% with 250 kg N ha^{-1} $year^{-1}$, 39% with 500 kg ha^{-1} and 75% with 900 kg N ha^{-1} (Barraclough *et al.*, 1984). In another lysimeter study, the nitrate leached from ryegrass on a sandy loam soil receiving 400 kg N ha^{-1} $year^{-1}$ amounted to 8–33 kg N ha^{-1} $year^{-1}$, of which the fertilizer contributed about 60–70% in the first winter following the application (Dowdell and Webster, 1980).

However, the proportions of the leached nitrate derived directly from the fertilizer (as assessed by ^{15}N) underestimate the effect of the fertilizer application on the amount of nitrate leached. The underestimation is due to the exchange that occurs between the ^{15}N in the fertilizer and the ^{14}N in the soil organic matter (see p. 109). For this reason, the overall effect of fertilizer N on nitrate leaching is best assessed from a comparison of the total amounts of N leached with and without the fertilizer application.

Influence of Slurry and Sewage Sludge on the Leaching of Nitrate

Usually, about half the N in cattle and pig slurry is present as ammonium and about half as organic N (see p. 79). When slurry is applied to grassland, there is some loss of N through the volatilization of ammonia (see Chapter 8) and sometimes an additional loss through surface runoff (Sherwood, 1986). Most of the remaining ammonium-N is nitrified to nitrate and, if this is not taken up by the sward, it is susceptible to loss through leaching and denitrification.

Measurements that have been made of nitrate leaching following the application of slurry to grassland have usually shown the amounts to be small. In a field study in which cow slurry was applied at rates up to 550 t ha^{-1}, providing about 2500 kg N ha^{-1}, less than 1% of the slurry N was found subsequently as nitrate in the field drains, though it is possible that some nitrate bypassed the drains (Burford *et al.*, 1976). With pig slurry applied three times per year for 3 years, at rates averaging about 1400 kg total N ha^{-1} $year^{-1}$, on two soils in Ireland, it was estimated that 15% of the N was leached on the well-drained loam soil but only 2% on the impermeable gley soil (Sherwood, 1986). Smaller proportions were leached at lower rates of application. In a 10-year lysimeter study in which cow slurry, providing the equivalent of 200 kg N ha^{-1}, was applied each spring to ryegrass growing on both a sandy loam and a chalk soil, only 10–14 kg N ha^{-1} was leached, an amount similar to that from a comparable sward receiving ammonium nitrate at 125 kg N ha^{-1} $year^{-1}$ (Unwin, 1986). Increasing the number of applications of slurry to four per season, supplying a total of 500 kg N ha^{-1}, increased the amount of N leached to about 40–44 kg N ha^{-1}, half the amount leached from 500 kg N ha^{-1} $year^{-1}$ supplied as ammonium nitrate (Unwin, 1986).

The timing of the application is an important factor in the leaching of nitrate from slurry. When diluted cow slurry was applied, at rates between 150 and 250 kg total N ha^{-1}, to plots of grassland on a shallow calcareous soil over chalk, at monthly intervals from September to January, the proportion of the N leached increased from 17% for the September application to 43% for the November application, but was 10% or less for the slurry applied in December and January (Froment *et al.*, 1992). The low leaching losses from applications in December and January were consistent with other results on a freely-drained soil (Thompson *et al.*, 1987). Factors that curtail the amount of nitrate leached from slurry applied to grassland include the occurrence of ammonia volatilization and denitrification, and the fact that the organic N mineralizes slowly. When slurry is injected into grassland rather than applied to the surface, nitrate leaching may be increased, particularly if the slurry is injected outside the growing season (Steenvoorden *et al.*, 1986). The amount of ammonia volatilization is reduced by injection, and more soluble N therefore remains in the soil. However, for moderate rates of slurry applied during the growing season, the amount of nitrate leached during the following winter may be less when the slurry is injected than when it is applied to the surface, due to a greater uptake of N by the grass (Steenvoorden, 1989). A possible means of reducing the leaching of nitrate from slurry is to incorporate a nitrification inhibitor (Pain, 1994), but such treatment is expensive. Although the N leached from slurry is normally in the form of nitrate, the application of large amounts of slurry during the autumn may result in some leaching of ammonium and organic N, as shown with pig slurry applied to a light sandy soil in the Netherlands (Steenvoorden *et al.*, 1986).

The application of sewage sludge may also increase the amount of nitrate leached from grassland, as reported from the Netherlands (De Haan, 1986) but the amount of nitrate is greatly influenced by the treatment, if any, previously applied to the sludge. 'De-watered' sludge cake, for example, may immobilize nitrate temporarily and thus minimize leaching (Smith *et al.*, 1992).

Loss of N by Surface Runoff

The surface runoff of water occurs when the intensity of rainfall (or the melting of snow) exceeds the capacity for infiltration into the soil. The water then moves laterally over the soil surface to ditches and streams, taking with it soluble, and sometimes particulate, material. Runoff is most likely to occur on soils that are poorly structured or have a high groundwater table, have a considerable slope or are frozen just below the surface. It occurs most commonly during the winter, and therefore affects slurry to a greater extent than fertilizer. Although fertilizer N is normally not applied at times when runoff is likely, applications in early spring to impermeable soils may be susceptible. When grazing animals are present, runoff may remove some of the N deposited in excreta. Much of the N in surface runoff from grazed or slurry-treated grassland may be in the form of ammonium (Heathwaite *et al.*, 1993).

The extent to which slurry or fertilizer N is lost by surface runoff increases with an increase in slope, but decreases with the length of time between the application and the rainfall event causing the runoff (Fig. 7.3). In some situations, runoff can be reduced by improving the porosity of the soil and, in other situations, it can be reduced by lowering the water table. With grassland, the extent of runoff is less when the density of tillers is relatively high and when there is a substantial layer of dead leaf litter at the soil surface.

In rotationally grazed paddocks receiving fertilizer N at a rate of 224 kg ha^{-1} $year^{-1}$, in a hilly area of Ohio, the loss of N by surface runoff represented 14–20% of the combined loss by runoff and leaching (Owens *et al.*, 1983). In Ireland, the application of pig slurry to grassland on two soils, at a rate of 3.7 t dry matter ha^{-1}, was found to produce a concentration of ammonium in the runoff water of more than 120 g N m^{-3} when runoff occurred within 2 days of the application, but much lower concentrations when the interval between application and runoff was more than 5 days (Sherwood and Fanning, 1981). A similar pattern of results was obtained with cattle slurry.

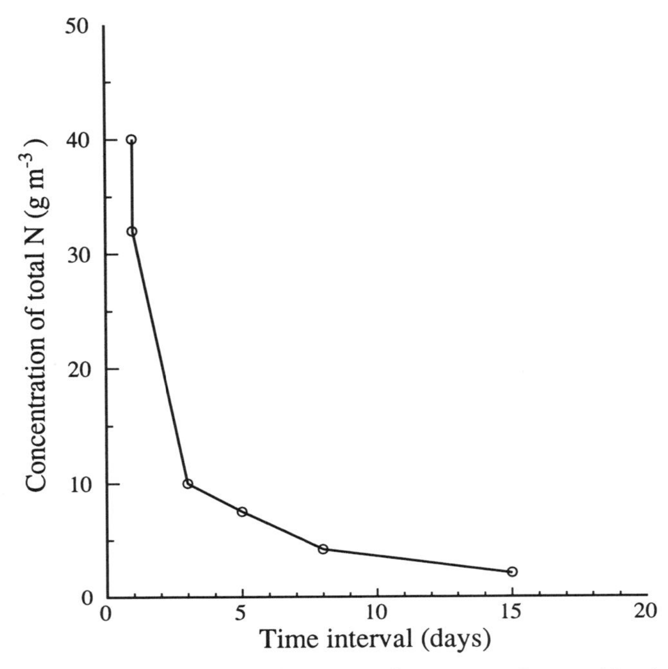

Fig. 7.3. The concentration of total N in runoff water as influenced by the time interval between the spreading of slurry (1 t DM ha^{-1}) and the rainfall event causing runoff. (Redrawn from Steenvoorden, 1988.)

Leaching and Surface Runoff of N from Grassland Catchments

Studies based on the analysis of drainage water from complete catchment areas have the advantage of avoiding the need to analyse numerous replicate samples of soil or leachate, but they have the disadvantage of being restricted to areas with an impermeable subsoil. In a catchment of 40 ha, with both cut and grazed grass receiving fertilizer N plus slurry and manure, in a high rainfall area in northwest England, the loss of nitrate in the stream draining the catchment increased from about 35 to 70 kg N ha^{-1} $year^{-1}$ during a 4-year period, as the input of N from fertilizer and slurry increased from 188 to 240 kg N ha^{-1} (see Garwood and Ryden, 1986). In addition, there was a small loss of N as ammonium (1–9 kg N ha^{-1}). Smaller amounts of N in the drainage water from a grassland catchment were reported in Northern Ireland, despite higher rates of fertilizer N. The area of 6 ha included both cut and grazed swards that received about 420 kg N ha^{-1} $year^{-1}$, about 75% as fertilizer and 25% as slurry, and the loss of nitrate in the drainage water was only 17 to 36 kg N ha^{-1} $year^{-1}$ (Jordan and Smith, 1985). These amounts are low in relation to the input of N and probably result from a combination of factors, e.g. the removal of some herbage as hay or silage, relatively high rates of denitrification and/or incomplete interception of water and nitrate by the drains.

In a study of nitrate in the stream flow from a 46 km^2 grassland catchment area in Devon during an 8-year period, the average amount of nitrate-N lost per year was 24 kg N ha^{-1}, but the amounts in individual years ranged from 8.3 kg in 1975/76 to 51.4 kg in 1976/77 (Webb and Walling, 1985). The actual fertilizer inputs were not recorded but the area was used mainly for dairy farming. The concentration of nitrate in the stream leaving the catchment was highest in December and lowest in early autumn. In a small grazed grassland catchment on a clay soil, there was a large difference in the amount of nitrate leached in two successive winters (Haigh and White, 1986). Several factors are thought to have contributed to this difference, but the timing of fertilizer applications in relation to weather conditions, and differences in the duration and intensity of grazing, were considered to be important.

Most of the nitrate derived from the various forms of N in surface runoff, together with some leached nitrate, adds to the concentration present in streams, rivers and lakes. The streams and rivers in lowland England usually have average nitrate concentrations of more than 5.5 mg nitrate-N l^{-1} but some have concentrations of more than 12.5 mg l^{-1} (Johnes and Burt, 1993).

Influence of the Ploughing of Grassland on the Leaching of Nitrate

The ploughing of grass swards often induces a large increase in the mineralization of N (see p. 125) and this may result in large amounts of nitrate being leached. A large proportion of the old grassland in lowland UK was ploughed during the period between 1940 and 1970, and this practice resulted in substantial increases in the concentration of nitrate in groundwater. Although some of the N mineralized as a result of ploughing is taken up by subsequent crops and some is denitrified, much of it is leached. In the first season after the ploughing of old grassland, water draining from the field may well contain up to 450 mg N l^{-1} (Whitmore *et al.*, 1992). When old permanent grassland at Rothamsted was ploughed, on a soil that contained 8100 kg N ha^{-1} to a depth of 25 cm at the time of ploughing, there was an average loss of 490 kg N ha^{-1} $year^{-1}$ during the next 3 years with most of the loss probably being due to leaching (Jenkinson, 1986).

In a study involving the ploughing of 6-year-old swards that had received 420 kg fertilizer N ha^{-1} each year, 275–310 kg N ha^{-1} were estimated to have been subject to leaching in the year following ploughing (Ryden *et al.*, 1984).

In another study, about 100 kg N ha^{-1} was estimated as being leached below 90 cm during two winters, following the ploughing of grass swards established 3 and 7 years previously, despite some uptake by crops of winter wheat (Cameron and Wild, 1984). About 42 kg N ha^{-1} was estimated as leached during 20 weeks following the ploughing of a 4-year meadow fescue ley that had received fertilizer N at 200 kg N ha^{-1} $year^{-1}$ (Bergström, 1987).

Substantial leaching may also occur following the ploughing of grass–clover leys that have received no fertilizer N. For example, it was estimated that about 70 kg N ha^{-1} was leached during the winter after the ploughing of a 4-year grass–clover ley in October (Watson *et al.*, 1993). Although grass–clover leys often form the basis of organic farming systems, the extent to which organic farming contributes to overall nitrate leaching is limited by two factors. First, the leys are normally ploughed only once in 3–4 years and, second, the arable crops in organic farming rotations do not receive fertilizer N, though they may receive organic manures and slurries. With good management, the total amounts of nitrate leached from an organic farm should not produce an average concentration in the drainage water of more than 11.3 mg NO_3-N l^{-1} (Watson *et al.*, 1993).

The profiles of nitrate concentration in the groundwater below two adjacent areas, one of continuous long-term grassland, and the other of previously long-term grassland that had been ploughed and then cropped for 12 years with cereals, are shown in Fig. 7.4. These results, and other similar data, provided the basis for a mathematical model developed to predict the amount of nitrate-N released following the ploughing of grassland of various ages (Oakes, 1982; Young, 1986). Some predictions from the model for the amount of nitrate leached in each of four successive years after ploughing are shown in Table 7.8. Another approach to predicting the amounts of nitrate that might be leached following ploughing has been to measure the quantities of the stubble, leaf litter, roots and macroorganic matter in swards of various types and ages, and the concentrations of N in these fractions (Whitehead *et al.*, 1990). In general, the predictions from these two approaches show good agreement, though the model implies that the amount of N mineralized reaches a maximum at a sward age of about 4 years whereas measurements of the N in

Table 7.8. Amounts of nitrate-N estimated as being susceptible to leaching after the ploughing of grass swards of various ages. (From Young, 1986.)

	Nitrate-N (kg ha^{-1}) released by ploughing			
Age of sward (years)	Total	Year of ploughing	Following year	Next following year
1	100–120	60–72	30–36	10–12
2	180–190	108–114	54–57	18–19
3	240–310	144–186	72–93	24–31
4 or more	280–380	168–228	84–114	28–38

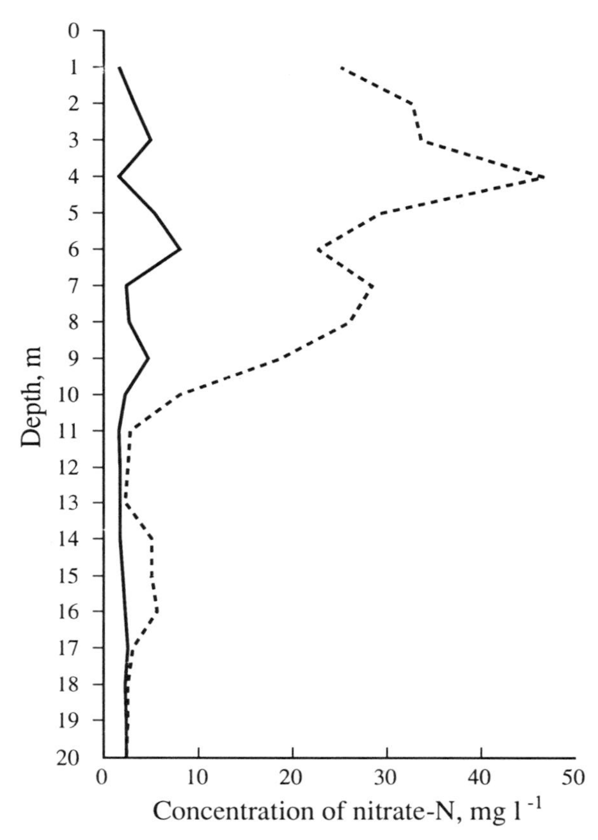

Fig. 7.4. Concentration of nitrate-N in the soil water at depths to 20 m below continuous long-term grassland (—) and an adjacent area ploughed 12 years previously and cropped with cereals in subsequent years (- - - -), both over chalk. (Redrawn from Foster *et al.*, 1982.)

stubble, roots etc indicate that the amount would increase substantially beyond this age (Whitehead *et al.*, 1990). In addition to the effect of sward age, the amounts of N in the stubble, roots etc. are influenced by the previous sward management, being increased by fertilizer N and by grazing as compared with cutting management. The amounts of nitrate leached after ploughing tend to reflect these differences. For example, when two plots of ryegrass that had received either nil or 400 kg fertilizer N ha^{-1} $year^{-1}$ for 4 years were ploughed and sown to wheat, the amounts of nitrate leached were 7 and 70 kg N ha^{-1} respectively (Webster and Dowdell, 1984b). Evidence of increased nitrate leaching due to higher rates of fertilizer N being applied to grass in the years before ploughing was also reported in a study of several farm sites (Lloyd, 1992a).

Environmental Impact of Nitrate Leaching

There are two main concerns about the leaching and runoff of nitrate from agricultural land. First the nitrate may give rise to appreciable concentrations in domestic water supplies and, second, it may contribute to the eutrophication of streams, rivers and lakes and encourage the excessive growth of algae. Although eutrophication usually reflects the supply of phosphate rather than nitrate, the effect of nitrate becomes apparent when substantial amounts of phosphate are also present. In recent years, the effects of eutrophication have been reported not only in rivers and estuaries but also in semi-enclosed areas of sea such as the Baltic Sea (Schrøder, 1990).

It is important to appreciate that a large proportion of agricultural land is used not only for agriculture but also for the catchment of water supplies, either by the re-charging of groundwater or by the collection of water from streams and rivers into reservoirs, and this additional function of land should be taken into account in agricultural management. As noted above, an important factor that affects the actual concentration of nitrate in groundwater (and in surface water) is the amount of rainfall. The concentration in groundwater reflects both the total amount of nitrate leached per year and the volume of water draining below the root zone, viz. the excess winter rainfall. With increasing excess winter rainfall, there is an increase in the amount of nitrate, expressed in terms of kg ha^{-1}, that can be leached before a concentration in the leachate of 11.3 mg nitrate-N l^{-1} is exceeded. In the relatively dry areas of eastern England, only about 30 kg N ha^{-1} can be leached without the average concentration for the winter drainage period exceeding 11.3 mg NO_3^--N l^{-1} (Greenwood, 1990). In the wetter western areas, the critical amount may be as much as 200–250 kg N ha^{-1}. However, the average concentration for a winter period obscures variation in time, and a concentration of 11.3 mg NO_3^--N l^{-1} may be exceeded during several leaching episodes despite the average concentration being much less. Variation with time is caused by factors such as pore-size distribution, the partitioning of nitrate between micropores and macropores, and the intensity and duration of rainfall.

Typically, less nitrate is leached from cut grassland than from arable cropping systems, but the amount leached from intensively grazed grassland may well exceed the amount from arable cropping. In all systems, high rates of fertilizer N increase the concentration of N in plant residues, which then release more inorganic N after mineralization. It is difficult to make a direct comparison between grassland and arable systems, partly because much grassland is cultivated from time to time. In a ley farming system, such cultivation may result in the N mineralized from grass residues contributing to the N leached from a subsequent arable crop.

There are a number of management practices that can achieve some reduction in nitrate leaching from grassland itself, by restricting the amount of inorganic N in the root zone at the end of the growing season (Garwood and Ryden, 1986; Steenvoorden, 1989). These include:

1. a greater integration of cutting and grazing, especially the cutting for silage late in the season of swards that have been grazed intensively earlier in the season;
2. adjusting the rate of successive fertilizer applications during the season to take account of weather conditions;
3. the greater use of grass–clover swards receiving little or no fertilizer N; and
4. the use of irrigation when necessary during the growing season to encourage the uptake of N.

In addition, where slurry is to be applied, less leaching is likely when the application is made in early spring than in late autumn though, with clay soils, application in early spring may not be possible.

In areas where concentrations of nitrate in groundwater exceed the limit of 11.3 mg NO_3^--N l^{-1} from time to time, measures may be taken to reduce the concentration in domestic water supplies. Long-term storage in reservoirs, a process that allows denitrification to occur, and the blending of high-nitrate water with supplies low in nitrate are two common examples (Royal Society, 1983). Treatments such as enhanced biological denitrification and ion exchange are also being developed (Croll and Hayes, 1988). However, the cost of removing 1 kg N from drinking water by a combination of ion exchange and denitrification is about 10 times greater than the cost of 1 kg fertilizer N (van der Meer and Wedin, 1989). The restrictions on agricultural management aimed at curtailing the amounts of nitrate being leached that are currently being introduced through the EC Nitrate Directive and subsequent EU regulations are described in Chapter 16.

8 Volatilization of Ammonia

Sources and Amounts of Ammonia in the Atmosphere

Large amounts of ammonia are volatilized to the atmosphere and the main sources are agricultural, with livestock excreta and fertilizers being most important. Other sources of ammonia including industrial processes, the combustion of fossil fuels and sewage treatment works are less important though their contributions are not well quantified (Buijsman *et al.*, 1987; Lee and Dollard, 1994). Once in the atmosphere, gaseous ammonia reacts progressively with acidic compounds to form ammonium aerosols (mainly ammonium sulphate and ammonium nitrate) and, depending on meteorological conditions, these may be transported over large distances. However, both gaseous ammonia and the aerosols are returned, sooner or later, to the land or sea surface by wet and dry deposition. The background concentration of ammoniacal N ($NH_3 + NH_4^+$) in the atmosphere is usually in the range 1–10 μg NH_3 m^{-3} (Dabney and Bouldin, 1990; Fangmeier *et al.*, 1994; Warneck, 1988) and is usually less than 3 μg NH_3 m^{-3} in areas that are not affected by intensive livestock production and/or the disposal of slurry (Allen *et al.*, 1988; Denmead *et al.*, 1976; Sutton *et al.*, 1993). The ratio of NH_3 to NH_4^+ in the atmosphere is usually less than 1 except in localized areas where large amounts of ammonia are volatilizing (Warneck, 1988). Residence times of NH_3 and NH_4^+ in the atmosphere depend on factors such as the rate of conversion of NH_3 to NH_4^+ and the rates of wet and dry deposition but, in general, are probably between 0.8 and 4 days for NH_3 and between 5 and 19 days for NH_4^+ (Fangmeier *et al.*, 1994; Sutton *et al.*, 1993). The dry deposition of ammoniacal N is relatively more important in areas where emissions are large, whereas wet deposition is more important where emissions are small (Fangmeier *et al.*, 1994).

The presence of intensively managed livestock has been shown to increase the concentration of ammoniacal N in the atmosphere close to the source. For example, concentrations, averaged over a period of 17 months were 20 and 29 μg NH_3 m^{-3} at two sites close to livestock farms in southeast England, whereas concentrations were less than 5 (and usually less than 3) μg m^{-3} at 17 other sites, both rural and urban (Allen *et al.*, 1988). At three sites in areas of particularly intensive livestock production in the Netherlands, mean concentrations expressed on a monthly basis ranged from 5 to 35 μg NH_3 m^{-3} and, on a few occasions, the daily average exceeded 250 μg N m^{-3} (Erisman *et al.*, 1987). In general, the highest concentrations of ammoniacal N occur in spring and summer, and are associated with the spreading of slurries, manures and fertilizers, and with increasing temperatures. Concentrations of more than 100 μg m^{-3} have been recorded immediately after the spreading of slurry (Klasink *et al.*, 1991). Locally high concentrations have also been reported near feedlots and intensive dairy farms in the USA (Hutchinson and Viets, 1969; Luebs *et al.*, 1974).

The effect of livestock production is due mainly to the volatilization of ammonia from urea, which usually accounts for between 60 and 90% of the N in urine (see p. 74). Urea is also increasingly used in synthetic form as a fertilizer, and its application in solid form to the sward surface of grassland allows ammonia to volatilize readily. The urea in both urine and fertilizer is hydrolysed by urease, a microbial enzyme that is widespread in soils (Bremner and Mulvaney, 1978) and on plants and plant litter (Freney and Black, 1988). On soils with a pH higher than about 6.3, the product of the hydrolysis is mainly ammonium bicarbonate:

$$CO(NH_2)_2 + H^+ + 2H_2O \rightarrow 2NH_4^+ + HCO_3^-$$

but, on soils with a lower pH, the reaction becomes increasingly:

$$CO(NH_2)_2 + 2H^+ + 2H_2O \rightarrow 2NH_4^+ + H_2O + CO_2$$

Except in dry or cold conditions, urea added to soils is hydrolysed completely within a few days (Holland and During, 1977; Stevens *et al.*, 1989c; Thomas *et al.*, 1988).

The urease activity of soils is associated with organic matter (Kissel and Cabrera, 1989; O'Toole *et al.*, 1982; Reynolds *et al.*, 1985). It therefore tends to decrease with depth, in parallel with the decrease in organic matter content (Bremner and Mulvaney, 1978; Mulvaney and Bremner, 1981), and is greater in grassland soils than in corresponding arable soils (O'Toole *et al.*, 1985b; Reynolds *et al.*, 1985; Whitehead and Raistrick, 1993b). Urease activity increases markedly with temperature over the range of 0–40°C (Vlek and Carter, 1983) though slight activity has been detected even below 0°C (Bremner and Mulvaney, 1978). The optimum pH is about 6–7 (Rachhpal-Singh and Nye,

1984). Anaerobic conditions appear not to influence the activity of existing urease, but they may inhibit the production of new urease (McCarty and Bremner, 1991).

Another precursor of ammonia is uric acid, which is the main form of N in poultry excreta, and which occurs in small amounts in the urine of cattle and sheep. It is hydrolysed by bacterial uricase enzymes, via allantoin and allantoic acid, to urea (Mahler and Cordes, 1971) which is then hydrolysed by urease. Uricase enzymes are present in soils (Martin-Smith, 1963) but there is little information on factors influencing their activity. Ammoniacal N is also produced during the decomposition of the protein in plant residues, but there is often little volatilization from this source because the resulting concentration at the soil surface is low, much lower than that resulting from the return of urine or the application of slurry or fertilizer N.

The ammonium produced in soils is subject to cation exchange, nitrification, microbial assimilation and plant uptake, and all of these processes tend to reduce the concentration in the soil solution. However, in solution, the ammonium is in equilibrium with free ammonia, and the ammonia in solution is in equilibrium with ammonia in the atmosphere:

$$\begin{array}{c} NH_3 \text{ (in atmosphere)} \\ \upharpoonleft\downharpoonright \\ NH_4^+ + OH^- \rightleftharpoons NH_3 + H_2O \text{ (in soil solution)} \end{array}$$

The volatilization of ammonia expressed as a proportion of the total ammoniacal N in the soil depends on these equilibria. In solution, the proportion of the ammoniacal N that is present as ammonia increases markedly with increasing pH (Fig. 8.1). When urea is hydrolysed in soil, H^+ ions are removed from solution (see p. 153) thus increasing the localized soil pH and promoting the volatilization of ammonia. For example, in a urine patch, there is a rapid rise in pH during the first 24 h, greatest near the soil surface and decreasing with depth (Ball *et al.*, 1979; Vallis *et al.*, 1982). A rise of 2.0 to 3.5 pH units is not uncommon in the surface 0.5 cm of soil (Sherlock and Goh, 1984; Vallis *et al.*, 1982). However, as ammonia volatilizes, the remaining ammonium ions effectively combine with hydroxyl ions to maintain equilibrium in the soil solution and this, together with the nitrification of ammonium, lowers the soil pH. The progressive decrease in the concentration of ammonia in the soil, and the lowering of soil pH, reduce the rate of volatilization, as illustrated by a typical cumulative volatilization curve (Fig. 8.2). The main influence on the second equilibrium, between ammonia in solution and ammonia in the atmosphere, is temperature. Increasing temperature tends to increase volatilization by decreasing the solubility of ammonia, and increasing its rate of diffusion (Freney *et al.*, 1981; Smith *et al.*, 1990). With soils, increasing temperature also tends to reduce the volume of the soil solution.

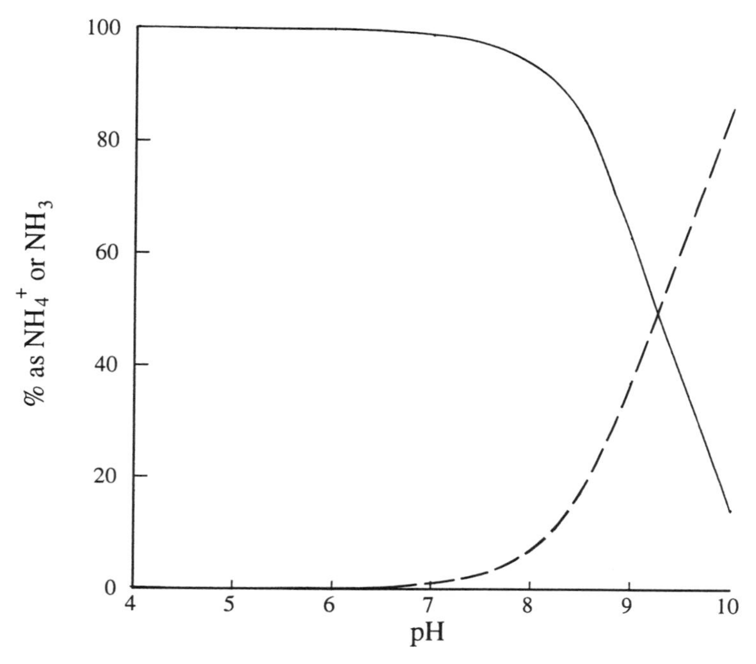

Fig. 8.1. Proportion (%) of ammoniacal N as NH_4^+ (—) and NH_3 (----) as influenced by solution pH. (From Court *et al.*, 1964.)

Methods for the Measurement of Ammonia Volatilization

In laboratory studies of ammonia volatilization, batches of soil amended with fertilizer or urine are usually incubated for between 1 and 3 weeks in a container supplied with a continuous flow of air, and the ammonia in the outgoing air is trapped in a dilute acid solution which is subsequently analysed for NH_4-N. Usually the acid is replenished at intervals during the incubation period. It is important, in studies of this type, that the flow of air should be sufficient to prevent the ammonia being reabsorbed by the soil (Fenn and Hossner, 1985), or dissolved in condensation in the containers or tubing.

In field studies of ammonia volatilization, two types of method have been used. One type involves trapping the ammonia volatilized under enclosures that are placed over treated areas of the field, while the second type, suitable for large plots or whole fields, involves measuring the ammonia concentration and the wind speed at various heights (or, in some circumstances, one height) above the surface of the field. Methods of this second type as known as micrometeorological methods.

The enclosure methods are relatively simple to operate and enable a number of different treatments to be compared but, as with laboratory

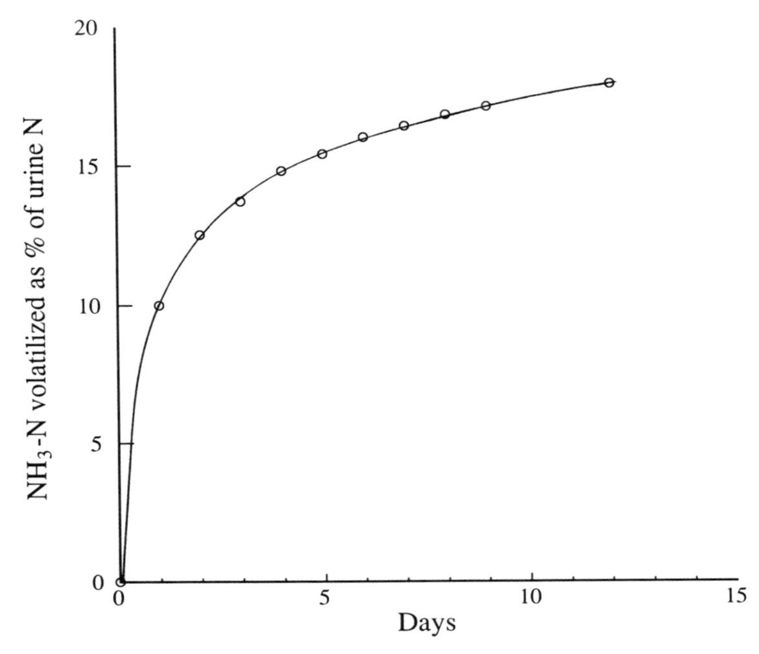

Fig. 8.2. Cumulative volatilization of ammonia from cattle urine applied to grassland: typical relationship for warm dry weather. (From Lockyer and Whitehead, 1990.)

methods, they require a continuous and substantial flow of air in order to produce satisfactory results. However, even with a continuous air flow, they have the disadvantage that the enclosures inevitably prevent rainfall reaching the soil, and modify wind speed and temperature, all factors which influence ammonia volatilization (Freney and Black, 1988). In addition, ammonia may be retained on the walls of the enclosure and fittings, and may dissolve in any water that condenses in the system. These problems can be minimized by using transparent 'wind tunnel' or similar enclosure systems which have relatively little effect on temperature and humidity of the air, and in which the rate of air flow is sufficient to prevent condensation (Kissel *et al.*, 1977; Lockyer, 1984; Ryden and Lockyer, 1985). When used to estimate volatilization from a whole field or a large plot, it is possible to move the enclosures at intervals to allow for the effect of rainfall, and to minimize the effect of spatial variation.

The micrometeorological methods are generally the most suitable for areas of 0.1 ha or more, as they have no effect on the factors that influence the rate of volatilization and they provide a value that is integrated over the whole area (Freney and Black, 1988). However, micrometeorological methods are not practical for comparisons of a number of adjacent experimental plots and, if they are to be effective, volatilization must be much greater from the area being assessed than from the surrounding area. Micrometeorological methods also require large numbers of measurements to be recorded and processed. The most widely used of the micrometeorological methods is the mass balance method which is based on the assumption that the amount of ammonia moving horizontally across a vertical plane of unit width on the downwind edge of a specific area is equal to the amount volatilizing from the surface of a strip of similar width upwind. The horizontal flux at any height can be determined from the horizontal wind speed and the concentration of ammonia at that height; and the total horizontal flux can be obtained by integrating over the height that is influenced by the ammonia volatilization (Denmead, 1983). The flux density can be expressed as:

$$F = \frac{1}{x}\int_0^z \overline{uc}\,\mathrm{d}z$$

where x is the distance travelled by the wind over the treated area, z is the height of the modified layer in the atmosphere, u is wind speed and c is the ammonia gas concentration in excess of the background. The profiles of u and c are obtained by measurements at a number of heights through the modified layer (Freney and Black, 1988).

It is also possible to make an assessment of the amount of ammonia volatilized from the measurement of ammonia concentration at a single height, often referred to as Z_{inst}, which varies with circumstances but is usually between 1 and 2 m (Wilson *et al.*, 1983). This modification greatly reduces the amount of equipment and labour, but the concentration at Z_{inst} is usually low, and measurement error is therefore much more significant than at lower heights. In an attempt to overcome this disadvantage, an intermediate type of procedure has been proposed (McInnes *et al.*, 1985).

In a comparison of an enclosure method and the mass balance micrometeorological method, the cumulative volatilization of ammonia was essentially the same, but the hourly rates differed considerably (Black *et al.*, 1985a). However, it has been suggested that micrometeorological methods may under-estimate actual volatilization, partly because it is difficult to obtain an accurate value for the background ammonia concentration and partly because the measured volatilization tends to be reduced by the deposition of ammonia from the atmosphere (Vertregt and Rutgers, 1991). Conversely, it is possible that enclosure methods may overestimate net volatilization if they increase temperature and/or the rate of air flow relative to the field situation.

The Influence of Weather Factors on Volatilization of Ammonia

The weather factors that are most important in influencing ammonia volatilization are temperature and rainfall, particularly the amount and timing of

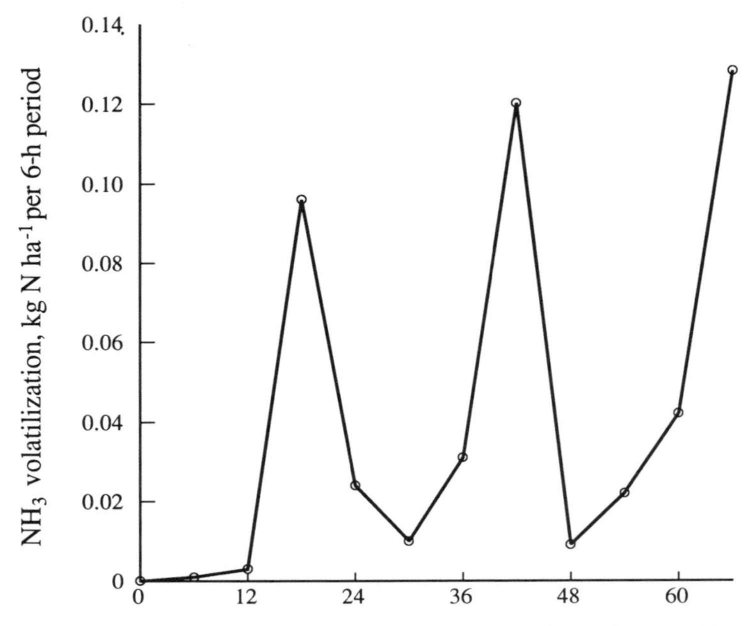

Fig. 8.3. Diurnal variation in ammonia volatilization from a grazed sward in southern England shown by 12 consecutive 6-hour periods in early autumn. (Redrawn from Hatch *et al.*, 1990a.)

rainfall after the application of urine or fertilizer. Other factors that sometimes have appreciable effects are the initial soil moisture status, the relative humidity of the air and wind speed.

The volatilization of ammonia from urea applied to soil generally increases with increasing temperature over the range 0–45°C (Ernst and Massey, 1960; Gasser, 1964; McGarry *et al.*, 1987), though the size and pattern of the increase have differed in different investigations. In some instances, there has been little effect of temperature (e.g. Holland and During, 1977) or little if any increase above 20°C (e.g. Whitehead and Raistrick, 1991). The increase in volatilization with temperature is due partly to increased urease activity, partly to a greater proportion of the ammoniacal N being present as ammonia gas, and partly to faster diffusion (Rachhpal-Singh and Nye, 1986a,b).

In the field, there is often a marked diurnal variation in the rate of ammonia volatilization, with the maximum rate occurring around midday and the minimum around midnight (Fig. 8.3). This pattern has been reported to occur following the application of fertilizer urea (Black *et al.*, 1985a; Harper *et al.*, 1983) and from grazed swards (Hatch *et al.*, 1990a). Following the application of slurry, diurnal variation is often small presumably because most of the ammoniacal N volatilizes quickly (Thompson *et al.*, 1990a). The diurnal pattern appears to be related mainly to fluctuation in temperature (Hargrove, 1989; Haynes and Sherlock, 1986) though differences in wind speed and turbulence may have some effect: both temperature and wind speed are generally lowest at night. In some weather conditions, ammonia may be retained in dew and volatilize as the dew evaporates (Harper *et al.*, 1983; Lightner *et al.*, 1990) thus accentuating the difference between day and night. The diurnal variation is greater in summer than in winter because the variation in both temperature and wind speed is usually greater in summer (Asman, 1992).

There is also a seasonal variation in ammonia volatilization, attributable mainly to differences in temperature. Thus, when cattle urine was applied to grass swards at different times of year between March and October, the proportion of the urinary N volatilized during a period of 15 days was found to be closely correlated with the average soil temperature measured at a depth of 3 cm on the three days following the application (Lockyer and Whitehead, 1990). However, in January, when volatilization occurred slowly, the total volatilization over 15 days was greater than predicted by this correlation.

With both urine and fertilizer urea, ammonia volatilization is reduced by heavy rain following the application. In field investigations, ammonia volatilization was greatly reduced if more than about 10 mm of rain fell within 2 days of fertilizer application (Black *et al.*, 1987) or the start of an intensive grazing period (Bussink, 1994). Volatilization was completely inhibited when 25 mm of rain fell immediately after the application of fertilizer urea to a bare soil (Bouwmeester *et al.*, 1985). However, volatilization was promoted when the topsoil was moist (Bouwmeester *et al.*, 1985) or when light rainfall followed the application of solid urea (Lightner *et al.*, 1990) or the start of a grazing period (Bussink, 1994). In a laboratory investigation, the volatilization of ammonia from urine applied to soil was progressively reduced as the amount of simulated rainfall was increased up to at least 16 mm, but the effect diminished as the time interval between the application of the urine and the rainfall was increased from 2 hours to 2 days (Whitehead and Raistrick, 1991).

The moisture content of the soil at the time of urea addition often has little effect on ammonia volatilization, unless the soil is so dry that urease activity is inhibited (Ferguson and Kissel, 1986; Rachhpal-Singh and Nye, 1986b; Vlek and Carter, 1983; Whitehead and Raistrick, 1991). Soil dryness has a greater effect with fertilizer urea than with urine as it restricts the rate at which the urea dissolves as well as curtailing hydrolysis (Black *et al.*, 1987; Reynolds and Wolf, 1987a; Vlek and Carter, 1983). When the soil is moist at the time that urea or urine is added but the weather is dry subsequently, the dry conditions are likely to increase volatilization (Burch and Fox, 1989). Conditions that favour the evaporation of water favour the volatilization of ammonia, as the volume of soil water decreases, and water moving upwards to the soil surface may be accompanied by ammonia in solution.

The relative humidity of the air generally has little effect on the volatilization of ammonia unless the soil is dry, when volatilization may be further reduced by a low relative humidity (Reynolds and Wolf, 1987a; Whitehead and Raistrick, 1991), probably due to restricted hydrolysis of urea. The effect of wind speed also depends on other conditions: increasing wind speed may increase volatilization if the soil moisture content is maintained but it may have a drying effect on the soil, thus reducing the hydrolysis of urea and ammonia volatilization (Bouwmeester *et al.*, 1985; Ferguson and Kissel, 1986; Reynolds and Wolf, 1987a).

The Influence of Soil Factors on the Volatilization of Ammonia

The amount of ammonia volatilized from urine or fertilizer N applied to soil normally increases with increasing ammonium concentration in the soil solution and with increasing soil pH. In the field, ammonia volatilization from urea (as a proportion of that applied in urine or fertilizer) has been found to increase with increasing rate of application (Hargrove *et al.*, 1977; Overrein and Moe, 1967; Whitehead *et al.*, 1989b) as well as with the factors that increase the rate of hydrolysis. For a given rate of application, the concentration of ammonium in the soil and the resultant pH are influenced by the cation exchange capacity (CEC) and buffering capacity of the soil. In general, the higher the CEC, the lower the concentration of ammonium in the soil solution and therefore the less ammonia is volatilized. Negative correlations between soil CEC and ammonia volatilization have been observed in several investigations (Lyster *et al.*, 1980; O'Toole *et al.*, 1985a; Whitehead and Raistrick, 1993b) with soils ranging in CEC from about 60 to 450 mmol kg^{-1}.

Although the initial soil pH does influence the extent of volatilization, a more important pH value is that attained by the zone of soil in contact with the urine or urea fertilizer after hydrolysis has occurred (Lyster *et al.*, 1980; O'Toole *et al.*, 1985a; Sherlock *et al.*, 1988). As noted above, the hydrolysis of urea results in a localized increase in pH, but the actual pH attained varies with the initial soil pH and the buffering capacity of the soil. The greater the buffering capacity, the smaller the increase in soil pH and the less the volatilization of ammonia.

In a study of the volatilization of ammonia from cattle urine applied to 22 soils, the proportion of the urinary N volatilized ranged from 7% to 41%, and the soil property most closely correlated ($r = -0.834$) with the extent of volatilization was CEC (Whitehead and Raistrick, 1993b). In this study, the titratable acidity of the soil was poorly correlated with ammonia volatilization whereas, in a similar study involving fertilizer urea applied to 36 soils, titratable acidity was more closely correlated with ammonia volatilization than was

CEC (Stevens *et al.*, 1989c). A possible reason for the greater effect of CEC in the investigation with urine is that urine infiltrates into the soil, and there is therefore a greater contact between the resulting NH_4-N and soil particles than occurs with fertilizer urea. The greater contact is likely to result in ammonium retention and therefore CEC being more important, relative to pH, with urine than with fertilizer urea.

Although urease is responsible for the conversion of urea to ammonium, there is often a negative correlation between urease activity and ammonia volatilization for a range of different soils (Reynolds and Wolf, 1987b; Whitehead and Raistrick, 1993b). The negative correlation arises because urease activity is positively correlated with contents of organic matter and clay, and hence with cation exchange and buffering capacities. Examples are provided by comparisons of grassland and arable soils on the same soil type. The grassland soils normally have a greater urease activity than corresponding arable soils, but there is more ammonia volatilization, following the addition of urea, from the arable soils which have lower cation exchange and buffering capacities (McGarry *et al.*, 1987; Whitehead and Raistrick, 1993b).

The soil factors that limit volatilization are likely to differ in different situations. In sandy soils with low cation exchange capacity, volatilization may be limited by the rate of urea hydrolysis whereas, in loam and clay soils, volatilization is more likely to be limited by high cation exchange and buffering capacities (Reynolds and Wolf, 1987b).

The Influence of Microbial Immobilization and Nitrification on Ammonia Volatilization

The microbial immobilization of ammonium reduces the amount in the soil solution and therefore the amount susceptible to ammonia volatilization. Studies with ^{15}N-labelled ammonium nitrate and urea have indicated that appreciable immobilization can occur within a few days, though some apparent immobilization may represent mineralization–immobilization turnover rather than net immobilization (see p. 119).

Nitrification also tends to reduce the volatilization of ammonia by removing ammonium from the soil solution and by lowering soil pH, but its effect is difficult to quantify. The development of nitrification often shows a sigmoidal pattern, with a lag period of a few days after the addition of ammonium followed by an increase in activity (see p. 116). With this pattern, nitrification is likely to have little effect on ammonia volatilization. In laboratory studies, the rate of nitrification was generally slower than that of urea hydrolysis, both when fertilizer urea was added to 36 soils (Stevens *et al.*, 1989a) and when cattle urine was applied to 22 soils (Whitehead and Raistrick, 1993b). However, when soil receives a second application of ammonium or urea, there may be no further lag period in nitrification, and nitrification might then curtail ammonia volatilization (Fleisher and Hagin, 1981). Despite the suggestion that ammonia volatilization might be minimized by applying a small amount of ammonium, before the main application of urea or ammonium, to stimulate the nitrifying bacteria (Fleisher and Hagin, 1981), this effect has not been confirmed. Thus, when two successive applications of urine were made to the same area (Sherlock and Goh, 1984), and when ammonium fertilizer was applied to a urine-affected area (Meyer and Jarvis, 1989), ammonia volatilization from the second application was increased, suggesting that any stimulation of nitrification had a negligible effect on volatilization.

Low temperatures restrict both urea hydrolysis and nitrification and, below about 10°C, the restriction appears to be greater with nitrification than with urea hydrolysis (Low and Piper, 1961; Haynes, 1986c, cf. Kissel and Cabrera, 1989). This suggests that, at low temperatures, ammonium derived from urea would remain susceptible to volatilization for a longer period than it would at higher temperatures.

The Influence of Plants on Ammonia Volatilization

When urine or fertilizer urea is partially retained by the leaves and dead leaf litter of grassland, there is an increase in the amount of ammonia volatilized (e.g. Nelson *et al.*, 1980). In particular, the presence of a substantial amount of dead leaf litter on the soil surface tends to increase volatilization (Fenn *et al.*, 1987; Nelson *et al.*, 1980; Whitehead and Raistrick, 1992). The litter provides a surface for urease activity (Hoult and McGarity, 1986), reduces contact between the urine or fertilizer and the soil, and has little cation exchange capacity to restrict volatilization. On the other hand, living plant leaves are able to assimilate gaseous ammonia released from the soil (see Chapter 2) and this process has been reported to reduce the amount of ammonia volatilizing from a grass–clover pasture in Australia (Denmead *et al.*, 1976). The uptake of ammonium-N by roots also reduces the potential for ammonia volatilization. Whether a grass sward enhances or curtails volatilization in comparison with a bare soil depends on factors such as the amount of leaf litter and the rate of grass growth. In several instances, the net effect has been to curtail ammonia volatilization (Hoult and McGarity, 1987; Kresge and Satchell, 1960; Whitehead and Raistrick, 1992) though, in one investigation, the removal of ryegrass vegetation had no significant effect (Simpson, 1968).

Volatilization from the Excreta of Grazing Animals

Urea is the major nitrogenous constituent in urine (see p. 74) and releases ammonia more readily than do other constituents such as allantoin, hippuric acid and creatinine (Whitehead *et al.*, 1989b). In studies of urine applied directly to grass swards, the proportion of the total N volatilized, as assessed from wind-tunnel measurements, has ranged from about 4% to 18% in the Netherlands (Vertregt and Rutgers, 1987, 1991), generally from 12% to 38% in New Zealand (Ball *et al.*, 1979; Carran *et al.*, 1982; Sherlock and Goh, 1984, 1985) and from 3% to 27% in southern England (Lockyer and Whitehead, 1990). A report that 66% of the urinary N was volatilized during warm dry conditions in New Zealand (Ball and Ryden, 1984) is exceptional. Volatilization from urine applied to grass swards is normally at a maximum during the first two days after application, and often becomes negligible after 8–14 days (see Fig. 8.2). However, volatilization may continue at a slow rate beyond 14 days, particularly in cool conditions when the initial rate of volatilization is low. As noted above, factors that favour volatilization include warm moist soil, low values for soil cation exchange and buffering capacity, high pH, and little or no rainfall during the first 2–3 days after the application of urine. The composition of the urine also affects the extent of volatilization. Studies with simulated urine have shown that hippuric acid, although itself releasing little ammonia, promotes the hydrolysis of urea and the volatilization of ammonia (Doak, 1952; Whitehead *et al.*, 1989b). The proportion of the N volatilized from simulated urine was also found to increase with increasing total concentration of N in the urine over the range 1 to 10 g N l^{-1}; but urine pH and salt concentration had little if any effect (Whitehead *et al.*, 1989b). When urine was applied to the same area on three successive occasions, at intervals of about 2 weeks, volatilization from the second and third applications was enhanced, the proportion of the urinary N volatilized increasing from 18% to 24% to 36% (Sherlock and Goh, 1984). On average over a grazing season in temperate regions, it is likely that about 15% of the N returned in urine would be volatilized as ammonia.

Of the N excreted in dung, most is insoluble in water and therefore not susceptible to volatilization as ammonia, at least in the short term. Few measurements of volatilization from dung have been made, and the amounts have usually been between 1% and 5% of the total N for measurement periods up to 14 days (Floate and Torrance, 1970; MacDiarmid and Watkin, 1972a; Ryden *et al.*, 1987). However, higher values within the range 2.8–8.1% were reported in a 60-day study in New Zealand (Sugimoto and Ball, 1989), and as much as 13% of the total N was reported as volatilized during 12 days in a study in the Netherlands (van der Molen *et al.*, 1989).

When ammonia volatilization has been assessed from grazed swards using the micrometeorological mass balance method, the results have usually

Table 8.1. Ammonia volatilization from three grazed swards: ryegrass receiving 420 kg N ha^{-1} $year^{-1}$, ryegrass receiving 210 kg N ha^{-1} $year^{-1}$ and ryegrass–white clover receiving no fertilizer N; mean values for two grazing seasons, assessed by a micrometeorological method. (From Jarvis *et al.*, 1989.)

	Sward		
	Ryegrass (420 kg N ha^{-1} $year^{-1}$)	Ryegrass (210 kg N ha^{-1} $year^{-1}$)	Ryegrass–white clover (No fertilizer N)
Ammonia-N volatilized, kg ha^{-1} $year^{-1}$	25.1	9.6	6.7
Proportion (%) of fertilizer N	6.0	4.6	–
Proportion (%) of estimated urinary N	12.1	11.2	9.8

indicated that rather less volatilization occurred from urinary N than would have been expected on the basis of the results obtained using wind-tunnels. For example, with grazed swards receiving 420 kg fertilizer N ha^{-1} $year^{-1}$, the average amount of ammonia volatilized was estimated at only 25 kg N ha^{-1} $year^{-1}$, representing approximately 10–12% of the urinary N (Table 8.1). Smaller proportions appeared to be volatilized from swards receiving 210 kg fertilizer N ha^{-1}, and from grass–clover swards receiving no fertilizer N. With the same type of method in the Netherlands, the proportions of the total excreted N apparently volatilized were only about 7–8% from swards receiving 400 or 550 kg fertilizer N ha^{-1} $year^{-1}$, and about 3% from a sward receiving 250 kg N (Bussink, 1992, 1994). If it assumed that most volatilization was from urinary N, these proportions would be equivalent to about 11% and 4% respectively, of the urinary N. Possible reasons for the difference between the results of micrometeorological and wind-tunnel investigations are suggested on p. 157.

Volatilization from the Excreta of Housed Animals

When animals are housed, the excreta produced are immediately subject to the action of enzymes, including urease, released by microorganisms present in the dung (Muck, 1981) and on the floors and other surfaces of the housings. The rate of urea hydrolysis is strongly influenced by temperature (Whitehead and Raistrick, 1993a) and, in the absence of soil to retain ammonium and to buffer any pH change, ammonia volatilization is often substantial. Recent work has shown urease activity on the floor of a cow barn to be associated with a layer of precipitated salts derived from constituents of urine and faeces, and that the urease activity could be largely removed by rinsing the floor with a dilute solution (0.5M) of hydrochloric acid (Ketelaars and Rap, 1994).

Table 8.2. Examples of estimated amounts of ammonia volatilized from livestock housings. (From Klarenbeek and Bruins, 1987.)

	Ammonia volatilization (kg N per animal per year)
Dairy cows	8.8*
Calves (0–6 months)	1.5
Sows (including litter)	8.1
Fattening pigs	2.2
Laying hens	0.04–0.30†

*October – May period only.
†Depending on housing and manure disposal systems.

Ammonia volatilization from buildings is difficult to quantify. With any particular housing system, the proportion of the urinary N volatilized will be influenced by the time interval between excretion and the collection, and possible dilution, of the excreta in a storage area or tank outside the housing. High concentrations of ammonia in the atmosphere have been reported in livestock housings, for example, several hundred $\mu g\ m^{-3}$ in cow sheds and several thousand $\mu g\ m^{-3}$ in pig sheds (Lee and Dollard, 1994). Some estimates of ammonia volatilized from housed cattle, pigs and poultry in the Netherlands are shown in Table 8.2. The estimates were based on the N : P ratios of the feedstuffs, animal products and excreta, and on the assumption that losses of N from the housings were due entirely to ammonia volatilization (Klarenbeek and Bruins, 1987). Although the use of bedding material, such as straw, tends to reduce the volatilization of ammonia from animal housings, the subsequent volatilization during storage is greater from farmyard manure (FYM) than from slurry (Hartung, 1991).

Losses of ammonia during the storage of liquid slurries are often small, partly due to dilution with additional water. The greater the dilution, the less the volatilization of ammonia (Hartung, 1991). A second factor is that a crust often forms on the surface of the slurry and this reduces volatilization. The formation of a crust occurs to a greater extent with cattle slurry than with pig slurry, and can be enhanced by the addition of chopped straw (De Bode, 1991; Wouters and Verboon, 1993). The volatilization of ammonia during storage can also be reduced by acidifying the daily production of slurry with nitric acid to about pH 4.5 (Wouters and Verboon, 1993).

In experiments conducted with slurry, stored without treatment for 26–36 weeks in above-ground tanks, the loss of ammonia accounted for 5–15% of the N in the slurry (De Bode, 1991). Rates of loss were two to three times greater from pig slurry than from cattle slurry (despite the two slurries being similar in pH and in their concentrations of ammoniacal N), and were greater in summer than in winter. The greater loss from pig slurry was probably due to its lower viscosity. Slurry from a beef fattening house that was stored for periods of 11–23 weeks lost, on average, only 6% of its N during storage (Gracey 1979).

The N in slurry that has been stored for more than a few weeks occurs in two main forms, ammonium-N and organically bound N, and often the two forms are present in approximately equal amounts (see p. 79). When slurry is spread in the field, little volatilization occurs during the spreading process (Phillips *et al.*, 1991) but a substantial proportion of the ammoniacal N, usually between 15 and 75%, volatilizes within a few days (Beauchamp *et al.*, 1982; Klarenbeek and Bruins, 1991; Lauer *et al.*, 1976; Pain *et al.*, 1989; Stevens and Logan, 1987). The proportion is relatively high (compared with the N in urine from grazing animals) because much of the liquid is retained by the solid fraction of the slurry and does not infiltrate into the soil. There is often more volatilization from slurry when it is applied to a grass sward rather than to bare soil, because the presence of herbage and leaf litter further reduces infiltration of the liquid fraction into the soil (Amberger, 1991; Thompson *et al.*, 1990a). When slurries contain relatively high contents of dry matter and ammoniacal N, the proportion of the ammoniacal N that volatilizes tends to decrease with increasing rate of application, but this relationship may not apply to dilute slurries (Frost, 1994; Thompson *et al.*, 1990b). Temperature has a major effect on the initial rate of ammonia volatilization, but less effect on the cumulative amount volatilized after several days. At temperatures above about 10°C, the rate of volatilization is usually highest during the first few hours after spreading and then declines rapidly (Pain *et al.*, 1989; Sommer *et al.*, 1991; Stevens and Logan, 1987; Thompson *et al.*, 1987) though a slow rate of volatilization may continue for 15 days or more (Thompson *et al.*, 1990a,b). Often more than half the volatilization from slurry occurs within 24 h of application (Pain *et al.*, 1989; Thomson *et al.*, 1987) though, at temperatures close to zero, a slow but fairly uniform rate of volatilization continues for several days (Sommer *et al.*, 1991). The difference between late summer and winter, in volatilization from cattle slurry applied to grassland, is illustrated in Fig. 8.4.

The decline in the rate of volatilization with time after spreading is due partly to the formation of a surface crust and, as a crust forms more readily at higher temperatures, it tends to counteract the direct effect of temperature on volatilization. Increasing wind speed, at least at speeds up to 2.5 m s^{-1}, tends to increase the rate of ammonia volatilization from slurry (Sommer *et al.*, 1991; Thompson *et al.*, 1990b) but rainfall shortly after spreading reduces volatilization (Klarenbeek and Bruins, 1991).

The extent of ammonia volatilization is also influenced by the composition of the slurry, especially its concentration of NH_4-N (Brunke *et al.*, 1988), its pH (Frost *et al.*, 1990) and moisture content (Pain *et al.*, 1989; Sommer and Christensen, 1991). In general, the greater the proportion of total N in the form of ammonium, the greater the proportion of the total N volatilized. And, for a constant amount of NH_4-N, the higher the content of solid material, the greater the volatilization (Sommer and Olesen, 1991). Some of the solid

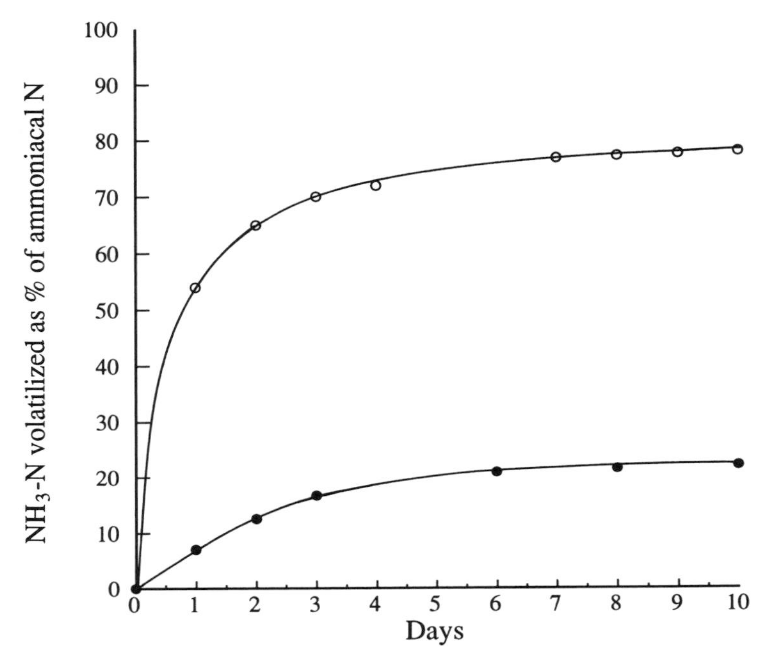

Fig. 8.4. Cumulative volatilization of ammonia from cattle slurry applied to grassland in southern Germany in late summer (○) and winter (●). (Redrawn from Amberger, 1991.)

material in slurry can be removed by mechanical separation (e.g. by the use of a mesh) and volatilization from the liquid fraction (which contains almost all the NH_4-N) is then reduced, mainly due to more of the liquid infiltrating into the soil. The reduction in volatilization is greater the finer the mesh size used to separate the solid material, and separation through a 0.1 mm mesh sieve was found to reduce volatilization by 50% (Frost *et al.*, 1990). Dilution of the separated liquid has an additional effect (Stevens *et al.*, 1992).

It is possible to reduce ammonia volatilization by acidifying slurry before spreading to about pH 5.5 or less (Frost *et al.*, 1990; Klasink *et al.*, 1991: Stevens *et al.*, 1989b; Stevens *et al.*, 1992) or, alternatively, by adding chemicals such as aluminium chloride or calcium chloride (Molloy and Tunney, 1983; Pain *et al.*, 1990). However, in general, the cost of these treatments outweighs the benefit of conserving N in the soil. Another means of reducing ammonia volatilization from slurry is to inject it into the soil, usually at a depth of between 6 and 15 cm or in a V-shaped slot in the soil (Frost, 1994; Pain, 1994; Phillips *et al.*, 1991; Thompson *et al.*, 1987; Wouters and Verboon, 1993). Much of the ammonium-N is then adsorbed by the soil colloids, and some is nitrified, before it is able to escape to the atmosphere. Although injection is effective in reducing volatilization, it also entails an appreciable cost, inevitably causes some damage to the sward and may increase denitrification (Thompson *et al.*, 1987).

When livestock excreta are collected as farmyard manure, ammonia volatilization during storage is greater than with slurry. Studies of straw-based cattle manure indicated that about 10–50% of the total N was lost after 4–7 months of storage, with most of the loss thought to have been due to ammonia volatilization (Kirchman, 1985). There was a clear relationship between the loss of N and the initial C : N ratio of the manure; about 40% of the total N being volatilized when the C : N ratio was 16–18 : 1 but only about 8% when the ratio was 40 : 1 (Kirchman, 1985). Ammonia volatilization from solid cattle manure (as well as from liquid pig slurry) applied to field plots was found to be related to conditions of temperature, wind speed and humidity: most volatilization occurred when conditions favoured rapid drying (Brunke *et al.*, 1988).

When large numbers of cattle are kept in feedlots, a system that is common in the USA, large amounts of excreta are deposited per unit area and ammonia volatilization is high, partly because the pH of the surface layer may increase to between 8.5 and 9.9 (National Research Council, 1978). Micrometeorological measurements over a feedlot in Colorado indicated that the average rate of ammonia volatilization was equivalent to 1.4 kg N ha^{-1} h^{-1}, with lower rates when the surface was wet and higher rates when it was drying rapidly (Hutchinson *et al.*, 1982).

Volatilization from Fertilizers and from Sewage Sludge Applied to Grassland

Ammonia is sometimes used directly as a fertilizer, either as the anhydrous form or as an aqueous solution. Both forms are injected into the soil and the extent to which ammonia volatilizes during this process varies with soil characteristics. In general, there is more volatilization from sandy soils than from loam and clay soils, especially under dry conditions. However, appreciable volatilization may occur from loam and clay soils when conditions are either so dry or so wet that the injection slits do not seal well (Nelson, 1982). Stoniness increases volatilization.

Of the various solid N fertilizers, urea is increasingly important despite the disadvantage of being susceptible to ammonia volatilization. When applied to grassland, the proportion volatilized is often between 10% and 30%, depending on the weather and soil factors discussed above, and on the rate of application. The proportion of urea N volatilized tends to increase with increasing rate of application (Black *et al.* 1985b, 1987; Ryden *et al.*, 1987b). Volatilization can be curtailed to some extent by the use of additives such as acids, inorganic salts and urease inhibitors (Fenn and Hossner, 1985; Lightner *et al.*, 1990; Rodgers *et al.*, 1987). The inhibitor NBPT (see p. 255) is generally effective (Watson *et al.*, 1990b) but the cost of treatment is rarely justified. The application of fertilizer urea in solution, rather than as solid granules, may influence the extent of volatilization but both an increase (Lightner *et al.*, 1990) and a decrease in volatilization (Titko *et al.*, 1987) have been reported.

Volatilization is usually, though not always, less from ammonium salts than from urea (Black *et al.*, 1985b; Fenn and Hossner, 1985; Sommer and Jensen, 1994; Whitehead and Raistrick, 1990). The reactions involved in volatilization from ammonium salts are chemical and depend mainly on the inorganic soil properties, though microbial immobilization and nitrification influence the time period during which the ammonium is susceptible to volatilization. The initial soil pH has a major effect on volatilization from ammonium salts, but chemical interactions between soil and fertilizer are also important, particularly with the ammonium phosphates and ammonium sulphate. With mono-ammonium phosphate, little volatilization ($< 2\%$ of the total N) occurs from most soils, but as much as 35% of the N may volatilize from highly calcareous soils (Whitehead and Raistrick, 1990). The reason for this difference appears to be that, when there is only a moderate amount of Ca in the soil, mono-ammonium phosphate reacts to form calcium ammonium phosphate which tends to inhibit volatilization whereas, when large amounts of $CaCO_3$ are present, it forms octacalcium phosphate and ammonium carbonate, thus increasing pH and the volatilization of ammonia (Terman and Hunt, 1964). Di-ammonium phosphate shows a similar difference in reaction products between soils containing little or no $CaCO_3$ and those containing large amounts: volatilization is much greater from highly calcareous soils due to the precipitation of $CaHPO_4$, octacalcium phosphate and, possibly, apatite (Terman and Hunt, 1964). However, even with slightly acid and neutral soils, volatilization is often greater from di-ammonium phosphate than from other ammonium fertilizers, though much less than from urea (Black *et al.* 1985b; Sommer and Jensen, 1994; Whitehead and Raistrick, 1990). With ammonium sulphate, there is also the possibility of reaction with soil Ca, to form insoluble calcium sulphate, a process which releases ammonium, tends to raise pH and therefore increases volatilization. On some calcareous soils, ammonia volatilization is greater from ammonium sulphate than from other forms of fertilizer N, including urea, (Hargrove *et al.*, 1977; Terman, 1979; Whitehead and Raistrick, 1990), and may exceed 40% of the N (Hargrove *et al.*, 1977). However, on non-calcareous soils, volatilization from ammonium sulphate often amounts to less than 5% of the N (Sommer and Jensen, 1994; Whitehead and Raistrick, 1990). With ammonium nitrate, there are no chemical reactions resulting in insoluble products and so the extent of volatilization is governed primarily by the initial pH of the soil. Volatilization rarely exceeds 20% of the ammonium-N (10% of the total N), and on many soils it is negligible (e.g. Lightner *et al.*, 1990).

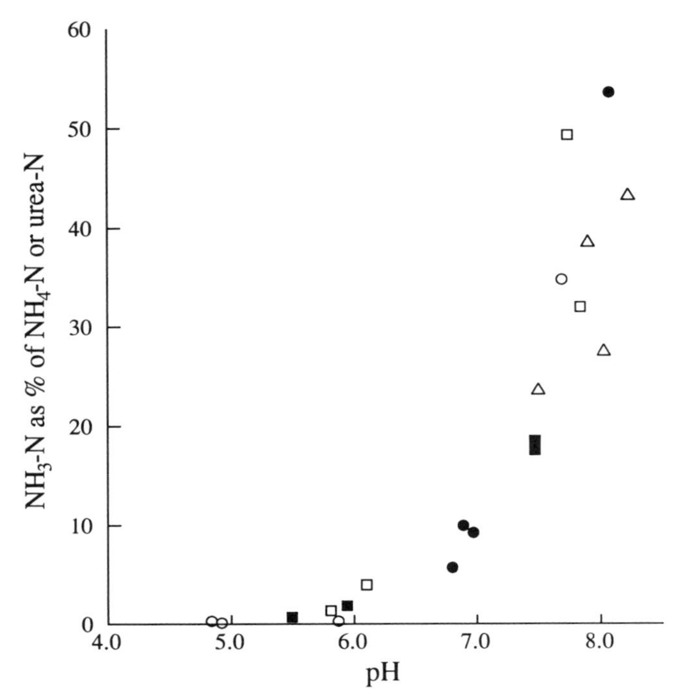

Fig. 8.5. Total volatilization of ammonia after 8 days in relation to the soil pH attained after 24 h in mixtures of mono-ammonium phosphate (○), di-ammonium phosphate (●), ammonium sulphate (□), ammonium nitrate (■), and urea (△) with four soils ranging in initial pH from 5.5 to 7.4. (From Whitehead and Raistrick, 1990.)

The interaction between fertilizer and soil may change the localized soil pH; and the extent of volatilization from fertilizers containing urea or ammonium-N is highly dependent on the pH attained in the immediate vicinity of the fertilizer granules (Lyster *et al.*, 1980; O'Toole *et al.*, 1985a; Sherlock *et al.*, 1988; Whitehead and Raistrick, 1990). A close relationship was found between total volatilization after 8 days, expressed as a percentage of the ammonium- or urea-N, and the pH attained after 24 hours by a corresponding mixture of fertilizer and soil, when five forms of solid fertilizer N were added to four soils under uniform conditions (Fig. 8.5). When urea or ammonium fertilizer is applied to grazed grassland, ammonia volatilization is enhanced by the presence of recent urine patches (Black *et al.*, 1984; Meyer and Jarvis, 1989), presumably due at least partly to their higher pH.

The results of the laboratory and field studies cited above suggest that, in temperate grassland areas, the proportions of fertilizer N likely to be volatilized as ammonia would typically be as follows:

mono-ammonium phosphate, 1–2%,
di-ammonium phosphate, 4–6%,

ammonium sulphate, 8–12%,
ammonium nitrate, 1–3%,
urea, 15–20%.

Variation in the extent of volatilization is largely due to differences in soil and weather conditions as described above.

Ammoniacal N constitutes up to half the total N in anaerobically-digested sewage sludge (though it is less in other types of sludge) and is susceptible to volatilization when the sludge is spread. The extent of volatilization varies with the factors described above, but was estimated to amount to 55–60% of the ammoniacal N during 5–7 days, in studies of sludge applied to the surface of bare soil in Canada (Beauchamp *et al.*, 1978).

Volatilization from Living Plant Material

As well as having the capacity to absorb gaseous ammonia from the atmosphere (see p. 21), plant leaves may also lose ammonia to the atmosphere. Volatilization from leaves occurs mainly during senescence (presumably because more proteolysis is occurring within the plant tissue at that time) but volatilization may occur at younger stages of growth if the supply of N exceeds requirements (Schjoerring, 1991; Sutton *et al.*, 1993). The process has been studied mainly in annual species such as wheat and maize and, in these species, ammonia volatilization was found to be eight to ten times greater from senescing lower leaves than from actively growing upper leaves (Farquhar *et al.*, 1979). With metabolically active leaves, there is more ammonia volatilization during the day than at night, because photorespiration stops and the stomata close during darkness (Schjoerring, 1991; Sutton *et al.*, 1993). Less ammonia is likely to volatilize from perennial grasses than from annual crops because some of the N not required in senescing leaves is translocated to the growing leaves or to the roots (Wetselaar and Farquhar, 1980). However, attempts to assess the extent of ammonia volatilization from crops, including grass, that have received fertilizer N are complicated by the difficulty of distinguishing between volatilization from the foliage and volatilization from the soil. Partly for this reason, the extent to which ammonia volatilizes from living plant material is still uncertain.

Volatilization from Decomposing Plant Material

Plant proteins undergoing microbial decomposition are hydrolysed to amino acids and these, in turn, are hydrolysed to release ammonium-N. There is often an increase in the pH of decomposing plant material and therefore the possibility of some volatilization of ammonia. Laboratory studies have shown that volatilization does occur from decomposing grass herbage, and that the rate of volatilization increases with temperature (Salt, 1965; Whitehead *et al.*, 1988). With two batches of grass herbage decomposing under wind tunnels in the field, the extent of volatilization was shown to depend also on the concentration of N in the herbage. Cut grass herbage containing 3% N lost 10% of its total N through ammonia volatilization during a 28-day period of simulated showery weather, whereas no volatilization was detected from herbage with a concentration of only 0.9% N (Whitehead and Lockyer, 1989). Volatilization from decomposing plant material becomes appreciable only after a lag period of several days, and plant material that is dried rapidly loses little if any ammonia by volatilization (Terman, 1979; Whitehead *et al.*, 1988). However, fertilized crops of grass intended for hay or pre-wilted silage, if cut and then left in the field due to showery weather, may lose appreciable amounts of ammonia. A slow rate of volatilization may also occur continually from dead leaves decomposing *in situ*, particularly in grazed swards where leaves are subject to physical damage by treading, and in recently cut swards when the herbage is not completely removed. However, such slow volatilization is difficult to detect and the extent to which decomposing plant material volatilizes ammonia in the field situation is uncertain. Micrometeorological measurements above a field of lucerne indicated that about 2.3 kg N ha^{-1} was volatilized from the fragments of decomposing herbage remaining after three successive hay cuts (Dabney and Bouldin, 1985).

Ammonia Emissions from Grassland and Livestock Production Systems in the UK

Information on the extent of volatilization of N from livestock excreta, from fertilizers and from plant materials, together with data on livestock numbers and fertilizer use, can be used to estimate the total emission of ammonia from specific geographic regions such as individual countries. This approach has been used by several authors to estimate the emission of ammonia from agricultural systems in the UK and other European countries as a whole, but estimates have varied rather widely (e.g. ApSimon *et al.*, 1987; Asman, 1992; Buijsman *et al.*, 1987; Jarvis and Pain, 1990; Kruse *et al.*, 1989).

An estimate for the UK, based on the information now available, is presented here. First, the amounts of ammonia volatilized from cattle, sheep, pig and poultry production have been estimated on the basis set out in Tables 8.3–8.6. The calculations involve a number of assumptions; on the average feed consumption per livestock unit, the average concentration of N in the diet, the

Table 8.3. Estimated ammonia volatilization from cattle in the UK (based on assumed average values for DM consumption, % N in diet, proportions of dietary N excreted in urine and dung, and proportions of urine-N and dung-N volatilized.)

	Cows	Heifers < 2 year	Beef cattle	Calves
DM consumption (kg per animal per year)	6200	2700	2700	1000
% N in DM consumed	3.0	3.0	2.4	3.0
Proportion (%) of dietary N excreted	80	80	90	80
N excreted (kg N per animal per year)	149	65	58	24
Ratio of grazing to housing (yearly basis)	50 : 50	65 : 35	80 : 20	30 : 70
Ratio of urine N : dung N	65 : 35	65 : 35	55 : 45	65 : 35
Proportion (%) of urine N volatilized, grazing	15	15	15	15
Proportion (%) of urine N volatilized, housed*	55	55	55	55
Proportion (%) of dung N volatilized, grazing	3	3	3	3
Proportion (%) of dung N volatilized, housed	3	3	3	3
Amount of N volatilized, grazing (kg per animal per year)	8.1	4.5	4.4	0.8
Amount of N volatilized, housed (kg per animal per year)	27.4	8.3	3.7	6.2
Total amount of N volatilized (kg per animal per year)	35.5	12.8	8.1	7.0
UK population (× 1000)†	4800	1000	3000	3000
Ammonia volatilized from cattle in UK (kt N $year^{-1}$)	170.4	12.8	24.3	21.0
Total = 228.5 kt N $year^{-1}$				

*Based on references cited by Ryden *et al.* (1987).
†Based on MAFF June census data for 1993.

ratio of N in the urine to that in the dung, the ratio between the time spent grazing and the time spent in housing for the population over a year, and the proportions of the N in urine and dung volatilized as ammonia in both grazing and housed systems. Of the assumed values indicated in each Table, the greatest uncertainty is in the proportion of the excreted N volatilized from housed animals: the proportions indicated are summations of estimates for volatilization from the animal house itself, plus volatilization during the storage of the slurry or manure, plus volatilization after spreading. The estimates of ammonia volatilization per year, for the UK as a whole, derived from these calculations amount to 229 kt N for cattle, 33 kt N for sheep, 27 kt N for pig and 61 kt N for poultry production. To estimate the total volatilization from UK agriculture, these values have been combined with estimates for volatilization from fertilizers and from decomposing plant material (Table 8.7). The ammonia volatilized directly from fertilizers is assumed to be, on average, 3.4% of the fertilizer N applied. This figure is based on the assumptions that about 70% of the fertilizer N is in the form of ammonium nitrate, 11% in the form of urea and 15% as mono- and di-ammonium phosphate; that

Table 8.4. Estimated ammonia volatilization from sheep in the UK (based on assumed average values for DM consumption, % N in diet, proportions of dietary N excreted in urine and dung, and proportions of urine-N and dung-N volatilized).

	Adult sheep	Lambs < 1 year
DM consumption (kg per animal per year)	550	150
% N in DM consumed	2.2	3.0
Proportion (%) of dietary N excreted	90	75
N excreted (kg N per animal per year)	10.9	3.4
Ratio of grazing to housing (yearly basis)	95 : 5	90 : 10
Ratio of urine N : dung N	55 : 45	65 : 35
Proportion (%) of urine N volatilized, grazing	15	15
Proportion (%) of urine N volatilized, housed	50	50
Proportion (%) of dung N volatilized, grazing	2	2
Proportion (%) of dung N volatilized, housed	2	2
Amount of N volatilized, grazing (kg per animal per year)	0.94	0.32
Amount of N volatilized, housed (kg per animal per year)	0.15	0.11
Total amount of N volatilized (kg per animal per year)	1.09	0.43
UK population (× 1000)*	21,800	22,100
Ammonia volatilized from cattle in UK (kt N $year^{-1}$)	23.8	9.5
Total = 33.3 kt N $year^{-1}$		

*Based on MAFF June census data for 1993.

the ratio of the use of fertilizer N on arable land to that on grassland is about 1.8 : 1 and that volatilization per unit area of arable land is 30% less than from grassland (due to some fertilizer being incorporated into the soil) (Whitehead and Ronstrich, 1990). The average volatilization from surface-applied fertilizers of different types is estimated as described on p. 169. As indicated above, the amounts of ammonia volatilized from living plants and from decomposing plant material are difficult to assess. In the present calculation, volatilization from living plants has been considered to be negligible, but a figure has been included for decomposing plant material in grassland. It is assumed that the 7.2 million ha of lowland grassland produce an average yield of 5.5 t DM ha^{-1} (Green and Baker, 1981), that an amount equivalent to 20% of this yield decomposes above the soil surface, that the average concentration of N in the herbage is 2.5% and that 5% of this N is volatilized. These assumptions result in an emission of ammonia from decomposing herbage of 10 kt N $year^{-1}$.

The resultant total for ammonia volatilization from UK agriculture is about 415 kt NH_3-N $year^{-1}$, a value that is similar to another recent estimate of 468 kt for volatilization from livestock production and fertilizer use in the UK (Asman, 1992). However, other recent calculations produced a much lower estimate for UK agriculture of < 200 kt NH_3-N (Jarvis and Pain, 1990). These calculations were based partly on micrometeorological measurements of ammonia volatilization from grazed swards, and partly on measurements

Table 8.5. Estimated ammonia volatilization from pigs in the UK (based on assumed average values for DM consumption, % N in diet, proportions of dietary N excreted in urine and dung, and proportions of urine-N and dung-N volatilized).

	< 20 kg	20–50 kg	50–110 kg	> 110 kg
DM consumption (kg per animal per year)	200	450	680	800
% N in DM consumed	2.8	2.6	2.4	2.6
Proportion (%) of dietary N excreted	80	85	85	85
N excreted (kg N per animal per year)	4.5	9.9	13.9	17.7
Ratio outdoors : housed (yearly basis)	5 : 95	10 : 90	10 : 90	10 : 90
Ratio of urine N : dung N	65 : 35	60 : 40	60 : 40	60 : 40
Proportion (%) of urine N volatilized, outdoors	15	15	15	15
Proportion (%) of urine N volatilized, housed*	60	60	60	60
Proportion (%) of dung N volatilized, outdoors	4	4	4	4
Proportion (%) of dung N volatilized, housed*	4	4	4	4
Amount of N volatilized, outdoors (kg per animal per year)	0.02	0.10	0.14	0.19
Amount of N volatilized, housed (kg per animal per year)	1.72	3.35	4.70	5.98
Total amount of N volatilized (kg per animal per year)	1.74	3.45	4.84	6.17
UK population (× 1000)[†]	2000	2180	2480	1000
Ammonia volatilized from pigs in UK (kt N $year^{-1}$)	3.5	7.5	12.0	4.2
Total = 27.2 kt N $year^{-1}$				

*Based on references cited by Ryden *et al.* (1987b).
[†]Based on MAFF June census data for 1993.

using enclosures for slurries applied to grassland. In common with the assessments in Tables 8.3–8.7, there was most uncertainty over the amounts of NH_3-N volatilized from livestock housings. The wide range in the recent estimates for ammonia volatilization in the UK, from < 200 to > 450 kt NH_3-N $year^{-1}$, highlights the need for more reliable data and the importance of refining methods of measurement. However, it is worth noting that a total emission of about 420 kt NH_3-N $year^{-1}$ would be equivalent to about 25% of the fertilizer N used in UK agriculture. In comparison with the UK estimate, mentioned above, of 468 kt NH_3-N, ammonia emissions from the Netherlands have been estimated at 222 kt, from France at 774 kt and from Europe as a whole 7638 kt $year^{-1}$ (Asman, 1992). In many European countries, changes in livestock numbers and in methods of housing have resulted in the annual amounts of ammonia volatilized from agricultural sources increasing by 50–100% since 1950 (ApSimon *et al.*, 1987; Lee and Dollard, 1994).

Table 8.6. Estimated ammonia volatilization from poultry in the UK (based on assumed average values for DM consumption, % N in diet, proportions of dietary N excreted and proportions of excreted N volatilized).

	Hens	Pullets etc.	Broilers etc.
DM consumption (kg per bird per year)	47	24	26
% N in DM consumed	2.7	2.5	3.2
Proportion (%) of dietary N excreted	80	78	78
N excreted (kg N per bird per year)	1.01	0.47	0.65
Proportion (%) of excreted N volatilized*	65	65	65
Amount of excreted N volatilized (kg per bird per year)	0.66	0.31	0.42
UK population (× 1000)†	39,000	12,500	75,700
Ammonia volatilized from poultry in UK (kt N year^{-1})	25.7	3.9	31.8
Total = 61.4 kt N year^{-1}			

*Based on references cited by Ryden *et al.* (1987b).
†Based on MAFF June census data for 1993.

Table 8.7. Estimated ammonia volatilization from UK agriculture (based on estimates in Tables 8.3–8.6 plus estimates for volatilization from fertilizers and decomposing plant material).

	Ammonia volatilization (kt N year^{-1})
Cattle production	229
Sheep production	33
Pig production	27
Poultry production	61
Fertilizers (direct volatilization)*	55
Decomposing plant material*	10
Total	415

*See text.

Environmental Impact of Ammonia Emissions

One consequence of ammonia volatilization is that potentially plant-available N is distributed over a wider area than would otherwise occur and, on a global scale, this involves a net transfer of ammonia from the land to the oceans (Warneck, 1988). Transfer to largely enclosed areas of sea, such as the Baltic Sea, contributes to the eutrophication of these areas (Schrøder, 1990). However, some of the ammonia volatilized from the soil surface is absorbed and assimilated by vegetation in the immediate vicinity, as observed with a grass–clover pasture in Australia (Denmead *et al.*, 1976), the amount being influenced by dew and humidity as well as by the concentration in the air (ApSimon and Kruse-Plass, 1991). Ammonia moving laterally from localized sources, such as urine patches or areas of land treated with slurry, may also be partially absorbed by vegetation close to the source. Using ryegrass grown in pots with ^{15}N-labelled nitrate it was shown that gaseous ammonia absorbed at various distances of up to 130 m from the yard of a dairy farm, during a 47-day period, declined from the equivalent of about 30 kg N ha^{-1} at a distance of 10 m from the yard to 7 kg N ha^{-1} at a distance of 130 m in the prevailing downwind direction (Sommer and Jensen, 1991). Other evidence of the absorption of atmospheric ammonia by unfertilized grassland or moorland at greater distances has been reported from the Netherlands (Heil *et al.*, 1988) and from the UK (Sutton *et al.*, 1992). With an isolated source of ammonia, such as intensive livestock housing, there is evidence that the concentration in the atmosphere may return to the background concentration at a distance of about 1500 m in winter and 3000 m in summer (Fangmeier *et al.*, 1994). The decline in concentration is due partly to deposition, partly to vertical mixing and partly to conversion to ammonium aerosols. The deposition of ammoniacal N at two moorland sites, with mean concentrations in the atmosphere of 0.55 and 2.6 $\mu g\ m^{-3}$, was estimated to be 7.4 and 16.5 kg N ha^{-1} $year^{-1}$ respectively: the dry deposition of ammonia was similar in magnitude to the wet deposition of ammonium (Sutton *et al.*, 1992). Deposition is increased by the presence of trees. Estimates of total N deposition in forested areas in the Netherlands include 64 kg N ha^{-1} $year^{-1}$, with 75% in the form of ammonium (van Breemen *et al.*, 1982), and 115 kg N ha^{-1} $year^{-1}$, with 83% in the form of ammonium (Draaijers *et al.*, 1989).

The absorption of ammoniacal N by vegetation and soils may have a beneficial effect in some situations by increasing the supply of N to crops and thus reducing the need for purchased fertilizer N: it will represent a major component of the 30–35 kg N ha^{-1} deposited in southeast England (see p. 88). However, with natural or semi-natural vegetation, such absorption may result in undesirable changes in botanical composition, with those species most responsive to N becoming increasingly dominant. For example, in areas of heathland in the Netherlands, a change from heather to grass dominance has been attributed to inputs of nitrogen from the atmosphere (Roelofs, 1986). The replacement of species such as *Calluna vulgaris* (L) by grasses e.g. *Molinia caerulea* (L) and *Deschampsia flexuosa* (L) (Roelofs and Houdijk, 1991) has been associated with the deposition of between 20 and 60 kg N ha^{-1} $year^{-1}$, with 60–90% of the N being in the form of ammonium sulphate. Ultimately, those species least able to withstand the increased competition resulting from enhanced N supply may disappear completely (Fangmeier *et al.*, 1994).

The deposition of ammoniacal N has also been implicated in the deterioration of forests in the Netherlands and elsewhere in Europe (Lee and Dollard, 1994; Roelofs and Houdijk, 1991). Part of the effect is thought to be due to the direct uptake of ammonia/ammonium by the leaves with a

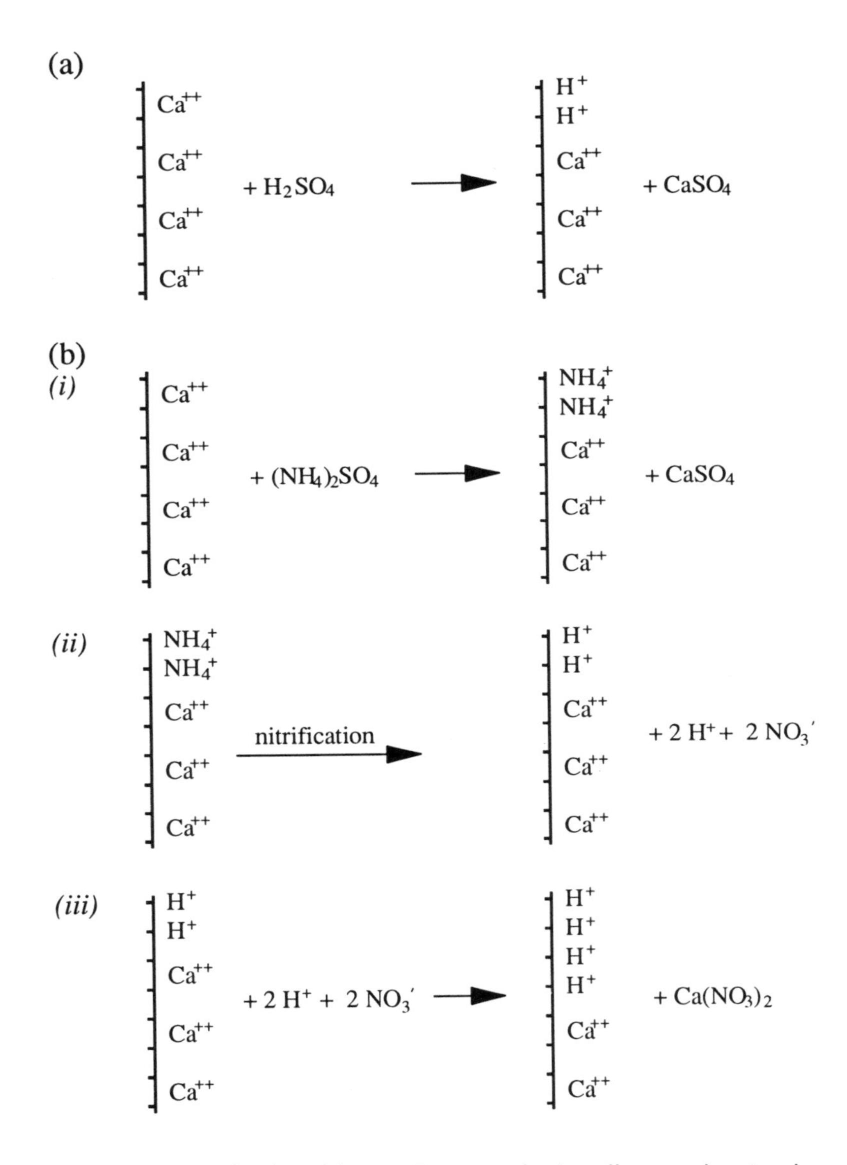

Fig. 8.6. Processes of soil acidification by atmospheric pollutants, showing the displacement of calcium ions from cation exchange sites and the formation of leachable salts: (**a**) by sulphuric acid, (**b**) by ammonium sulphate.

consequent excretion, and possibly induced deficiency, of cations such as K, Ca and Mg. Part of the effect is thought to be due to soil acidification which, at pH < 4.5, inhibits nitrification and therefore promotes the accumulation of exchangeable ammonium in the soil and the depletion of K, Ca and Mg by leaching (Roelofs and Houdijk, 1991; Sutton *et al.*, 1993). Deposition in forests is greatest at the forest edge.

Most of the ammoniacal N deposited on the land surface is either taken up by plants or is nitrified, and both these processes tend to reduce the soil pH (Sutton *et al.*, 1993). Assuming that all the ammonium is nitrified, the acidifying effect of ammonium sulphate is twice that of sulphuric acid (Fig. 8.6), and the deposition of ammonia (and its oxidation to nitric acid) is thought to be a major source of soil acidification in some areas (van Breemen and van Dijk, 1988). Nitrification is largely responsible for the paradox that, although volatilized ammonia tends to neutralize nitric and sulphuric acids in the atmosphere, it accentuates the acidifying effects of acid rain on soils.

With point sources of ammonia, such as the ventilation outlets of intensive animal housings, the ammonia may have more obvious effects on nearby vegetation. For example, ammonia from poultry housing has been reported to damage nearby pine trees (Kaupenjohann *et al.*, 1989; Roelofs *et al.*, 1985), probably due to induced nutrient imbalances in the foliage or soil acidification. The deposition of ammoniacal N from poultry housing over an 18-year period has also been reported to reduce crop yields in a field adjacent to the housing as a result of soil acidification (Spiers and Frost, 1987).

Ammonia not absorbed close to the source of emission is quickly dispersed and diluted in the atmosphere. Thus, with a 600-cow intensive dairy farm in California, the average concentration of ammonia in the air at a height of 1.2 m was 540 μg NH_3-N m^{-3} at the perimeter fence compared with 16 μg NH_3-N m^{-3} at a distance of 0.8 km downwind (Luebs *et al.*, 1974). Ammonia in the atmosphere dissolves rapidly in water droplets and, depending on the concentration of acidic compounds such as sulphuric and nitric acids, reacts to form aerosols. Although some of the ammonium aerosol will be deposited within a few kilometres of the source of the ammonia, some will persist in the atmosphere for several days or more, and will be transported over much larger distances (ApSimon and Kruse-Plass, 1991; Warneck, 1988). The average distance travelled is difficult to assess but will clearly depend on various factors such as weather conditions, topography and the nature of the vegetation in the vicinity of the source. Studies in the Netherlands showed that, with point sources of ammonia, the effect on the amount of deposited ammoniacal N was particularly noticeable within a distance of 1 km (Schuurkes *et al.*, 1988). However, the aerosols may travel much further. Assuming a residence time in the atmosphere of 6 days, and an average wind speed of 5 m s^{-1}, the average transport distance of ammonium sulphate and ammonium nitrate would be 2500 km (Irvin and Williams, 1988).

9 Volatilization of Gaseous Nitrogen and Nitrogen Oxides through Denitrification and Nitrification

Introduction

Emissions of gaseous N_2, nitrous oxide (N_2O) and nitric oxide (NO) from soil occur through a combination of the microbial processes of denitrification and nitrification, and purely chemical reactions (chemo-denitrification). Denitrification produces a mixture of N_2, N_2O and NO; and nitrification produces (in addition to nitrate) a mixture of N_2O and NO. Denitrification tends to occur intensively over short time periods, whereas the release of N_2O and NO during nitrification occurs much more slowly but more continuously. Although denitrification is the main pathway for the release of N_2 from the soil, both denitrification and nitrification are important as sources of the nitrogen oxides. The relative importance of denitrification and nitrification as sources of N_2O is uncertain, but more N_2O is produced from soils treated with ammonium (and therefore undergoing nitrification) than from soils treated with nitrate (Warneck, 1988).

Estimates of the amounts of soil N lost through denitrification, nitrification and chemo-denitrification have a greater uncertainty than estimates of the amounts lost through leaching and ammonia volatilization, mainly because the methods of measurement are rather complex and are unable to provide accurate values for periods of more than a few days. An additional difficulty arises because the rate of denitrification, in particular, varies widely from time to time at the same location, and also from point to point in the soil (Parkin, 1987; Parsons *et al.*, 1991b). However, evidence from short-term measurements and long-term N balance studies (see Chapter 15) suggests that in many regions the amount of soil N lost through the combined effects of denitrification,

nitrification and chemo-denitrification is likely to be broadly similar to the amount lost through nitrate leaching.

Denitrification

Denitrification is the process by which nitrate is converted, by bacterial enzymes, to the gaseous compounds, N_2, N_2O and NO, which then diffuse into the atmosphere. In soils, denitrification is brought about mainly by certain types of soil bacteria which have the ability to grow in anaerobic conditions if nitrate and/or nitrite are present. Such bacteria are numerous in most soils and often account for between 1% and 5% of the total heterotrophic population (Steele, 1987). The denitrifying bacteria use nitrate and nitrite as electron acceptors in anaerobic conditions, though in aerobic conditions they use molecular oxygen. However, the various denitrifying bacteria vary greatly in their sensitivity to oxygen and in the concentration of oxygen at which denitrification in induced (Lloyd, 1993). In general, in anaerobic conditions, they produce enzymes that catalyse the reactions in the sequence:

$$NO_3^- \rightarrow NO_2^- \rightarrow NO \rightarrow N_2O \rightarrow N_2$$

Most denitrifying bacteria produce all the enzymes in the sequence, but some are unable to produce nitrate reductase while others are unable to produce nitrous oxide reductase (Firestone, 1982; Knowles, 1981, 1982; Lloyd, 1993). In general, nitrate reductase and nitrous oxide reductase are the enzymes that are most easily inhibited by oxygen and, consequently, when conditions are only marginally anaerobic, the proportion of N_2O increases. It is thought that only small amounts of NO are produced by denitrification, as the activity of nitric oxide reductase is higher than that of nitrite reductase (Lloyd, 1993). The ratio of $N_2 : N_2O$ is very variable.

To the extent that the main product of denitrification is N_2, the process reverses the effect of N_2 fixation by returning N from the biosphere to the atmosphere as N_2. The release of N_2O has the same effect, as most N_2O is converted eventually to N_2 and O atoms by UV radiation (Warneck, 1988). However, NO differs from N_2 and N_2O in being oxidized to HNO_3 which returns to the earth's surface through wet and dry deposition. Of the three gaseous compounds, N_2 has no environmental impact but N_2O and NO are potentially harmful (see p. 198).

In addition to the denitrification pathway described above, a small amount of denitrification may occur as a result of some fungi and bacteria, other than the denitrifying bacteria, having the ability to respire anaerobically by reducing nitrate to nitrite: some fungi and bacteria may also have the ability to convert nitrite to ammonium with the release of a small amount of N_2O (Haynes and Sherlock, 1986).

Although most of the N_2O produced in soils is thought to diffuse into the atmosphere, some may be leached from the soil in drainage water during the winter. This fraction may sometimes amount to a few kg N ha^{-1}, comparable to the amount diffusing to the atmosphere (Colbourn and Dowdell, 1984). Some of the N_2O produced may also be adsorbed by soils (Hauck, 1984b) and may be reduced further to N_2 (National Research Council, 1978).

Little denitrification occurs in soils in which the concentration of nitrate is consistently low, as in grassland receiving little or no fertilizer N and carrying few livestock. In more intensively managed grassland, high concentrations of nitrate inevitably occur in the soil from time to time, due to the application of fertilizer and/or the return of urine N. However, the presence of a high nitrate concentration does not necessarily result in denitrification. The process occurs mainly when conditions are anaerobic, at least in localized zones of the soil, and when there is a supply of available carbon. Most denitrification occurs during periods when conditions are particularly favourable, usually beginning a few hours after the onset of rainfall or irrigation (Hutchinson and Davidson, 1993; Smith and Tiedje, 1979). Rates of denitrification, measured in the field, have often shown a poor correlation with environmental factors such as, soil moisture, temperature and nitrate concentration (Burton and Beauchamp, 1985) though peaks in denitrification have been associated with times when the water-filled pore space was more than 60–80% of total pore space (Aulakh *et al.*, 1992).

An important factor in the environmental impact of denitrification is the ratio of N_2O to N_2, and this ratio tends to be increased by high concentrations of nitrate and by acid conditions, and to be decreased by increasingly poor aeration, by increasing availability of organic C (Aulakh *et al.*, 1992; Firestone, 1982; Granli and Bockman, 1994; Haynes and Sherlock, 1986) and by increasing temperature (Granli and Bockman, 1994). The effect of nitrate supply is illustrated by measurements of denitrification before and after the addition of ammonium nitrate, which indicated that N_2O accounted for 60% of total denitrification from soil plus fertilizer, but only 17% from soil alone (Colbourn *et al.*, 1984). In another investigation, the proportion of N_2O was greatest at times when total denitrification was high (Eggington and Smith, 1986) and nitrate presumably plentiful.

Methods for the Measurement of Denitrification

A major problem in assessing denitrification in the field is caused by the difficulty of measuring changes in the concentration of N_2 in the atmosphere. It is not feasible to assess denitrification by measuring the loss of nitrate because this can occur through leaching and plant uptake, as well as denitrification. The difficulty in measuring changes in the concentration of N_2 is due to the fact that N_2 comprises 78% of the earth's atmosphere. The small changes caused by denitrification can therefore be measured only if sealed containers are used, and the natural atmosphere is replaced by an artificial atmosphere, usually with most of the N_2 being substituted by an inert gas, e.g. Ar (Limmer *et al.*, 1982; Scholefield *et al.*, 1990). This type of procedure restricts the length of time over which measurements can be made but it does enable short-term emissions of N_2 to be quantified. It is simpler to measure N_2O and NO than to measure N_2 because their background concentrations are low, but measurements of N_2O and NO do not provide an assessment of total denitrification. The use of ^{15}N-labelled nitrate enables denitrification from specific sources to be measured but the sensitivity of this type of procedure is low unless the soil is placed in a container with a relatively small volume of atmospheric N_2, or with He or Ar in place of atmospheric N_2. In addition, N_2O interferes with the measurement of N_2 by mass spectrometry, though a mixture of ^{15}N-labelled N_2 and N_2O can be analysed for both constituents by appropriate mass spectrometry techniques (Stevens *et al.*, 1993).

In view of these various difficulties, the methods that are now most widely used for assessing denitrification are based on the use of acetylene to inhibit the reduction of N_2O to N_2. With this type of procedure it is assumed that, when a soil is incubated in an atmosphere containing acetylene, the amount of N_2O produced is equivalent to the amount of N_2O plus N_2 that would have been produced in the absence of acetylene. And, since N_2O has a low background concentration in the atmosphere (about 300 ppb), the increase is relatively easy to measure by gas chromatography. The concentration of acetylene introduced into the atmosphere is normally between 0.1 and 10.0%, and often about 1% (v/v). Incubation of soil in the presence of acetylene can be carried out in the field using small plots under enclosures (Ryden and Dawson, 1982; Smith and Arah, 1990) or using soil cores placed in a container in the field (Ryden *et al.*, 1987a). Alternatively, the incubation may be carried out using soil cores in the laboratory (e.g. Parkin *et al.*, 1984). With small enclosures in the field, acetylene is injected into the soils through probes inserted to various depths. The time required to establish an adequate concentration of acetylene in soils in the field varies with soil porosity, and ideally the acetylene concentration should be monitored (Smith, 1987). When soil cores are used and these are incubated in contact with the air, denitrification may be less than in undisturbed soil, but the difference can be minimized by surrounding the cores with a plastic sheath (Tiedje *et al.*, 1989). In addition to the difficulties of ensuring that acetylene is present in adequate amounts, and that the aeration of the soil is changed as little as possible, there are other problems in using acetylene. One is that acetylene also inhibits nitrification and it therefore reduces the amount of nitrate becoming available for denitrification from the soil (Mosier, 1980). A second disadvantage is that acetylene tends to become less effective with time due to its being oxidized slowly by microorganisms in the soil: this process accelerates after the microorganisms have been exposed to acetylene for more than a few days (Tiedje *et al.*, 1989). Although potentially serious, these disadvantages can be minimized by making repeated measurements at intervals of 1 to 3 days (Aulakh *et al.*, 1992; Ryden and Dawson, 1982) and by statistical treatment of the data to allow, so far as possible, for variability in time and space (Tiedje *et al.*, 1989).

Micrometeorological methods for N_2O, similar to those used for ammonia (see p. 155) have also been examined but have had only limited success because the small concentration gradients are difficult to measure with the required precision. However, new and more sensitive long-path methods based on tunable lasers are being developed and are likely to produce improved accuracy in the future (Granli and Bockman, 1994; Smith and Arah, 1990).

Influence of Soil Aeration and Water Status on Denitrification

Most denitrification occurs when conditions are anaerobic, and consequently when the supply of molecular oxygen is insufficient for the microorganisms decomposing organic matter, at least at some microsites within the soil. In this situation, the denitrifying bacteria utilize nitrate or nitrite as electron acceptors, and use some of the oxygen in the nitrate or nitrite to produce CO_2 from the organic matter. Anaerobic conditions throughout a soil are usually due to excessive amounts of water, since the rate of diffusion of oxygen is about 10,000 times faster through air than through water. Although the entire soil may become anaerobic in completely waterlogged soils, most wet soils have some macropores that contain air and provide aerobic zones. Soils vary widely in porosity, reflecting differences in texture and structure, and they therefore vary in the soil water content, or tension, at which denitrification becomes appreciable (Smith and Arah, 1990). A small amount of denitrification may occur, even when the soil water content is low, if anaerobic microsites are present, and these are most likely to occur in clay soils and in aggregates with a radius of about 9 mm or more (Greenwood, 1975). Also, some denitrifying bacteria may denitrify to some extent under aerobic conditions (Lloyd, 1993). However, for appreciable denitrification to occur, it is usually necessary for the soil moisture content to be greater than 60% of the waterholding capacity (Aulakh *et al.*, 1992) or for the soil water tension to be below 2–20 kPa (20–200 mbar). The greatest increase in denitrification normally occurs between 100% and 200% of water holding capacity, as air is displaced from the larger soil pores (Knowles, 1981).

Water-filled pore space (or its converse, air-filled pore space), expressed as a percentage of total pore space is generally more closely related to denitri-

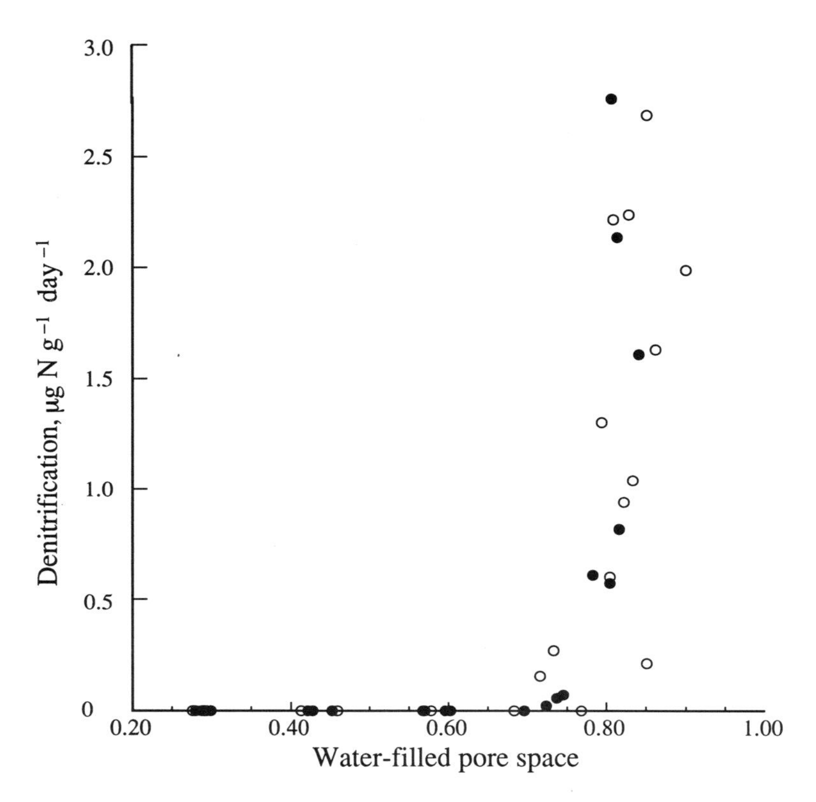

Fig. 9.1. Denitrification in relation to water-filled pore space (as a fraction of total pore space) in several soils of medium-fine (○) and coarse (●) texture. (Redrawn from Aulakh *et al.*, 1992.)

fication than is soil water content alone (Aulakh *et al.*, 1992; Sexstone *et al.*, 1985). This is especially so when soils of different textures are being compared. Critical values for water-filled pore space of between 65% and 90% have been reported (Aulakh *et al.*, 1992; Smith and Arah, 1990): an example is shown in Fig. 9.1. The concentration of oxygen in the soil atmosphere below which denitrification becomes appreciable has been reported to range from 4% to 17% (Smith and Arah, 1990). However, these critical values, for both water-filled pore space and oxygen concentration, are influenced by the average size of soil aggregates, by soil water content and by soil respiration rate, and therefore by the supply of available C (Smith and Arah, 1990). For a particular water content, denitrification increases with decreasing O_2 concentration and, for a particular aeration status, it increases with increasing water content (Knowles, 1981).

When a period of intense rainfall or irrigation causes a soil to become saturated at the surface, it may induce intense but short-term denitrification (Aulakh *et al.*, 1992). However, in the longer term, if rainfall or irrigation exceeds the field capacity of the soil, it will tend to leach the nitrate to a greater depth where denitrification is less likely to occur. Heavy rainfall on a clay soil is likely to increase denitrification more than leaching, but heavy rainfall on a sandy soil is likely to increase leaching more than denitrification. In the field, the installation of a drainage system tends to reduce denitrification by improving aeration and by increasing the downward movement of nitrate (Scholefield *et al.*, 1988). Conversely, in cattle feedlots, the anaerobic conditions that result from soil compaction and the abundance of excreta tend to promote denitrification (Ellis *et al.*, 1975).

Influence of Temperature and Soil pH on Denitrification

Denitrification occurs at temperatures between about 0°C and 75°C but there is a pronounced decline below 10°C, and this temperature has occasionally been reported as the minimum for denitrification to occur (Firestone, 1982). Between 10°C and 35°C, the rate is very dependent on temperature, often doubling with a 10°C increase over this range (Knowles, 1982). Denitrification often reaches a maximum at around 65°C and then declines to zero as microorganisms are destroyed at higher temperatures (Aulakh *et al.*, 1992; Hauck, 1984b). Where there is an abundant source of available C, such as that provided by slurry, the rate of loss can be substantial at temperatures of less than 10°C (Thompson, 1989). An increase in the availability of C may also explain why freeze/thaw cycles in soils tend to increase subsequent denitrification (Christensen and Tiedje, 1990; Edwards and Killham, 1986).

Denitrification is inhibited by strongly acid conditions (pH < 4) but soil pH has little effect on denitrification in the range 6–8 (Firestone, 1982). Appreciable denitrification can occur in moderately acid soils, and there is evidence that the denitrifying population adapts to acid conditions (Aulakh *et al.*, 1992). The proportion of N_2O increases with increasing acidity even though there may be no decrease in the total production of $N_2 + N_2O$ (Aulakh *et al.*, 1992; Firestone, 1982).

The seasonal pattern of denitrification reflects the combined effects of changes in soil water status and temperature. Denitrification tends to increase during early spring with increasing temperature, but it then decreases as the soil becomes drier. During the autumn, there is often an increase in denitrification as the soil becomes wetter, followed by a decrease as temperature declines (Ryden, 1986).

Influence of Available Organic Carbon on Denitrification

Even if the concentration of nitrate is high and the concentration of oxygen is low, denitrification will occur only when there is a supply of available organic C as a substrate for the growth of the denitrifying bacteria, and as a source of electrons for their respiration. Organic C also has an indirect effect on denitrification, through encouraging the growth of other microorganisms that consume oxygen and therefore reduce soil aeration (Firestone, 1982).

The effect of organic matter has been shown most clearly in studies of denitrification potential, i.e. the rate of denitrification that occurs when nitrate is non-limiting and the soil is incubated under anaerobic conditions at a constant temperature. Ideally in such studies, the soil should not be dried before incubation, as drying may have a considerable effect on the microbial activity that develops after re-wetting. In several studies, each with a range of soils, denitrification potential has been positively correlated either with total soil organic matter (e.g. Beauchamp *et al.*, 1980; Reddy *et al.*, 1982), or with water-soluble organic matter (Bijay-Singh *et al.*, 1988; Burford and Bremner, 1975), or with mineralizable carbon (Bijay-Singh *et al.*, 1988; Burford and Bremner, 1975) or with extractable reducing sugars (Stanford *et al.*, 1975). In general, the relationship is closer with some measure of readily available organic matter than with the total soil organic matter (Granli and Bockman, 1994). A relationship with carbon mineralized during anaerobic incubation, in a study of 27 soils, is shown in Fig. 9.2. Denitrification potential is usually greater in grassland soils than in similar arable soils and, in any one soil, tends to decline with depth (Aulakh *et al.*, 1992; Colbourn *et al.*, 1984) reflecting differences in the content of soil organic matter. Within the root zone, denitrification is concentrated in the rhizosphere (Smith and Tiedje, 1979), probably also due to a greater supply of available organic matter. Although denitrification normally decreases with depth in the soil, it sometimes occurs at depths of more than 60 cm, possibly as a result of soluble organic matter being leached from the surface soil (Aulakh *et al.*, 1992; Jarvis and Hatch, 1994).

It has been suggested that, because grassland soils have a high root density and are therefore well supplied with exudates and decomposing root material, the rate of denitrification is unlikely to be limited by organic matter (Clark and Paul, 1970; McGill *et al.*, 1981). However, the correlations between denitrification potential and organic matter indicate that this is not necessarily so, especially when fertilizer nitrate is applied. It is possible that the correlation between denitrification and organic matter reflects, at least partly, the relationship between organic matter and the size of the microbial biomass (Drury *et al.*, 1991).

In addition to influencing the general pattern of denitrification, organic matter is largely responsible for the variation that occurs from point to point in the soil. Fragments of easily decomposed plant material may induce intense

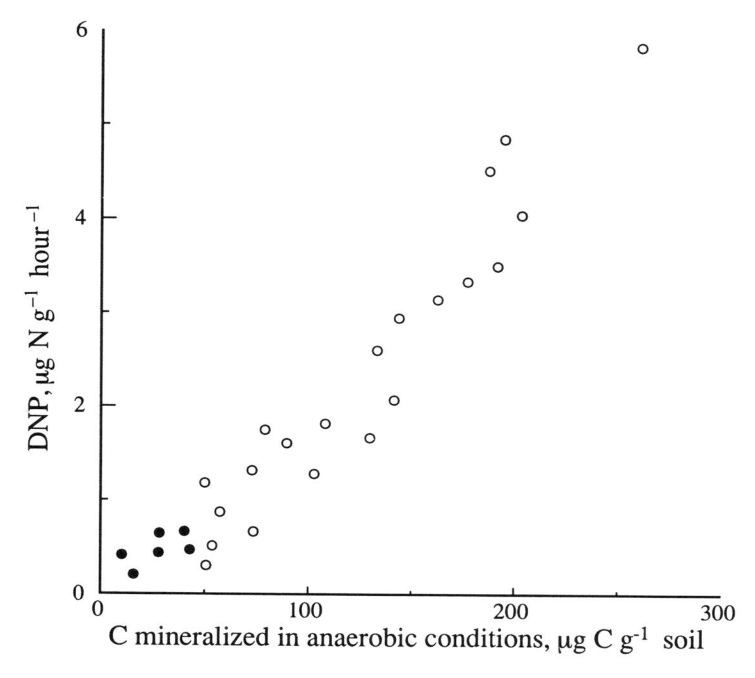

Fig. 9.2. The denitrification potential (DNP) of 21 grassland soils (○) and 6 arable soils (●), assessed in the field-moist condition in spring, in relation to the amount of C mineralized under anaerobic conditions. (Redrawn from Bijay-Singh *et al.*, 1988.)

but localized denitrification in so-called 'hot spots' (Parkin, 1987). As earthworm casts are enriched in both inorganic N and soluble organic matter, they often contain a large number of such hot spots (Elliott *et al.*, 1990; Knight *et al.*, 1989). The influence of earthworms on denitrification depends, however, on the nature of their food source and is greatest when this has a low C : N ratio (Parkin and Berry, 1994). In plots of long-term grassland, with and without fertilizer N, worm casts were estimated to contribute between 10% and 22% of the total denitrification, which was assessed at 29–83 kg N ha^{-1} $year^{-1}$ (Elliott *et al.*, 1991). Average rates of denitrification from earthworm casts and bulk soil, measured during a 9-week period in autumn are compared in Fig. 9.3. However, in contrast to their effect of promoting denitrification in worm casts, earthworms may reduce denitrification in the bulk of the soil by forming burrows that enhance drainage and aeration.

The incorporation of readily decomposable organic matter into soil (e.g. by the addition of organic manures or the ploughing of leys) can increase subsequent denitrification, but the amount of the more humified organic matter appears to have little effect (Webster and Goulding, 1989). In a study

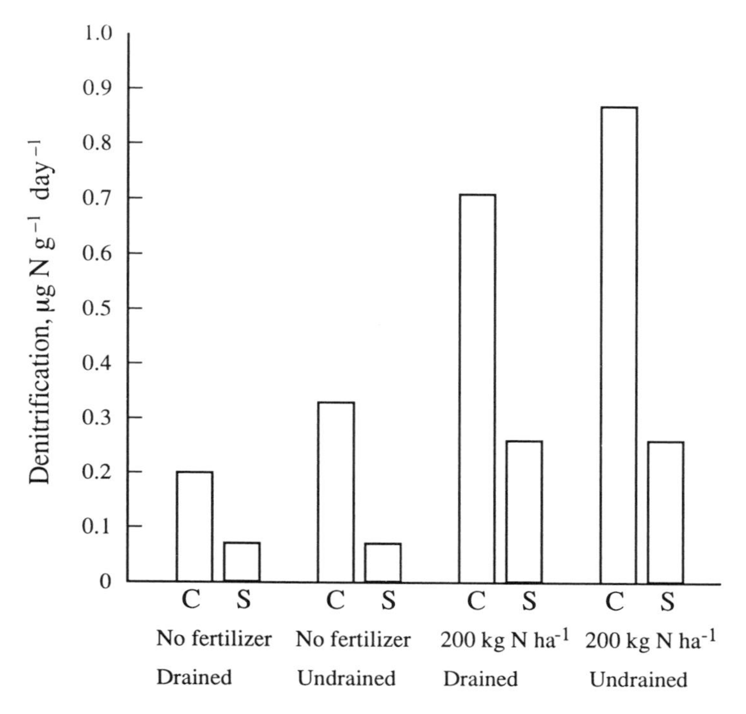

Fig. 9.3. Rates of denitrification during autumn in earthworm casts (C) and bulk soil (S) from plots of long-term grassland receiving either no fertilizer N or 200 kg N ha^{-1} $year^{-1}$ and either drained or undrained. (Data from Elliot *et al.*, 1991.)

of several slurries and solid manures added to soil, the rate of denitrification was closely correlated with the concentration of water-soluble carbohydrate and, in general, with the concentration of volatile fatty acids (Paul and Beauchamp, 1989). Because water-soluble organic matter is susceptible to leaching, the addition of slurries and manures may promote denitrification in the subsoil.

Influence of Plants on Denitrification

The presence of plants can both promote and inhibit denitrification. Thus, plant roots tend to promote denitrification by providing organic matter in the forms of root exudates and dead cell material, and also by reducing the amount of oxygen in the rhizosphere through respiration. On the other hand, growing plants absorb nitrate and therefore reduce the amount available for denitrification. They also absorb and transpire water, and thus improve aeration and the rate of oxygen diffusion in the soil (Stevenson, 1986). As these various effects tend to counteract one another, the net effect of the presence of plants may be either to increase or decrease denitrification, or it may be negligible (Beauchamp *et al.* 1989; Firestone, 1982; Granli and Bockman, 1994; Haynes and Sherlock, 1986).

Denitrification from Grassland in Relation to Fertilizer Use

There is little denitrification in grassland soils unless fertilizer or slurry is applied, or the area is grazed intensively, because nitrate concentrations are generally low. Although nitrate is released by the mineralization of soil organic matter, the process occurs progressively during the growing season, and the nitrate once released, is taken up rapidly by the grass. However, when high concentrations of nitrate occur following the addition of fertilizer N or slurry, or locally in urine patches, denitrification may be substantial. But, even with such additions, denitrification in long-term grassland soils may be restricted by the presence of numerous macropores which will prevent the soil from becoming generally anaerobic. The average loss of N as N_2O from lightly grazed shortgrass prairie in the USA has been estimated to be only 0.1 kg N ha^{-1} $year^{-1}$ (Parton *et al.*, 1988b) and so, even if the N_2 : N_2O ratio were 15 : 1, the total denitrification loss would be less than 2 kg N ha^{-1} $year^{-1}$.

In a study involving cut grassland receiving fertilizer N at 300 kg N ha^{-1} $year^{-1}$, denitrification was greater from a clay soil with impeded drainage than from a more freely drained clay-loam soil, and was greater from calcium ammonium nitrate than from urea (Table 9.1). Denitrification was negligible (< 2 kg N ha^{-1} $year^{-1}$) from plots that received no fertilizer (Jordan, 1989). In another study with cut ryegrass swards, but on a loam soil that received either 250 or 500 kg N ha^{-1} $year^{-1}$ as ammonium nitrate, it was found that, for most of the year, the rate of denitrification was less than 0.15 kg N ha^{-1} day^{-1}, and that higher denitrification rates of between 0.2 and 2.0 kg N ha^{-1} day^{-1} occurred only when the fertilizer was applied at a soil water content of more than 20% w/w (equivalent to a water-filled porosity of more than 62%) in spring or summer (Ryden, 1983). The annual denitrification loss, based on a

Table 9.1. Denitrification (kg N ha^{-1} $year^{-1}$) from cut grassland on two soil types receiving 300 kg N ha^{-1} $year^{-1}$, as either calcium ammonium nitrate (CAN) or urea, or receiving no fertilizer. (From Jordan, 1989.)

	CAN	Urea	No fertilizer
Freely-drained clay loam	29	10	< 1
Poorly-drained clay	79	31	< 1

Table 9.2. Denitrification potential assessed in the laboratory (means of seven determinations), and actual denitrification in the field, of grassland and arable plots both receiving fertilizer N at 56 kg N ha^{-1} $year^{-1}$ (means of five assessments on each of 4 days during an 8-day period). (From Bijay-Singh *et al.*, 1989.)

	Denitrification potential of soil (μg N g^{-1} h^{-1})	Field denitrification (g N ha^{-1} day^{-1})
Grassland	1.46	30
Arable	0.31	192

succession of measurements, was estimated to amount to about 10 kg N ha^{-1} when fertilizer was applied at 250 kg N ha^{-1}, and to about 30 kg N ha^{-1} when fertilizer was applied at 500 kg N ha^{-1}. Over the year as a whole, a major factor limiting denitrification was the low concentration of nitrate in the soil, caused largely by plant uptake (Ryden, 1983). When calcium nitrate was applied to cut grassland on an imperfectly drained soil, at rates of 100 or 200 kg N ha^{-1} per application at various dates, denitrification ranged from less than 1% to 21% of the N applied, with the highest proportions being denitrified after applications in the spring (Eggington and Smith, 1986).

In a comparison of denitrification potential (with nitrate non-limiting) in soil from two adjoining plots maintained for 31 years under either continuous grassland or arable cultivation, both receiving 56 kg N ha^{-1} $year^{-1}$ as fertilizer, the rate was five times greater in the grassland than in the arable soil, presumably due to a greater supply of available C in the grassland soil. However, the actual rate of denitrification in the field was often much lower in the grassland plot (Table 9.2), probably due to its greater porosity which resulted in it returning to a soil moisture tension of more than 5.5 kPa much more quickly after irrigation or rainfall than did the arable plot (Bijay-Singh *et al.*, 1989).

With grazed grassland, measurements made at three sites, at intervals of between 3 and 10 days during two autumn–spring periods, indicated that annual denitrification increased substantially as the rate of ammonium nitrate fertilizer was increased over the range 100–750 kg N ha^{-1} $year^{-1}$ (Barraclough *et al.*, 1992; Jarvis *et al.*, 1991). On the site with most denitrification, a poorly drained fine loam soil, the mean value increased from about 18 kg N ha^{-1} $year^{-1}$ at a rate of fertilizer N of 100 kg N ha^{-1} to about 48 kg with fertilizer N at 450 kg N ha^{-1} $year^{-1}$ (Barraclough *et al.*, 1992).

Influence of Grazing Animals on Denitrification from Grassland

Denitrification is generally greater from grazed swards than from cut swards. The difference is most apparent in the autumn when there is often an accumulation of nitrate in grazed swards, but may occur at other times if rainfall or irrigation are sufficient to promote denitrification. With swards that were irrigated during the summer, daily rates of denitrification from a grazed sward ranged from 0.05 to 0.35 kg N ha^{-1} day^{-1}, rates which were 1.7 to 16 times greater than those from a comparable cut sward (Ryden, 1984). The return of urine by grazing animals increases not only the amount of urea (and hence of nitrate) in the soil, but also the content of soil water in those areas affected by urine. It also increases the amount of soluble organic matter, partly through the addition of urine constituents and partly through an increased solubility of soil organic matter resulting from the temporarily higher soil pH (Monaghan and Barraclough, 1993). Although the addition of urine has sometimes promoted denitrification, as noted in relation to N_2O emission following the application of sheep urine (Sherlock and Goh, 1983) and of simulated cattle urine (De Klein, 1994), this has not always been so (e.g. Limmer and Steele, 1983). In a laboratory study involving undisturbed cores of a clay-loam soil under long-term grassland, treated with cattle urine and incubated in a sealed chamber with an artificial atmosphere, 30–65% of the urine N plus the N mineralized from soil organic matter was emitted as N_2, and 1–5% was emitted as N_2O during a period of 30 days (Monaghan and Barraclough, 1993). Although conditions in this experiment favoured denitrification and curtailed the volatilization of ammonia (due to restricted air flow), the results illustrate the potential loss from urine returned to grassland soils with plentiful organic matter.

Denitrification from drained plots of grazed grassland on a relatively free-draining clay loam soil in Northern Ireland, with calcium ammonium nitrate applied at rates from 100 to 500 kg N ha^{-1} $year^{-1}$, increased with increasing rate of fertilizer (Fig. 9.4).

Influence of Slurry Application on Denitrification

The application of slurry tends to promote denitrification, partly by supplying both inorganic N and organic C, and partly by promoting anaerobic conditions in the soil. The anaerobic conditions result partly from the higher soil water content, partly from the slurry blocking macropores, and partly from the available organic matter increasing the consumption of oxygen (Burford, 1976). In three treatments involving the injection into grassland of (i) a solution of ammonium chloride, (ii) a mixture of ammonium chloride and cattle slurry and (iii) cattle slurry alone, each providing 96 kg N ha^{-1}, the amount of N denitrified during a 12-week period increased with increasing addition of C in the form of slurry (Table 9.3): the loss from treatment (iii) represented about half of the ammonium-N added in the slurry (Thompson, 1989). It is likely that the volatile fatty acids (mainly acetic, propionic and butyric acids)

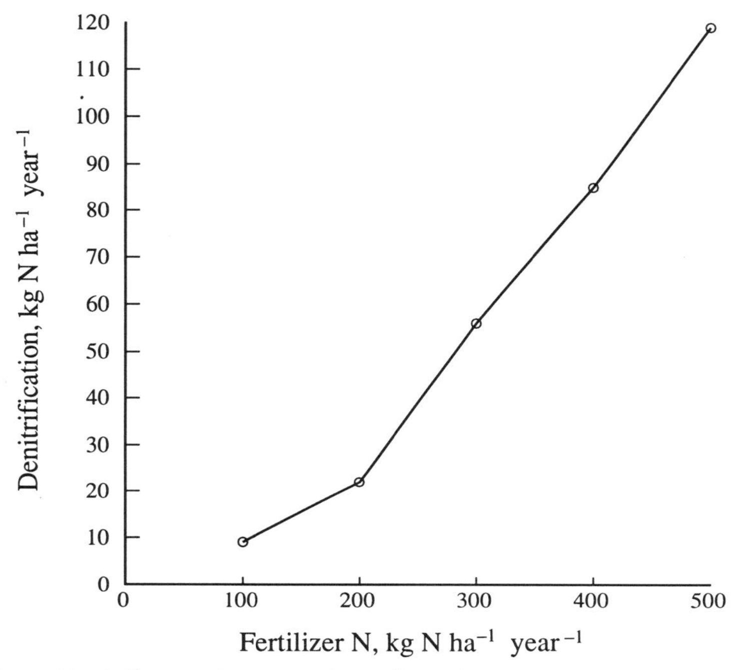

Fig. 9.4. Denitrification from grazed grassland plots, on a clay loam soil, receiving fertilizer N, as calcium ammonium nitrate, at five rates (mean values for 2 years based on sampling at approximately 10-day intervals). (Data from Watson *et al.*, 1992.)

produced during the storage of slurry are readily available sources of C for denitrifying bacteria in the soil (Beauchamp *et al.*, 1989). However, although slurry tends to increase denitrification, especially when injected into grassland (Comfort *et al.*, 1990; Thompson *et al.*, 1987), there is sometimes little effect when it is applied to the surface (Eggington and Smith, 1986; Thompson *et al.*, 1987).

When cattle slurry was applied at various times to two soils at a rate which provided about 100 kg NH_4-N ha^{-1}, the extent of denitrification varied widely (from nil to 54% of the NH_4-N) depending on soil type, and the timing and method of application (Thompson and Pain, 1989). Most denitrification occurred from applications in autumn and winter, and when the slurry was either injected into the soil or was acidified to reduce the volatilization of ammonia. The effect of soil type was surprising in that, when slurry was applied in the autumn, denitrification was much greater from a freely-drained than from a poorly-drained soil. This was attributed to a lack of nitrification of NH_4-N in the poorly-drained soil that was already almost saturated at the time that the slurry was applied. When slurry was applied to the surface of grassland on the freely-drained soil, denitrification accounted for about 30% of the NH_4-N

Table 9.3. Denitrification from grassland plots injected in late January with 80 m^3 of (1) a solution of NH_4Cl, (2) a mixture of NH_4Cl and cattle slurry and (3) cattle slurry alone, each providing 96 kg N ha^{-1} but differing in the addition of C; measurements of denitrification made at weekly intervals for 12 weeks. (From Thompson, 1989.)

	Control	NH_4Cl	NH_4Cl + slurry	Slurry
C added in slurry (kg C ha^{-1})	0	0	860	1720
NH_4-N from NH_4Cl (kg N ha^{-1})	0	96	48	0
NH_4-N from slurry (kg N ha^{-1})	0	0	48	96
Denitrification (kg N ha^{-1})	2	7	11	49

from applications in autumn and winter and about 6% of the NH_4-N from applications in spring. Denitrification was increased to more than 40% of the NH_4-N when the slurry was injected into the soil or was acidified before surface application, treatments that were intended to reduce the volatilization of ammonia. Conversely, denitrification was reduced by mixing a nitrification inhibitor (dicyandiamide or nitrapyrin) with the slurry, though nitrapyrin was effective only when the slurry was injected (Thompson and Pain, 1989). The incorporation of dicyandiamide into an autumn application of slurry providing 110 kg ammoniacal N ha^{-1} reduced denitrification from 24 to 12 kg N ha^{-1} (Pain *et al.*, 1994).

Emissions of N_2O and NO during Nitrification

There is increasing evidence that nitrification may be a major source of the N_2O and NO emitted by soils. Several studies have shown N_2O to be produced in soils under aerobic conditions in the presence of ammonium-N (Hutchinson and Davidson, 1993; Sahrawat and Keeney, 1986) and there is clear evidence that nitrifying bacteria are involved. For example, N_2O was evolved when sterilized soils were treated with ammonium sulphate, inoculated with nitrifying bacteria and incubated under aerobic conditions (Blackmer *et al.*, 1980). Also the production of N_2O from well-aerated soils was greatly curtailed by treatment with a nitrification inhibitor (Sahrawat and Keeney, 1986). Nitrification was considered to account for more than 60% of the N_2O produced from a semi-arid grassland soil in USA (except when the soil was saturated), though the estimated total N_2O production was < 0.2 kg N ha^{-1} $year^{-1}$ (Mummey *et al.*, 1994).

It is possible that some of the N_2O emission associated with nitrification may result from nitrite acting as a substrate for denitrifying organisms. This may occur if conditions generally become less aerobic, or if the nitrite moves into less aerobic microsites in the soil (Poth and Focht, 1985). The possibility that several reaction pathways may be involved makes it difficult to quantify the direct contribution of nitrification to the production of N_2O in the field situation. Nitrous oxide produced through nitrification is more likely to volatilize than that produced through typical denitrification, since the conditions of aeration associated with nitrification allow more to diffuse to the atmosphere rather than dissolving in the soil water.

It has been suggested that urine patches might provide a major source of N_2O released by nitrification in grazed grassland (Haynes and Williams, 1993). Although studies with cores of a clay-loam grassland soil treated with urine indicated that this pathway was of minor significance in comparison with normal denitrification (Monaghan and Barraclough, 1993), in other studies with both a sandy and a loam soil, there was evidence of substantial N_2O emission due to nitrification (De Klein, 1994; Velthof and Oenema, 1994). The relative importance of nitrification and denitrification as sources of N_2O will depend on soil conditions but, because both processes are influenced by a variety of soil factors, it is difficult to predict total N_2O emission (Mosier, 1994). A generalized relationship between the water-filled pore space of the soil and the emissions of N_2O and N_2 from both nitrification and denitrification has been postulated, as shown in Fig. 9.5. When soil water content is low, there is little emission of N_2O (and none of N_2) because O_2 is plentiful and nitrification proceeds to nitrate. As the soil water content increases, mineralization increases, and nitrification increasingly produces N_2O rather than nitrate. Over the range of moderate soil water contents, denitrification becomes increasingly important, initially with a high ratio of N_2O to N_2. When the soil water content is high, nitrification stops and denitrification proceeds increasingly to N_2 (Granli and Bockman, 1994). The relationship shown in Fig. 9.5 suggests that maximum production of N_2O is likely when the soil water content is at about field capacity and both nitrification and denitrification are occurring.

An assessment of measured N_2O emissions (but excluding N_2) from numerous experiments, including 65 comparisons between fertilized and unfertilized sites, indicated that, in all but one of the comparisons, higher emissions occurred from the fertilized sites (Eichner, 1990). For example, with long-term grassland, applications of ammonium nitrate and cow slurry greatly increased the emission of N_2O (Christensen, 1983). The highest emission rates were from experiments in which anhydrous ammonia was used as the N source, supporting the view that nitrification is important in the production of N_2O.

With NO, volatilization from soils appears to be due mainly to nitrification. Incubation studies have shown the addition of ammonia to enhance NO emission under aerobic, but not anaerobic, conditions and more of the added ammonia to be volatilized as NO than as N_2O. The proportion of added ammoniacal N released as NO by nitrification has been reported to range from 0.1% to 10% and is often greater than the proportion released as N_2O (Hutchinson and Davidson, 1993).

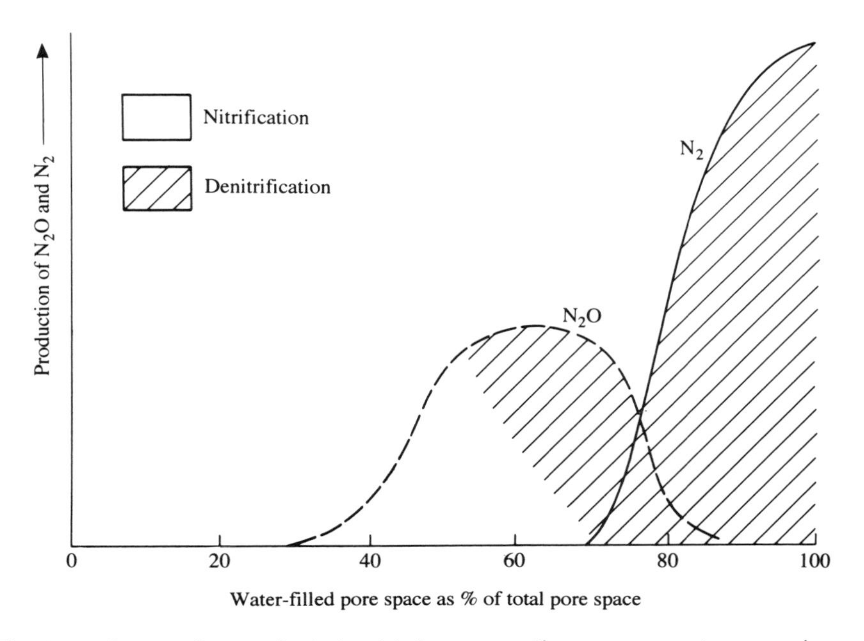

Fig. 9.5. Proposed general relationship between soil water content (expressed as water-filled pore space) and amounts of N_2O and N_2 produced in the soil. (From Granli and Bockman, 1994.)

In a study of the emissions of both NO and N_2O from plots of ryegrass, emissions were largest shortly after the application of ammonium nitrate, and the balance between NO and N_2O was related to soil aeration (Skiba *et al.*, 1992). In periods of low rainfall, the ratio of NO to N_2O was more than 100 : 1, suggesting that nitrification was the major source of the NO, whereas during periods of heavy rainfall on a clay loam soil the ratio was less than 0.001 : 1, a ratio typical of denitrification.

Chemo-denitrification

Chemo-denitrification occurs when nitrite reacts with other soil constituents to release gaseous NO_2, NO, N_2O and/or N_2. Nitrite does not normally accumulate in soils but it may do so if ammoniacal fertilizers or urea are applied at high rates (Morrill and Dawson, 1967; Vinten and Smith, 1993). The high concentrations of ammonia that result from such applications, combined with high pH, tend to inhibit the oxidation of nitrite to nitrate by *Nitrobacter* and therefore promote the accumulation of nitrite. As the ammonia is progressively nitrified, the soil pH declines and acid conditions may develop, at least in localized microsites, even in neutral or slightly alkaline soils. In acid conditions, nitrite forms nitrous acid and this may then decompose spontaneously:

$$2HNO_2 \rightarrow NO + NO_2 + H_2O$$

Normally the NO, at least, will volatilize. However, if volatilization is restricted while the soil is largely aerobic, the NO is converted to NO_2 which dissolves in soil water to form HNO_3, thus reducing the volatile loss (Nelson, 1982).

Other possible mechanisms for the release of gaseous products from nitrite are the reaction between nitrous acid and amino compounds in the soil:

$$HNO_2 + R\text{-}NH_2 \rightarrow R\text{-}OH + H_2O + N_2$$

and the reaction of nitrous acid with ammonium:

$$HNO_2 + NH_4^+ \rightarrow N_2 + 2H_2O$$

The total and relative amounts of N_2 and the nitrogen oxides released from acid soils through the decomposition of nitrite depend to a large extent on the pH and physical conditions of the soil (Hauck, 1984b).

Although chemo-denitrification has not been quantified, several observations have indicated its potential importance. Thus, substantial gaseous losses of N have been observed from soils in conditions that were not conducive to microbial denitrification or to the volatilization of ammonia, but were conducive to the accumulation of nitrite, e.g. after the application of high rates of ammonium or ammonium-forming fertilizers (Smith and Arah, 1990). Also, nitrite has been observed to decompose rapidly, with the formation of gaseous forms of N, when added to acidic soils, both sterilized and unsterilized (Nelson, 1982). Chemo-denitrification, possibly involving ferrous iron as a chemical reducing agent, may be significant in some subsoils (Smith and Arah, 1990).

Environmental Impact of the Gaseous Products of Denitrification and Nitrification

One consequence of denitrification is that plant-available N is lost from the soil and, in an agricultural context, this is clearly undesirable. On the other hand, with the increasing concern over the presence of nitrate in water supplies, denitrification may be regarded as beneficial in some situations, since it reduces the movement of nitrate into streams and rivers, and into underground aquifers.

Whether or not the gaseous products of denitrification, nitrification and chemo-denitrification have further environmental impact depends on the balance between N_2, N_2O and NO. The loss of N_2 from the soil has no impact on the wider environment, but N_2O contributes to two environmental concerns, climate change and depletion of the ozone layer. One reason for N_2O having an appreciable environmental impact is that it has a long residence time in the atmosphere, probably between 100 and 200 years (Granli and Bockman, 1994; Wayne, 1993). It has a low solubility in water and therefore only a small proportion is removed by precipitation and a large proportion moves into the stratosphere. Although N_2O is chemically inert in the lower atmosphere, it absorbs radiation in the infra-red band and therefore has a warming effect on the earth's surface. The effect of N_2O, per unit weight, is 180 times greater than that of CO_2 and is thought to account for about 5% of the total greenhouse effect (Bouwman, 1989). About 70% of the N_2O released to the atmosphere is derived from soil (Mosier, 1994) and the concentration in the atmosphere is increasing, probably by about 0.2% per year (Skiba *et al.*, 1992; Warneck, 1988). The increase is thought to be due partly to the widespread conversion of forest and grassland to arable cultivation and partly to the increased use of fertilizer N (Mosier, 1994). A second environmental effect of N_2O arises when it reaches the stratosphere and reacts progressively with electronically excited O atoms, produced by the photolysis of ozone. Most of the N_2O is converted to N_2 and O_2, with perhaps 5–10% being converted to NO (Jenkinson, 1990; Warneck, 1988). Reactions of N_2O and NO in the stratosphere contribute to the destruction of ozone, thus allowing more UV radiation to reach the earth's surface.

As indicated above, both denitrification and nitrification contribute to the production of N_2O and, in the field, the rate of these transformations is highly variable (Hauck, 1984b; Mosier, 1994). With denitrification, the conditions that favour N_2O relative to N_2 (high concentrations of nitrate, low temperature, acid microsites, restricted availability of organic C and marginal anaerobiosis) tend to restrict total denitrification and this suggests that the amounts of N_2O released may be relatively small (Hauck, 1984b). With nitrification, the factors that influence the production of N_2O are not well understood and the quantities of N_2O released are uncertain.

Although it is impossible to provide any quantitative estimate of total N_2O emission from grassland on a national or global basis (IPCC, 1992), it is clear that emissions increase with increasing fertilizer use and, almost certainly, with stocking density. Soil and weather factors are of major importance, but the adoption of good agricultural practices can help to minimize N_2O emissions. It is particularly important to avoid the application of excessive N in fertilizers and slurries (Granli and Bockman, 1994).

Nitric oxide that is released from the soil (or from other sources) to the lower atmosphere differs in its effects from NO formed in the stratosphere. Its main environmental impact is in causing soil acidification. Most NO is quickly converted via NO_2 to nitric acid, which is then removed from the atmosphere by wet and dry deposition (Warneck, 1988). Because NO and

NO_2 are in photochemical equilibrium in the atmosphere, they are usually designated together as NO_x. Their residence times in the atmosphere are similar at 35–36 days (Hauck, 1984b). Although most NO_x in the atmosphere results from the combustion of fossil fuels, reactions in the soil result in the release of appreciable amounts of NO but more work is needed on the factors that influence these amounts.

10

Use of Fertilizer Nitrogen and Slurry Nitrogen on Grassland: Recovery and Response

Introduction

The yield of herbage from intensively managed grassland in temperate regions is typically between 8,000 and 15,000 kg dry weight ha^{-1} $year^{-1}$. This yield may be harvested either by a sequence of cuts, or by rotational or continuous grazing, or by a combination of cutting and grazing. In addition, the annual production of unharvested stubble and roots may well amount to between 6,000 and 12,000 kg dry weight ha^{-1} $year^{-1}$. The concentration of N in the herbage, either of grass receiving fertilizer or of a productive grass–clover mixture, is typically between 2.5 and 3.5%, and the concentrations in stubble and roots are typically between 1.5 and 2.0%. The amount of N contained in the annual production of plant material by intensively managed grassland is therefore generally between 300 and 700 kg N ha^{-1}. Such an amount is considerably greater than that made available for plant uptake through the net mineralization of soil organic matter plus inputs from the atmosphere. In order to achieve high yields of herbage, it is therefore necessary to provide additional N, either by fertilizer application, or by the use of legumes to fix atmospheric N_2, and/or by the application of slurries or manures. Each of these sources of N has advantages and disadvantages.

Fertilizer N has several advantages. First, it is possible to predict quite accurately the increase in herbage yield that will result from its use under specific soil and weather conditions. Second, it is relatively cheap in comparison with the value of grass herbage for livestock production. Third, it is easily applied and, fourth, advice on its use is readily available. The main disadvantages of fertilizer N are that it may result in environmental pollution, and that substantial energy costs are incurred in its manufacture and distribution.

Pollution may arise through the leaching of nitrate into groundwaters, through the volatilization of ammonia and/or through the volatilization of nitrous oxide, three processes all of which are all enhanced by high rates of fertilizer N. In addition, the risk of water in streams and ditches being polluted by silage effluent or slurry is greater in situations where high yields of herbage are produced by the use of fertilizer N. In relation to the energy costs of fertilizer manufacture, the monetary component is reflected in the price paid by the farmer, but the environmental component, as with the cost of other forms of pollution, is currently borne mainly by the wider community.

When legumes are used as a source of herbage N, the expense of fertilizer N is avoided and the risks of pollution through nitrate leaching, ammonia volatilization and nitrous oxide production, are less than with high rates of fertilizer N. However, some risk of pollution remains and, for a grazed grass–clover sward, the risk is similar to that from grass fertilized with a moderate rate of fertilizer N, perhaps 250 kg N ha^{-1} $year^{-1}$. Also, if silage is produced from grass–clover swards and fed to housed cattle, there is some risk of the pollution of streams and ditches by silage effluent and/or slurry. Grass–clover swards also have some definite disadvantages in comparison with fertilized grass. The growing season of clover is shorter than that of grass, mainly because growth starts later in the spring, and the herbage yield produced in spring is therefore lower than that from fertilized grass. It is also more difficult to predict the total annual yield of a grass–clover sward at the beginning of the season, partly because the proportion of clover in the sward sometimes declines markedly from one year to the next (see Chapter 3). A further disadvantage is that, if herbage containing too large proportion of clover is fed to cattle, there is a tendency for them to develop bloat (see Chapter 14). Despite these various problems, if an optimum proportion of clover in the sward (around 30–40% of the herbage produced) can be maintained from year to year, the advantages of grass–clover swards may well outweigh the disadvantages.

The application of slurry or manure as a source of N has the advantage that use is made of material that would otherwise be wasted but would have to be disposed of. Slurry is the main form in which excreta from housed livestock are applied to grassland, though farmyard manure (FYM) is used in some areas, generally where management is fairly extensive. In addition to the slurries and manures derived from livestock excreta, sewage sludges are often applied to grassland, though usually as a means of disposal rather than a source of nutrients. Slurry has a number of disadvantages as a source of N. It is bulky, much of it is produced during the winter when wet or cold conditions often make application difficult, and the response in terms of herbage yield is much less predictable than that to fertilizer N. The use of slurry may also result in an unpleasant odour over a wide area during and after the application. There will almost certainly be some loss of N through the volatilization of ammonia, and there may be some pollution of water in ditches and streams if surface runoff occurs after application. Some of these disadvantages can be minimized by management practices, such as the installation of storage tanks (to allow slurry to be applied when soil conditions are suitable) and the separation of slurry into predominantly liquid and solid fractions (see p. 217). Farmyard manure and sewage sludge have disadvantages similar to those of slurry, and sewage sludge has the additional disadvantage that it often contains appreciable amounts of heavy metals, and sometimes other pollutants, that tend to accumulate in the soil.

Forms of Fertilizer N

Almost all forms of fertilizer N are derived from ammonia which is synthesized from nitrogen in the atmosphere (see p. 2). On a world scale, the most important fertilizers are urea, ammonium nitrate, the mono- and di-ammonium phosphates and ammonia itself, the last being either anhydrous or in solution. The nitric acid required for the production of ammonium nitrate is also obtained from ammonia (by oxidation), while the phosphoric acid required for the production of the ammonium phosphates is obtained from the acid treatment of phosphate rock.

Urea is the most concentrated solid N fertilizer (about 46% N) and, if the cost of construction of new manufacturing plant is taken into account, it is less expensive to manufacture than are ammonium nitrate and the ammonium phosphates. It is also cheaper to transport and is therefore generally less expensive to the farmer than other solid forms of fertilizer N. Urea is soluble in water and deliquescent and, in order to overcome the deliquescence, it is produced in the form of prills or granules that are usually coated with a resin. However, some sources of prilled urea are susceptible to crushing and, if this occurs, the fertilizer is likely to be applied unevenly. In addition to being used as a straight fertilizer, urea is included in some compound fertilizers, sometimes as urea phosphate and sometimes with other forms of N, such as ammonium sulphate or calcium nitrate. Urea is the most widely-used form of fertilizer N on a world scale though, in countries with a long history of fertilizer manufacture, much of the manufacturing plant is designed for ammonium nitrate and this form is also produced in large amounts.

Ammonium nitrate (35% N) is widely used in Europe and is the major form of fertilizer N in the UK. Until the 1960s, ammonium nitrate was available only in forms that included calcium carbonate, and it was sold in the UK as Nitro-chalk or Nitra-shell, which contained between 16% and 21% N. Such mixtures are now known as 'calcium ammonium nitrate'. The addition of calcium carbonate to the ammonium nitrate results in easier handling, and greatly reduces the risk of fire and explosion: it also counteracts the tendency of ammonium nitrate to acidify the soil and therefore reduces the need for liming. In the mid 1960s, the problems of handling and fire risk were overcome by a process that involves prilling and subsequent treatment of the granules with a water-repellent material, such as calcium stearate, and a conditioning material, such as diatomaceous earth. The resultant prilled form of ammonium nitrate, which contains 33–34% N, is stable during storage and spreading, costs less per kilogram N than does calcium ammonium nitrate, and is now widely used.

The ammonium phosphates (mono-ammonium phosphate and di-ammonium phosphate, which contain 12% and 21% N respectively) are often included as constituents of compound fertilizers. Di-ammonium phosphate is also available as a straight fertilizer. Ammonium sulphate (21% N), once widely available as a by-product from the manufacture of coal gas, is now used to only a small extent. Calcium, potassium and sodium nitrates are available as fertilizers in small amounts, but are rarely used on agricultural grassland.

As ammonia itself is the initial product of chemical N_2 fixation, it is cheaper, per unit of N, than are ammonium and nitrate salts and urea. Ammonia is available as a fertilizer as anhydrous ammonia (82% N) and as an aqueous solution (21–29% N) but, with both forms, injection to a depth of about 10 cm in the soil is necessary in order to minimize volatilization (van Burg *et al.*, 1967). Anhydrous ammonia requires the use of pressurized containers and special injection equipment whereas aqueous ammonia, which is more bulky, is under only slight pressure and therefore requires less expensive equipment.

The various N fertilizers can be divided into three groups based on the pH of their aqueous solutions (Hauck, 1984a). The first group comprises salts that are the product of a weak base and a strong acid (e.g. ammonium nitrate, ammonium sulphate and mono-ammonium phosphate) and these dissolve to form an acid solution which affects the soil around each fertilizer granule. The second group comprises salts that are products of a strong base and a strong acid (e.g. calcium nitrate, potassium nitrate and sodium nitrate) and these are neutral and have no direct effect on soil pH. The third group comprises salts, or compounds that form salts, of a strong base with a weak acid, and these form an alkaline solution in the soil close to the fertilizer. Examples of this group include urea, anhydrous ammonia and di-ammonium phosphate. The effect on the localized soil pH in the vicinity of fertilizer granules (or site of injection) influences subsequent transformations of the fertilizer N, including the extent of ammonia volatilization (see p. 169), the rate of nitrification (see p. 115) and the accumulation of nitrite which is then susceptible to chemo-denitrification (see p. 196). However, in the long term, almost all ammonium fertilizers and urea tend to acidify the soil (see p. 263) and cause a loss of exchangeable cations. These effects are due partly to the nitrification of ammonium to nitrate and partly to the uptake of ammonium (see p. 18). The reactions involved in soil acidification resulting from the nitrification of ammonium sulphate are illustrated in Fig. 8.6. Soil acidification, due to the repeated application of fertilizer over many years, reduces the population of earthworms and induces partially decomposed plant material to accumulate at the soil surface (Ma *et al.*, 1990).

In addition to the soluble forms of fertilizer N, various synthetic slow-release fertilizers have been developed with the aim of providing a continuous supply of plant-available N over a period of several months, but the high cost of such fertilizers generally precludes their use on agricultural grassland (see Chapter 13).

Response of Grassland to Fertilizer N

The response of grass swards to fertilizer N, applied at a range of different rates, has been examined in numerous field trials. Many of the trials have involved ryegrass, and most have been harvested by cutting rather than grazing. Normally, the fertilizer has been applied at intervals during the growing season, often in three to seven equal amounts, with the first application being made at the beginning of the season and the others after each harvest except the last. This pattern of application, which reflects the continuing need for N during the season, produces a greater annual yield and better quality herbage than does a single application in spring (Castle and Reid, 1968).

The results of trials with cut swards have shown that the annual yield of herbage increases substantially with increasing rate of fertilizer N, and a typical response curve, based on a number of trials in the UK, is shown in Fig. 10.1. With no fertilizer N (and with little or no clover in the sward) the annual yield of grassland herbage in lowland areas is usually between 1 and 5 tonnes dry matter (DM) ha^{-1}. The actual yield depends on soil and weather conditions, and on soil N supply as influenced by the previous management of the field. As the rate of fertilizer N is increased, there is often an almost linear increase in herbage yield, of between 20 and 30 kg DM per kg N, until the rate of application is at some point between 250 and 400 kg N ha^{-1} $year^{-1}$ (Morrison *et al.*, 1980; Reid, 1972; van Burg *et al.*, 1981; van der Meer and van Uum-van Lohuyzen, 1986). At higher rates, the response per kilogram of additional fertilizer N declines until the maximum yield is attained, at which point the response to additional fertilizer N is zero. At even higher rates of application, there is usually no significant effect on yield though eventually the response may become negative with some decrease in yield. The field trials that have included rates of fertilizer N of more than 500 kg N ha^{-1} $year^{-1}$, have generally shown maximum yield to be attained at a point between 600 and 1200 kg N ha^{-1} $year^{-1}$ (Armitage and Templeman, 1964; Hunt, 1966b; Lee *et al.*, 1977; Morrison *et al.*, 1980; Prins and Neetson, 1982; Reid, 1966; Richards 1977; Schmidt and Tenpas, 1965). Responses broadly similar to that shown in

Fig. 10.1 for ryegrass in the UK have been reported for timothy, smooth bromegrass and cocksfoot in Indiana, USA (George *et al.*, 1973), and for Bermudagrass and Bahiagrass in Florida, USA (Overman *et al.*, 1992). Three examples of actual response curves involving high rates of fertilizer in the UK are shown in Fig. 10.2. For any particular location, the rate of fertilizer that produces maximum yield is highest when the sward is defoliated frequently, and when soil and weather conditions are most favourable for grass growth (see Chapters 11 and 12).

Provided that fertilizer N is applied on several occasions during the growing season, differences in the relative amounts applied on the successive occasions have little effect on the response in terms of total annual yield (Morrison *et al.*, 1980), though there is an influence on the distribution of herbage production through the season (see p. 232).

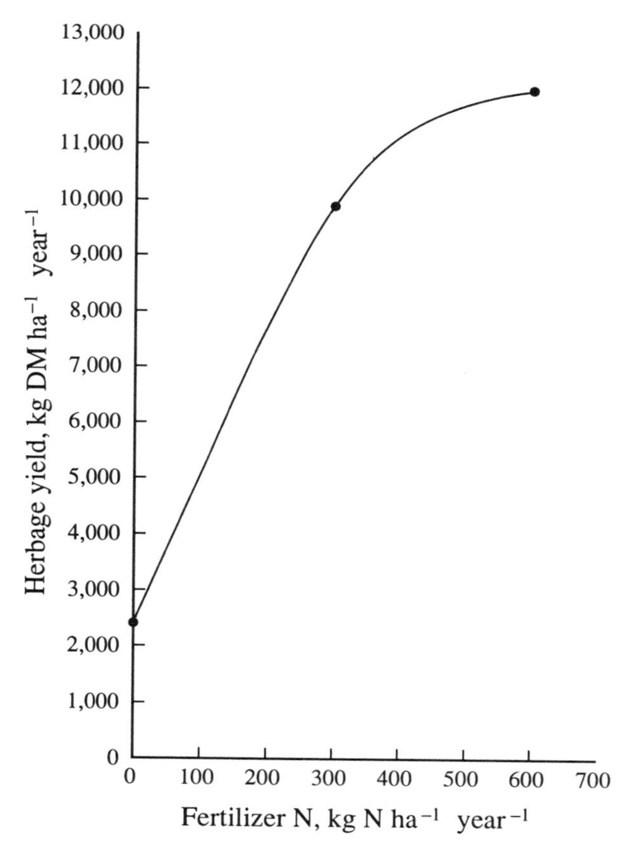

Fig. 10.1. Generalized response curve of productive grassland to the application of fertilizer N. (Based on data from Morrison *et al.*, 1980.)

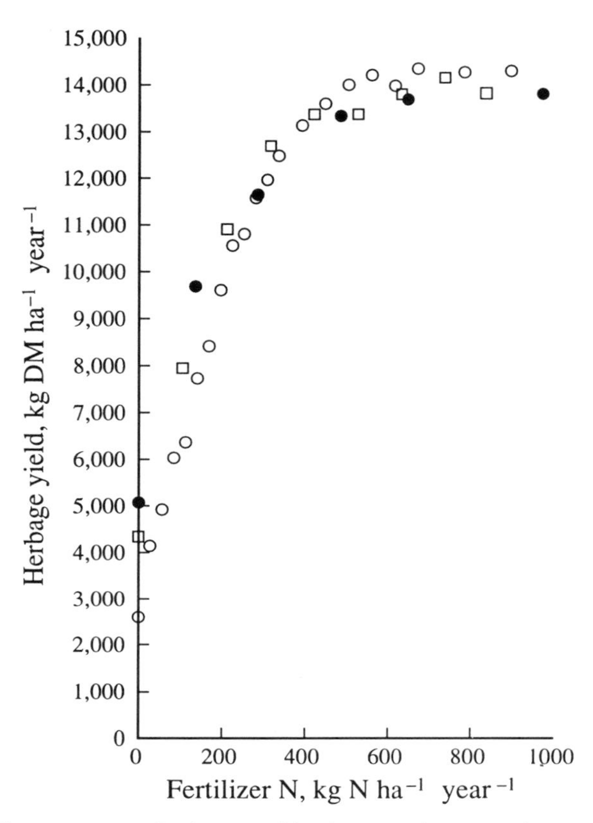

Fig. 10.2. The response to high rates of fertilizer N of perennial ryegrass in southwest Scotland (○) (Reid, 1970, 1972), of perennial ryegrass in southwest England (●) (Armitage and Templeman, 1964) and of Italian ryegrass in southeast England (□) (ICI, 1966).

The supply of N from the soil often has little influence on the response to fertilizer N (in terms of kg DM per kg N) at rates up to about 300 kg N ha^{-1} year^{-1}, though it does influence the actual yield at any specific rate of fertilizer. Also, other factors being equal, the higher the supply of N from the soil, the lower the optimum rate of fertilizer application. Although the supply of N from the soil usually has little effect on response, there may be an effect of fertilizer N applied in the previous year. High rates of fertilizer N (> 500–600 kg N ha^{-1} year^{-1}) often result in decreases in the number of tillers, the weight of roots, the depth of rooting and the carbohydrate reserves of the grass, and these factors all tend to curtail herbage production in the following year (Prins and Neeteson, 1982).

Despite soil N status having little effect on the response to fertilizer N, there is often considerable site-to-site variation in response (e.g. Hopkins *et al.*, 1995; Morrison *et al.*, 1980) due to factors such as the available water capacity of the soil and the rainfall during the growing season. For an individual site, a response curve based on average results for several years, provides a useful means of estimating the economically optimum rate of fertilizer application, i.e. the rate at which the marginal increase in the value of the herbage is equal to the marginal increase in the cost of fertilizer. Such a calculation cannot be precise, because weather varies from year to year and because the value of the herbage is influenced by the livestock system in use, by the nutritional quality of the herbage and by the degree to which it is actually utilized by the livestock. Nevertheless, it is possible to assume a monetary value for herbage of a certain digestibility, in the context of a particular livestock system. A calculation of this type indicated that a minimum response of 5.7 kg DM kg^{-1} N was necessary for an economic return from grass fed to dairy cows in the UK in 1970 (Reid, 1970) while a similar calculation in the Netherlands indicated that a minimum response of 7.5 kg DM kg^{-1} N was necessary (Prins *et al.*, 1980; van der Meer and van Uum-van Lohuyzen, 1986). In assessing the results of a multi-centre trial in England and Wales, the optimum rate of fertilizer application for a particular site was decided on a different basis, viz. the rate that produced a marginal response of 10 kg DM per kg N (Morrison *et al.*, 1980). There were two reasons for this choice of optimum rate, referred to as N_{10}. First, it produced a yield that was about 90% of the maximum while requiring only about 60% of the fertilizer N needed for maximum yield and, second, at rates higher than this calculated optimum, the response to additional N usually showed a marked decline.

In a few investigations, the response to fertilizer N has been assessed in terms of milk or meat production. From a response expressed in these terms, it is possible to calculate the optimum rate of fertilizer application from the ratio between the price received for milk or meat and the cost of fertilizer N (Gordon, 1974). If the herbage is utilized effectively, the response to fertilizer N in terms of milk or meat production is closely correlated with the response in terms of herbage yield (Gordon, 1974) and is due almost entirely to increased stocking rate rather than increased production per animal. Targets for the production of milk and meat from dairy and beef cattle grazing swards receiving a range of rates of fertilizer N have been proposed (Holmes, 1968, 1974) and are summarized in Table 10.1.

In terms of milk production, the targets suggest an almost linear increase up to a fertilizer rate of 450 kg N ha^{-1} $year^{-1}$, with each kilogram N yielding about 15 kg of milk. This is not often achieved in practice. An assessment of data from 845 herd-years within the ICI Recorded Farms Scheme (UK) indicated that the average response in terms of milk yield was 9.9 l of milk kg^{-1} fertilizer N (Hawkins and Rose, 1979). However, responses of about 15 kg milk kg^{-1} N have been obtained in experiments in Germany (van Burg *et al.*, 1981). Also, in a trial in the Netherlands in which a ryegrass sward, grazed continuously by dairy cattle, received fertilizer N at rates ranging from 250 to

700 kg N ha^{-1} $year^{-1}$, the response in terms of milk yield ha^{-1} to an increase in fertilizer from 250 to 400 kg N ha^{-1} was equivalent to 16.3 kg milk kg^{-1} fertilizer N (Fig. 10.3). Neither the mean daily milk yield per cow nor

Table 10.1. Targets for dairy and beef production from grass. (From Holmes, 1968, with modifications for beef cattle from Holmes, 1974.)

		Target production in 180-day grazing season			
		Dairy cows		Beef cattle	
Fertilizer applied (kg N ha^{-1})	Herbage yield (kg DM ha^{-1})	Cows ha^{-1}	Milk (kg ha^{-1})	Cattle ha^{-1}	Liveweight gain (kg ha^{-1})
0	5,500	2.2	5,800	3.2	570
150	9,800	3.9	10,500	5.7	1,020
300	11,700	4.6	12,500	6.8	1,220
450	13,200	5.2	14,100	7.6	1,380

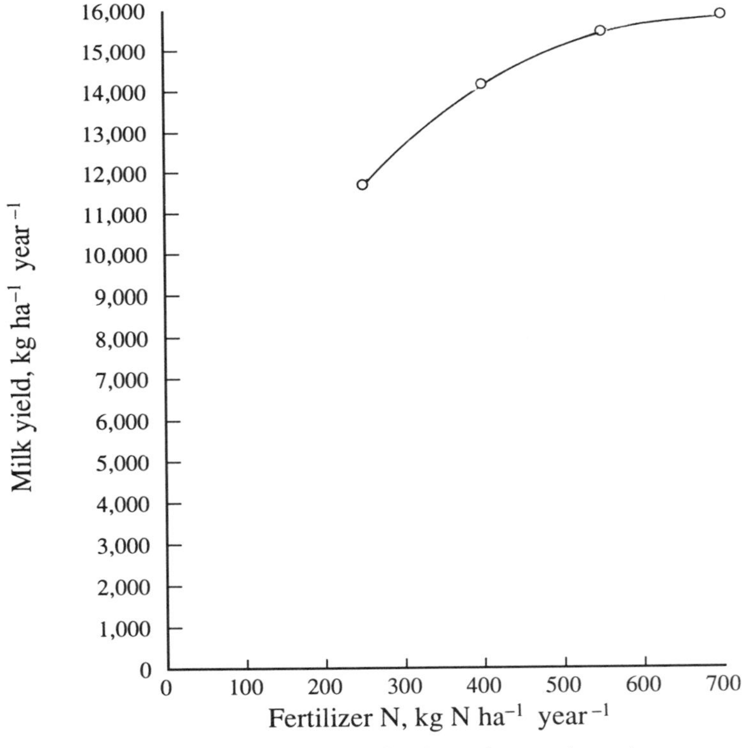

Fig. 10.3. Response in terms of milk yield (adjusted to 4% fat and 3.3% protein) ha^{-1} $year^{-1}$ from perennial ryegrass receiving four rates of fertilizer N; means of 3 years. (Data from Deenen and Lantinga, 1993.)

liveweight gain per cow was influenced by the rate of fertilizer N (Deenen and Lantinga, 1993).

For beef production, the response per hectare is very much influenced by the initial size of animal, since liveweight gain per animal per day is similar for animals ranging in weight from 150 to 450 kg, while stocking rate per hectare is greater for the smaller animals. The targets in Table 10.1 imply that response should amount to about 2 kg liveweight gain per kg fertilizer N for a rate of 300 kg N ha^{-1} $year^{-1}$, and this target is attainable by young animals up to an age of about 18 months. However in several experiments, which probably included animals of a wider size range, average responses of about 1 kg liveweight gain per kg N were obtained (van Burg *et al.*, 1981).

Recovery of Fertilizer N

The percentage recovery of fertilizer N in harvested herbage is usually expressed in terms of the 'apparent' recovery, calculated from the total amounts of N in the herbage and fertilizer. (To assess the 'actual' recovery of N from the fertilizer itself requires the use of ^{15}N-labelled fertilizer.) The apparent recovery is defined as the amount of N in the herbage of a fertilized sward, minus the amount in the herbage of a comparable unfertilized sward, expressed as a percentage of the fertilizer N applied. The amount of N in the herbage of the unfertilized sward provides an estimate of the supply from the soil and the atmosphere. Assessed in this way, the apparent recovery of fertilizer N by grassland is usually between 50% and 80%, and often about 65–70% (Dilz, 1988; Morrison *et al.*, 1980). Apparent recoveries of more than 80% were obtained in a trial in the Netherlands, in which perennial ryegrass, receiving fertilizer N at rates between nil and 700 kg N ha^{-1} $year^{-1}$, was cut at 4-weekly intervals (Deenen and Lantinga, 1993). A possible source of error in the calculation of apparent recovery is the assumption that the amount of N supplied by the soil is the same in fertilized and unfertilized plots: this may not be so if the fertilizer affects the mineralization of soil N and/or the distribution and activity of roots. However, discounting this possibility, a 67% recovery in the harvested herbage would represent complete uptake of the fertilizer N, if the N were partitioned between the herbage, and the stubble plus roots, in the ratio of 2 : 1. Evidence of a ratio close to this has been provided by investigations with ^{15}N (e.g. Bristow *et al.*, 1987; van der Meer and van Uum-van Lohuyzen, 1986). Apparent recoveries higher than 67% may indicate that the allocation of N to stubble and roots has been low, possibly due to frequent defoliation, or that the fertilizer has increased the mineralization of soil N. The apparent recovery of fertilizer N is generally highest at application rates of 200–350 kg N ha^{-1} $year^{-1}$, as illustrated in Table 10.2 and by the results of other investigations (e.g. Morrison *et al.*, 1980). With

Table 10.2. Herbage N yield, and apparent recovery of fertilizer N (applied as calcium ammonium nitrate) in Italian ryegrass at a wide range of fertilizer application rates. (From ICI, 1966.)

	Fertilizer N (kg ha^{-1} year^{-1})								
	0	105	210	316	420	526	632	738	843
Herbage yield, kg DM ha^{-1} year^{-1}	4,340	7,950	10,910	12,690	13,350	13,350	13,780	14,130	13,810
N yield, kg ha^{-1} year^{-1}	68	118	192	265	308	345	384	400	419
Apparent recovery (%)	–	47	59	62	57	53	50	45	42

application rates of less than about 200 kg N ha^{-1} year^{-1}, there is a tendency for a larger proportion of the N taken up to remain in the roots and stubble and, with application rates of more than about 350 kg N ha^{-1}, uptake may be incomplete. However, at high rates of fertilizer, there is a tendency for more of the N taken up to be allocated to the herbage rather than to roots (Deenen and Lantinga, 1993; Ennik *et al.*, 1980). There is no consistent difference in recovery between ammonium and nitrate forms of fertilizer N (see p. 253).

With a single application of fertilizer, the recovery of N at the next harvest increases with increasing time interval between the application and harvest (Table 10.3). A recovery of less than about 45% from a single application, and with a growth period of between 3 and 6 weeks and a cutting height of 3–6 cm, is likely to be due to a factor such as drought restricting growth, or to abnormal weather conditions causing a large leaching or gaseous loss of N.

As with response to fertilizer N, there is considerable site-to-site variation in recovery (Devine and Holmes, 1963; Morrison *et al.*, 1980; Stevens, 1988), reflecting differences in weather and soil conditions. For example, the apparent recovery of fertilizer N was found to be highest in dry years and lowest in wet years on poorly drained sites in the Netherlands, whereas it was best in normal years and low in dry years on sites with good drainage (van der Meer, 1983). In most field experiments in which the recovery of fertilizer N has been assessed for the season as a whole, equal applications of fertilizer N have been

Table 10.3. Apparent recovery (%) of fertilizer N by Italian ryegrass at various intervals after application, calculated from a basic application of 28 kg N ha^{-1}. (From Wilman, 1965.)

N applied (in excess of basic rate)	Weeks after application					
	1	2	3	4	5	6
56	2	34	70	78	86	76
112	1	20	59	67	72	74

applied for each growth period during the season. In an investigation in which the distribution of fertilizer N over six successive growth periods was 1/4, 1/4, 1/8, 1/8, 1/8, 1/8, a slightly higher recovery of N was achieved than with six equal applications (Morrison *et al.*, 1980).

In investigations in which the actual recovery of the N from a specific application of fertilizer has been assessed using ^{15}N-labelled fertilizer, the actual recovery has consistently been less than the apparent recovery, and the uptake of N derived from the soil has been increased by the application of fertilizer (Dawson and Ryden, 1985; Dowdell and Webster, 1980; Jansson and Persson, 1982). However, these differences are due mainly to mineralization/immobilization turnover (see p. 119) and do not indicate that the fertilizer has increased the net mineralization of soil N (though this may occur to a small extent).

The Fate of Fertilizer N Not Taken up by the Grass

Studies in which ^{15}N-labelled fertilizer has been applied to microplots (enclosed in metal or plastic casings) in field swards have provided information on the fate of the fertilizer not taken up by the grass, though few have been able to account for all the fertilizer N. One complicating factor is that some of the fertilizer applied to enclosed microplots is often leached rapidly through cracks at the edges of the microplots, caused at least partly by shrinkage of the soil during dry periods. Nevertheless, studies with ^{15}N have often shown a considerable amount of fertilizer N to be retained in the grass roots (Power, 1981; Triboi, 1987) with about half the maximum root content persisting in the roots for a year or more (Bristow *et al.*, 1987; Power, 1983). In addition, there is often a rapid and substantial immobilization of fertilizer N into soil organic matter, more so with ammonium and urea than with nitrate (Keeney and Macgregor, 1978; Stevens and Laughlin, 1989). Much of the immobilization appears to be due to incorporation into the microbial biomass (see p. 120). Some of the immobilized N some may be re-converted into inorganic forms within a few weeks (Bristow *et al.*, 1987) but some remains in organic forms for several years. However, on average, the recently immobilized N is mineralized much more readily than the N in the older soil organic matter (Smith and Power, 1985).

The Use of Fertilizer N on Grassland

The amounts of fertilizer N currently recommended for productive grassland in humid temperate regions such as western Europe are generally between 200 and 400 kg N ha^{-1} $year^{-1}$, the total amount being divided among several applications during the growing season. The amount recommended in a particular situation is less than that required for maximum herbage yield, because the efficiency of the fertilizer N (in terms of uptake and herbage yield) declines as maximum yield is approached. Actual recommendations for individual fields should take into account the soil type, the average weather conditions, whether slurry or manure is applied in addition to fertilizer, and the type of livestock enterprise on the farm.

Surveys of the use of fertilizer N in the UK have been carried out since the 1940s and these show that the average rate of application increased steadily until 1986, since when the average rate has fluctuated (see Fig. 1.4). Throughout the period during which surveys have been made, there has been a wide variation from farm to farm in the rate of application to grassland, a wider variation than with arable crops. Some of the variation reflects the location of the farm (e.g. whether it is lowland or upland) and the type of livestock enterprise, some reflects differences in the extent to which reliance is placed on N_2 fixation by clover, and some reflects differences in the use of slurry or farmyard manure. The rate of application is generally greater in lowland than in upland areas, and is generally greater with dairy farming than with beef and sheep production. An additional factor that contributes to farm-to-farm variation is the difficulty that farmers have in estimating the monetary value of grassland herbage, a difficulty that is largely due to the value being realized only after the herbage has been fed to livestock.

The current recommendations for the application of fertilizer N to grassland in England and Wales (MAFF, 1994) are based on the results of numerous field trials, carried out at many different sites and in years of differing weather conditions. Separate recommendations are made for (i) grazing, continuous and rotational, by cattle, (ii) grazing by sheep and (iii) cutting for silage, of both high and medium digestibility. With each of these management systems, a maximum rate of fertilizer N is recommended for each of three categories of available soil N as illustrated in Tables 10.4–10.6. The categories of available soil N are based on previous cropping, manuring and management. The 'low

Table 10.4. Recommended maximum rates for fertilizer N (kg ha^{-1}) for intensive continuous grazing by cattle. (From MAFF, 1994.)

	Soil N supply		
Time of application	Low	Moderate	High
Mid February – early March	60	60	60
Late March – early April	75	60	60
Late April – early May	75	60	50
Late May – early June	60	60	50
Late June – mid July	60	60	40
Late July – mid August	50	40	40
Annual total	380	340	300

Table 10.5. Recommended maximum rates of fertilizer N (kg ha^{-1}) for intensive grazing by sheep. (From MAFF, 1994.)

Time of application	Soil N supply		
	Low	Moderate	High
Mid February – early March	60	60	60
Late March – early April	60	60	50
Late April – early May	60	50	40
Late May – early June	60	50	40
Late June – mid July	50	40	30
Late July – mid August	40	30	30
Annual total	330	290	250

Table 10.6. Recommended maximum rates of fertilizer N (kg ha^{-1}) for silage of high digestibility (N to be applied at least six weeks before anticipated cutting date). (From MAFF, 1994.)

	Soil N supply		
	Low	Moderate	High
First cut	150	120	120
Second cut	110	100	100
Third cut	80	80	60
Fourth cut	80	80	60
Annual total	420	380	340

soil N' category includes fields that have recently been sown to grass after arable cropping, while the 'medium soil N' category includes fields of established grassland that have received moderate rates of fertilizer N in previous years, and the 'high soil N' category includes fields that have been intensively fertilized and grazed for several years or have been re-seeded from such swards. The rates of fertilizer N indicated by Tables 10.4–10.6 apply to highly intensive systems, and are reduced if the system is of moderate or low intensity. The rates for all systems are reduced if it is expected that growth will be restricted by a shortage of water during the summer, and are reduced further if organic manures or slurries are to be applied. This last reduction is based on the typical contents of available N in organic manures and slurries of which examples are shown in Table 10.7. In those areas designated as 'Nitrate Sensitive Areas', lower rates of fertilizer N than those indicated above are recommended. Recommendations for fertilizer N in Scotland, provided by the Scottish Agricultural Colleges, differ to some extent from the recommendations for England and Wales made by MAFF, though they are based on similar criteria. In particular, they assume a greater supply of N from the soil (Unwin and Vellinga, 1994).

Table 10.7. Examples of typical amounts of N available during the following growth period in slurries and farmyard manure applied at various times. (From Ministry of Agriculture and Food, 1994.)

Time of application	Cattle slurry (kg N m^{-3})	Pig slurry (kg N m^{-3})	Cattle FYM (kg t^{-1})
Winter	0.6	1.3	0.9
Early spring	0.9	1.8	1.2
Summer	0.6	1.0	n.a.

In the Netherlands, there has been a recent change in the basis of recommendations. Until recently, a standard recommendation of 400 kg N ha^{-1} $year^{-1}$ (from a combination of fertilizer and organic manures) was made for all soil types except well-drained peat, for which the recommendation was 250 kg N ha^{-1} $year^{-1}$. Since 1994, recommmendations have taken into account the need to reduce emissions of N, as well as agronomic and economic criteria. The main criteria used in the current recommendations are (i) the N delivery capacity of the soil (including deposition from the atmosphere), (ii) the apparent recovery of fertilizer N by grass on the soil type involved, (iii) the planned dry matter yield per cut or per grazing and (iv) the time of year ('t Mannetje, 1994). Lower rates are recommended for a particular cut or grazing if the weather is dry. However for high yields of grass on most mineral soils, in a year of normal rainfall, the recommendation for fertilizer plus slurry/manure N still amounts to about 400 kg N ha^{-1} $year^{-1}$ ('t Mannetje, 1994). The N delivery capacity of the soil is considered to range from 75 kg N ha^{-1} $year^{-1}$ on soils low in organic matter, containing less than 5 t total N ha^{-1}, and with an available water capacity of < 100 mm, to 200 kg N ha^{-1} $year^{-1}$ on peaty soils containing 10–15 t total N ha^{-1} and with an available water capacity of > 150 mm (Unwin and Vellinga, 1994). No allowance is made for the return of N in excreta during grazing because the effects of such N are considered to be highly variable.

The Use of N in Slurries and Manures

Of the organic manures applied to grassland, cattle and pig slurries are the most widely used. Most farmyard manure is applied to arable crops, but some is used on grassland in areas where there is little or no arable farming. All types of slurry and manure vary widely in N concentration and this variation makes it more difficult to use them effectively. With slurries, the concentration of N varies with the amount of water incorporated into the slurry (from the washing of floors and sometimes from rainfall) and also with the extent of ammonia volatilization during storage. With solid manures, much of the variation reflects losses due to ammonia volatilization and nitrate leaching during storage. Average values for concentrations of dry matter and total N in the main types of slurry and manure are shown in Table 4.9. However, if the approximate volume of the slurry is known, it is possible to calculate its N concentration from the number of animals involved, the time period during which they were housed and the typical amount of N excreted per animal which, for a dairy cow, would normally be about 40–50 kg N during a 6-month winter feeding period.

Not all the N contained in slurries and manures is readily available and the overall availability is influenced by factors such as C : N ratio and the ratio between organic N and ammoniacal N (Beauchamp and Paul, 1989). The availability is higher when the C : N ratio is low, and when the ratio of ammoniacal N to organic N is high. The N in organic manures can be regarded as being in three fractions: (i) ammoniacal N, all of which is readily available, (ii) organic N that becomes available during the first year after application and (iii) organic N that is relatively resistant to decomposition and becomes available slowly after the first year. The ammoniacal N, which often comprises 40–50% of the total N in slurries, as well as being readily available for uptake by plants, is also susceptible to ammonia volatilization, and to leaching and/or denitrification once it has been nitrified to nitrate. Estimates of the organic N mineralized during the first growing season, based on incubation with soil, have ranged from considerably less than 20% with cattle manures to as much as 70% or more with poultry manures (Beauchamp and Paul, 1989). Although only small amounts of N are mineralized each year after the first year, repeated applications of slurries and manures have a cumulative effect. The size of this effect can be estimated using the concept of the 'decay series', in which the amounts of N mineralized in successive years are expressed as a series of proportions (Smith and Peterson, 1982). For example, the decay series, 0.30, 0.10, 0.05 indicates that for a given application, 30% is mineralized in the first year, 10% of the residual organic N is mineralized in the second year, and 5% of the residual N is mineralized in the third and subsequent years. For applications repeated in successive years, the same series is applied to each yearly application. Clearly, if the amount of available N required each year remains constant, the amount of slurry or manure (or the amount of fertilizer N) should be decreased from year to year. Thus, with a manure containing 1.5% N and 30% moisture, and with a decay series of 0.35, 0.15, 0.10, 0.05, the amount of manure required to supply 200 kg of available N ha^{-1} $year^{-1}$ would be 54 t in the first year and decrease to less than 27 t by the 20th year of annual applications (Smith and Peterson, 1982). A further complication with organic manures is that the mineralized N is not necessarily taken up by grass to the same extent as inorganic N supplied directly by fertilizers. Factors such as the volatilization of ammonia, and the timing of the mineralization of organic N during the growing season, may result in a lower uptake from the organic manures.

Various chemical indices have been proposed for assessing the amount of N likely to become available from organic manures including slurries, but none has shown sufficient reliability to have become generally adopted. Examples include N extracted by digestion with pepsin (Castellanos and Pratt, 1981) and the N dissolved by a mixture of potassium dichromate and sulphuric and phosphoric acids (Douglas and Magdoff, 1991). One problem is that no method for predicting availability can predict the losses that may occur through ammonia volatilization etc.

To obtain the best results from the surface application of slurry to grassland, the slurry should be applied in amounts of less than 20 $m^3 ha^{-1}$, the grass should be short and applications should be made during, or just before, the growing season (Vetter *et al.*, 1987). High rates of application may result in scorch and smothering of the herbage, and the soil surface may become sealed, increasing the risk of surface runoff (Prins and Snijders, 1987). The surface runoff of slurry is most likely to occur when the soil has a low permeability and/or is saturated with water (Sherwood and Fanning, 1981) and also when frozen soil thaws at the surface but remains frozen at depth.

Response of Grassland to Slurry N

Even when the slurry is applied as evenly as possible, it produces highly variable responses. As indicated above, the response is influenced by the composition of the slurry, by weather conditions and by the time and rate of application. The efficiency of slurry on grassland is often curtailed by its effect of smothering the grass. The response per kg of total slurry N is normally less than to fertilizer N, but the ammonium N fraction may produce a response similar to that of fertilizer N (Adams, 1973a), provided that ammonia volatilization is minimal. The generally poor response to slurry N, in comparison with fertilizer N, is due partly to ammonia volatilization, partly to the slow availability of the organic N, and partly to the negative effect of particulate material that adheres to the leaves.

In an investigation involving 27 sites in England and Wales over 3 years, the efficiency of slurry N applied to grass cut for silage was found to be influenced mainly by the rate and timing of the application. On average, in comparison with N applied as ammonium nitrate, the efficiency of total slurry N was 38% for a low rate of slurry application in spring, and 17% for a high rate of application in autumn. Where both slurry and fertilizer N were applied, the effects were additive (Pain *et al.*, 1986). In another investigation, the average response of Italian ryegrass to slurry was 66% of that to ammonium sulphate and, when calculated on the basis of soluble N, the responses to slurry were similar to those from fertilizers (McAllister, 1966a,b). Responses to cattle slurry in Northern Ireland were generally greatest when it was applied in

February/March for the first harvest of silage, but were often poor when it was applied in May/June and July/August for the second and third harvests (Long and Gracey, 1990b).

Various techniques have been proposed to improve the efficiency of slurry N by reducing ammonia volatilization. These include: (i) the removal of fibrous solid material by mechanical separation, (ii) dilution of the slurry with water, (iii) acidification of the slurry, (iv) the incorporation of a nitrification inhibitor into the slurry before application, and (v) the injection of slurry into the soil. The removal of fibrous material and injection also prevent slurry particles adhering to grass leaves.

The removal of fibrous material can result in 30–80% of the solid material being removed, depending on the amount of solid material present initially and the method of separation. The liquid fraction resulting from the separation contains the soluble N (mainly ammonium) and almost all the K, while the solid fraction contains most of the organically bound N and almost all the P (Vetter *et al.*, 1987). An example of the effect of slurry separation on the partitioning of organic matter and N is shown in Table 10.8. The liquid fraction has the advantage over whole slurry that, when it is applied to the surface of grassland, more of the liquid infiltrates into the soil. When compared with a whole cattle slurry, the liquid fraction separated through a 1 mm mesh showed a reduction in ammonia volatilization of about 36% and an increase in herbage yield of about 24% (Frost and Stevens, 1991). In general, the finer the mesh through which the liquid slurry is separated, the less ammonia is volatilized and the higher the yield of herbage. However, fine mesh sizes (< *c.* 1 mm) are easily blocked and have slow rates of throughput (Frost *et al.*, 1990). Although the response per kg N is normally greater with the liquid fraction than with the unseparated slurry, the separation may be cost effective only if the solid fraction is utilized, e.g. by the sale of composted material for horticultural use (Vetter *et al.*, 1987). Slurry fibre produced by separation typically amounts to about 20% of the original slurry weight and, depending on the type of separator, has a content of dry matter of between 15% and 35% (Frost and Stevens, 1991). In some situations, the separation of a liquid fraction does not improve the response to slurry N (e.g. Long and Gracey, 1990b).

Table 10.8. Changes in organic matter and N after slurry separation, in which 1 m^3 slurry yielded 0.8–0.9 m^3 liquid fraction. (From Vetter *et al.*,1987.)

	Cattle slurry			Pig slurry		
	Original slurry	Liquid fraction	Solid fraction	Original slurry	Liquid fraction	Solid fraction
Dry matter, %	6.85	3.33	21.30	7.73	4.69	29.0
Organic matter, %	4.58	2.36	14.20	4.48	3.21	19.91
Total N, %	0.27	0.24	0.35	0.54	0.50	0.78

In the absence of separation of the fibrous material, the dilution of slurry with water often improves the response, particularly when the content of solid material is greater than about 10% and when the slurry is applied in warm dry weather. Although dilution increases the costs of transport and spreading, a dilution of 30% may increase yield sufficiently to be economic (Frost and Stevens, 1991; Vetter *et al.*, 1987). On intensively fertilized swards, the N efficiency of diluted cattle slurry, in comparison with fertilizer N, was about 35% for the first cut, 50% on an annual basis and 65% on a long-term basis (Schechtner *et al.*, 1980). The aeration of slurry during storage usually has little effect on response compared with normal storage, but the anaerobic digestion of cattle slurry (though not of pig slurry) often improves response, especially at the first harvest after application (Vetter *et al.*, 1987).

The acidification of slurry is effective in reducing the volatilization of ammonia and increasing the response in terms of herbage yield, but it has the disadvantages of cost and possible side-effects of the acids used (Frost *et al.*, 1990). At present, the cost of acidification exceeds the value of the additional response (Frost and Stevens, 1991).

In recent years, nitrification inhibitors have been studied as a means of improving the availability and uptake of slurry N but the yield benefits have been variable (Pain *et al.*, 1987). Trials in Norway showed that addition of dicyandiamide (DCD) to cattle slurry applied to grassland in autumn increased the yields of grass in the following spring by an average of 63% (Tveitnes and Haland, 1989). The period during which DCD remains effective depends on temperature: the concentration in the soil has been found to remain unchanged for more than 80 days at 8°C, but to start declining after 40–60 days at 14°C and after 20–40 days at 20°C (van Dijk and Sturm, 1983).

The injection of slurry into the soil at a depth of 7.5–15 cm also increases the efficiency of utilization (van der Meer *et al.*, 1987) because ammonia volatilization is largely prevented (Thompson *et al.*, 1987; Tunney and Molloy, 1986). However, on some soils, conditions may be suitable for the injection of slurry only during short periods in spring and autumn. Conditions may be too wet in winter and early spring to prevent damage to the soil structure by the tractor wheels and injection equipment, and may be too dry in summer to prevent serious damage to the grass roots (van der Meer *et al.*, 1987). Whenever slurry is injected, there may be some mechanical damage to the sward and some restriction of root activity due to anaerobic conditions close to the slurry but, with good soil conditions, these effects are short-lived (Prins and Snijders, 1987; Tunney and Molloy, 1986). The loss of yield from mechanical damage is influenced by the design of the tines and by their spacing, as well as by soil type and the water status of the soil (Hall, 1986). However, over a whole season, the injection of slurry rather than spreading it on the surface of grassland usually results in a greater uptake of N and a larger yield of herbage (De La Lande Cremer, 1986; Tunney and Molloy, 1986).

Recovery of Slurry N

The apparent recovery of total slurry N by grass in the year of application is usually less than 35% (Pain *et al.*, 1986; Unwin *et al.*, 1986) and is often less than 20%. Recovery is improved by appreciable rainfall within 72 h of the application (Long and Gracey, 1990b). Also, the procedures outlined above for improving response tends to improve recovery. The mineralization of the organic N in slurry contributes towards plant-available N in subsequent seasons, as described on p. 215 and it has been estimated with slurry that between 3% and 10% of the N applied in one season is taken up in the following season (Unwin *et al.*, 1986). In a field experiment in which pig and cow slurry were applied at three rates to a ryegrass sward over a period of 16 years, the recovery of N from the pig slurry was greater than that from the cattle slurry, and the recovery from both slurries increased from the lowest to the intermediate rate and then decreased at the highest rate of application (Christie, 1987).

The N in diluted urine is sometimes, though not always, as effective as fertilizer N in increasing yield. However, work over several years in southwest Scotland, showed that diluted urine applied to grass–clover swards resulted in a higher proportion of clover in the herbage than did the application of the equivalent amounts of N and K as fertilizer (Castle and Reid, 1987; Drysdale, 1965). The factor responsible for the higher clover content was not identified but the effect appeared not to be due to the higher soil pH nor to the N being supplied mainly as urea (Castle and Reid, 1987).

Use of Slurry and Fertilizer N in Combination

When slurry or FYM is applied to fields that are scheduled to receive fertilizer, allowance should be made for the nutrients supplied by the slurry. The average amount of N that is available for uptake in the first year after the application is shown in Table 10.7, though there is wide variation depending on factors such as the degree of dilution and the system of storage. In general, the amount is greatest when the slurry or FYM is applied in the spring, as losses by leaching and ammonia volatilization are usually small at that time, and some N may become available during the season from the mineralization of organic N. In areas where leaching normally occurs during the winter, slurry applied during the autumn or winter can be disregarded as a source of N for the first year (Adams, 1973a).

Response to the N in Sewage Sludge

In the UK about one-third of the sewage sludge applied to agricultural land is applied to grassland. It is applied mainly in the liquid form and usually at a rate that provides about 2 t dry matter ha^{-1} $year^{-1}$ (Hutchings, 1984). Grass is regarded as particularly suitable for the utilization of sludge because it can take up nutrients over the whole growing season, and because soils under grass are less likely than arable soils to have their structure damaged by vehicles distributing the sludge.

Sewage sludges vary considerably in their concentration of total N and in the proportion present in inorganic form, depending partly on the treatment they have received. Liquid undigested sludge usually contains 4–8% solid material, and between 1 and 4 kg N m^{-3}, of which only 5–15% is present as ammoniacal N (Coker *et al.*, 1987). Field trials have shown the response to N to be about 10–25% of that to fertilizer N in the first year but, in the second and subsequent years of successive treatments, the relative response has often increased to about 40% (Coker *et al.*, 1987). Sludge that is undigested comes mainly from small rural sewage works and is therefore unlikely to be contaminated by potentially toxic elements.

Much of the urban sewage sludge is digested either anaerobically or aerobically. The anaerobic digestion of sludge results in the decomposition of up to 40% of the original organic matter and this releases much of the organic N as ammonium. However, the residual organic matter decomposes slowly in the soil and therefore releases little inorganic N (Hutchings, 1984). In a 2-year trial, the availability to grass of the N in liquid digested sludge was found to average 44% on a freely-draining sandy soil and 34% on a poorly drained clay soil (Hutchings, 1984). Lower values of 18–20% even with repeated applications were reported from Ireland (O'Riordan *et al.*, 1987). When sludge is digested aerobically, most of the N remains in organic forms but the subsequent decomposition of the organic matter in the soil occurs more readily, and more of the residual N is mineralized soon after application. The availability of the N in aerobically-digested sludge in a 2-year trial was rather higher than that of anaerobically digested sludge, averaging 56% on a freely drained sandy soil, and 44% on a poorly drained clay soil (Hutchings, 1984). Ammoniacal N in sewage sludge is susceptible to volatilization (see p. 171) and this factor may contribute to availability being lower for anaerobically-digested sludge than for aerobically-digested sludge. Both types of sludge can be treated by pressing, or centrifugation, to produce a sludge cake containing 25–30% dry matter instead of the usual 4–8% dry matter. The removal of water is accompanied by the removal of much of the inorganic N and the supply of plant-available N from the cake then depends on mineralization of the organic N.

In a survey of sewage sludge from about 150 treatment plants in the USA, average concentrations of N were 4.2% (dry matter basis) in anaerobically digested sludge and 4.8% in aerobically treated sludge: average concentrations of NH_4-N were 0.16% and 0.04% respectively (Sommers, 1977). For a number of sludges in Ireland, the mean concentration of N in the dry matter of activated sludge (treated aerobically in an oxidation ditch) was 5.2%, while the concentration in anaerobically digested sludge was 3.2% (O'Riordan *et al.*, 1986). In the activated sludge, 11% of the N was present as ammonium whereas the corresponding figure for the anaerobically digested sludge was 14%. The mean concentration of N in sewage sludge in the UK has been estimated at 4.1% of dry matter (Jollans, 1981).

It is almost certain that increased amounts of sewage sludge will be applied to agricultural land in the future as greater restrictions are imposed on disposal at sea, but the amounts of sludge per unit area will be limited by regulations on the maximum permissible concentrations of heavy metals in sludge-amended soils (MAFF/DoE, 1993; O'Riordan *et al.*, 1994).

11

Response to Fertilizer Nitrogen: Influence of Sward Type, Pattern of Fertilizer Application and Method of Harvesting

Numerous factors influence the response of grassland to fertilizer N and among them are the botanical composition and age of the sward, the seasonal distribution of the fertilizer N and the method of harvesting the herbage (i.e. the frequency and height of defoliation, and whether by cutting or grazing). These factors are considered in this chapter.

The Influence of Species and Variety of Grass

Most of the species and varieties of grass that have been selected or bred for cultivation are similar in their response to fertilizer N. In the UK (and elsewhere with similar conditions), varieties of ryegrass and cocksfoot often produce the highest yields and the best responses, particularly at high rates of fertilizer N (Cowling and Lockyer, 1965; Hunt, 1966b; Reid, 1985). However, perennial ryegrass is much more widely sown than cocksfoot because, in addition to producing a high yield, it has a better nutritional quality in terms of digestibility and shows a greater persistence under various types of management. In the years before about 1960, it was common to sow mixtures of grass species. These usually included ryegrass and, in addition, cocksfoot, timothy and/or meadow fescue and, sometimes, white clover, red clover and/or herb species such as chicory. More recently, the use of species other than ryegrass has declined. The similarity in response of varieties of ryegrass, cocksfoot, timothy and meadow fescue, in lowland UK conditions, is shown in Fig. 11.1.

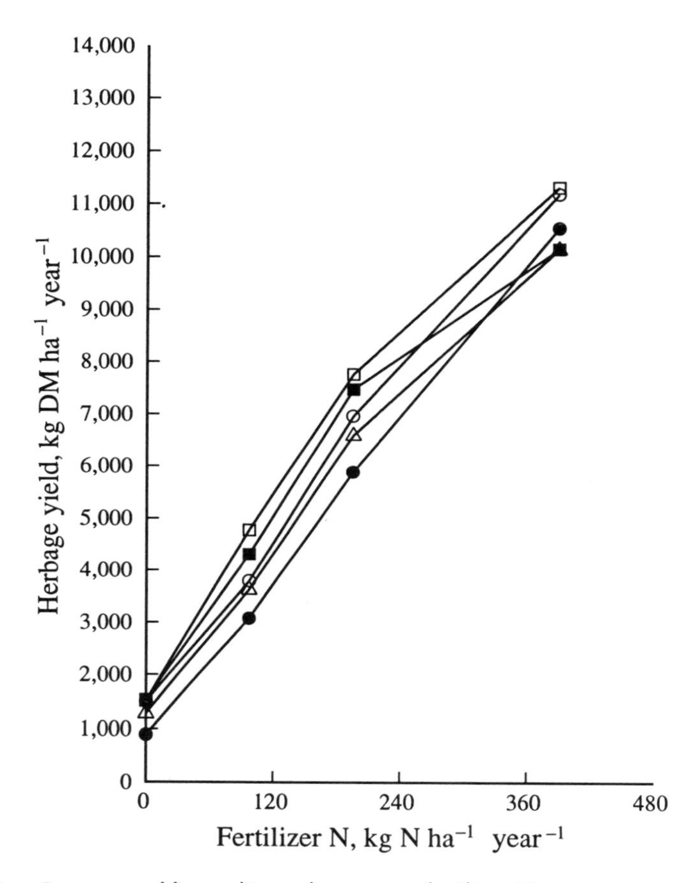

Fig. 11.1. Response of five cultivated grasses to fertilizer N: perennial ryegrass, S24 (○); perennial ryegrass, S23 (●); cocksfoot, S37 (□); timothy, S48 (■); meadow fescue, S215 (△). (Data from Cowling and Lockyer, 1965.)

In upland areas, species other than ryegrass often dominate and produce higher yields. Red fescue, for example, has been shown to produce higher yields than ryegrass at various rates of fertilizer N, though its establishment after sowing is relatively slow (Davies *et al.*, 1984). In drier regions of the world, particularly those with high summer temperatures such as parts of the USA, drought-resistant species such as tall fescue and smooth bromegrass generally produce the highest yields (e.g. Collins, 1991; Cooper *et al.*, 1962; Heath *et al.* 1985) whereas in cooler but rather dry regions, such as eastern Canada and northeastern USA, timothy produces high yields.

Within a species, the varieties that have been bred for high yields under intensive management generally respond more to fertilizer N than do strains of the same species taken from natural or seminatural grasslands. This difference was illustrated by various strains of perennial ryegrass, including the variety S24, grown in sand culture (Antonovics *et al.*, 1967). Different clones of perennial ryegrass, although not belonging to specific varieties, also varied considerably in their efficiency of N utilization, and required widely different amounts of fertilizer N to achieve similar yields when grown as spaced plants in the field (Lazenby and Rogers, 1965). In a comparison of seven bred varieties of perennial ryegrass (S24, Premo, Houba, Barlenna, Mellee and Scempter Pasture), the variety Melle produced the greatest response to high rates of fertilizer N in the field, in terms of both dry matter yield and N uptake into the herbage (Lee *et al.*, 1977).

Semi-natural grasslands usually contain a number of indigenous grass species, sometimes including one or more of the cultivated species mentioned above. The indigenous species generally show less response to moderate to high rates of fertilizer N than do the cultivated species but they may produce higher yields at low rates of fertilizer N. For example, in a comparison of *Agrostis stolonifera*, *Festuca rubra*, and *Holcus lanatus* with perennial ryegrass (cv. S23), *Festuca rubra* and *Holcus lanatus* responded well to 60 kg N ha^{-1} $year^{-1}$, though less well than perennial ryegrass to 180 kg N ha^{-1} $year^{-1}$; *Agrostis stolonifera* showed less response than did the other three species at each rate of fertilizer N (Haggar, 1976). A similar comparison of six secondary grasses with perennial ryegrass and timothy produced the results shown in Fig. 11.2. *Holcus lanatus*, *Festuca rubra* and *Agrostis stolonifera* were more productive than perennial ryegrass (cv. Perma) with no fertilizer N and equally as productive with 120 kg N ha^{-1} $year^{-1}$ but, at higher rates of fertilizer N up to 480 kg N ha^{-1} $year^{-1}$, they were less productive than perennial ryegrass (Frame, 1991). However, in a 3-year study (on a low fertility soil) in which an existing grass sward dominated by *Agrostis stolonifera* was compared with re-seeded perennial ryegrass, the *Agrostis* showed a consistently lower response to fertilizer N at rates up to 400 kg N ha^{-1} $year^{-1}$ (Sheldrick *et al.*, 1990).

When the need arises to intensify production from an area of semi-natural grassland, the decision on whether or not to cultivate and re-seed is influenced by the expected response of the existing sward to fertilizer application. The cost of re-seeding is high and, to be worthwhile, substantially greater yields must be obtained from the re-seeded sward (Dibb and Haggar, 1979). As indicated above, areas of semi-natural grassland can produce good yields of herbage with inputs of fertilizer N of up to about 250 kg N ha^{-1} $year^{-1}$, if soil and weather conditions are favourable. However, in terms of digestibility of the herbage, perennial ryegrass is superior to the secondary grasses at all rates of fertilizer N.

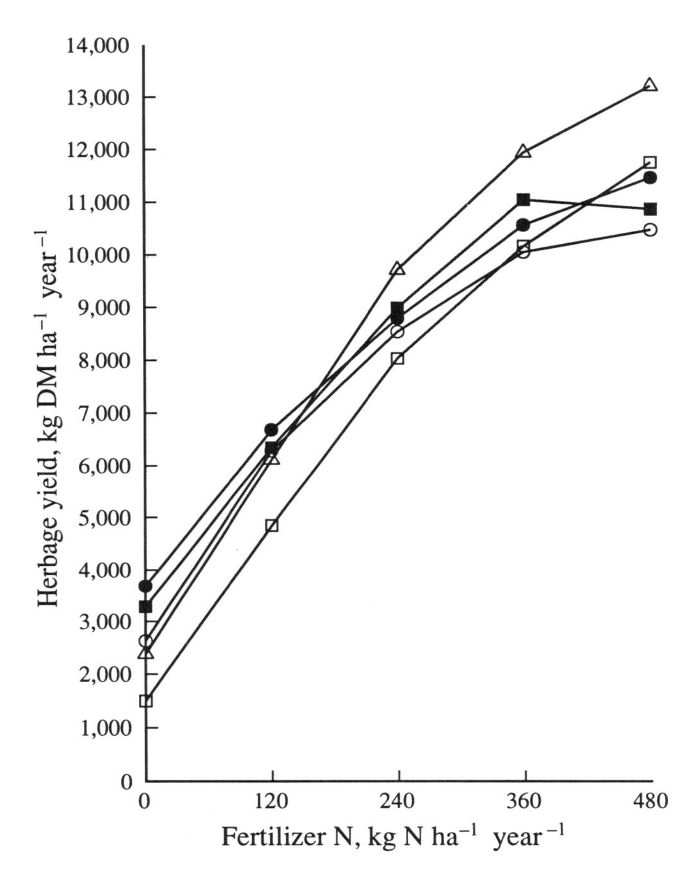

Fig. 11.2. Response of four secondary grasses and perennial ryegrass to fertilizer N: *Agrostis stolonifera* (○); *Holcus lanatus* (●); *Poa pratensis* (□); *Festuca rubra* (□); perennial ryegrass (△). (Data from Frame, 1991.)

The Response of Grass–Clover Compared with All-grass Swards

In the absence of fertilizer N, a higher yield of herbage is obtained from a grass–clover sward than from an all-grass sward at the same location, a difference that reflects the ability of the clover to fix N_2 from the atmosphere. However, when fertilizer N is applied to a grass–clover sward, N_2 fixation is reduced (see Chapter 3), and the yield response per unit of fertilizer N is therefore less than with a comparable all-grass sward. Typical responses from all-grass and grass–clover swards in the UK, based on the results of several investigations reported during the period 1960/72 are shown in Fig. 11.3.

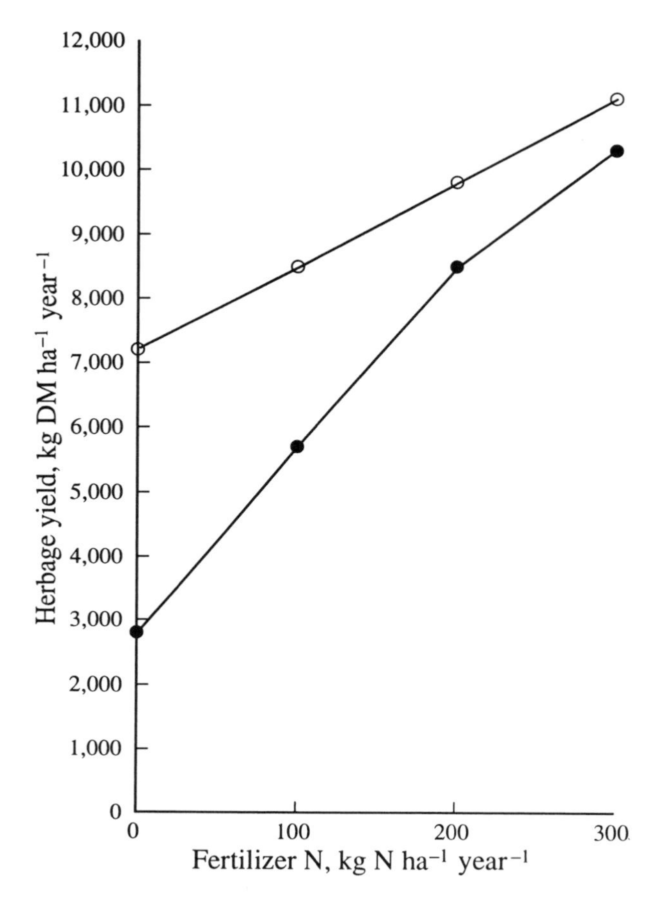

Fig. 11.3. Response of grass–clover (○) to fertilizer N in comparison with response of grass alone (●): means of data from several experiments in the UK. (Redrawn from Cowling, 1982.)

Remarkably similar average results were obtained in a subsequent study at 22 sites in the UK, carried out during the period 1978/81 (Fig. 11.4). The response to fertilizer N was examined with three sward types, perennial ryegrass (S23) alone and perennial ryegrass plus white clover (both cv. Blanca and cv. S100), all cut at 4-weekly intervals. On average over the 22 sites, higher yields were obtained from both grass–clover swards than from the pure grass swards at rates of fertilizer N up to 400 kg N ha^{-1} (Fig. 11.4). The results of both sets of experiments indicate that, whereas the average response from all-grass swards to rates of fertilizer N up to 300 kg N ha^{-1} $year^{-1}$ is about 20–25 kg DM per kg N, the average response from grass–clover swards is about 10–12 kg DM

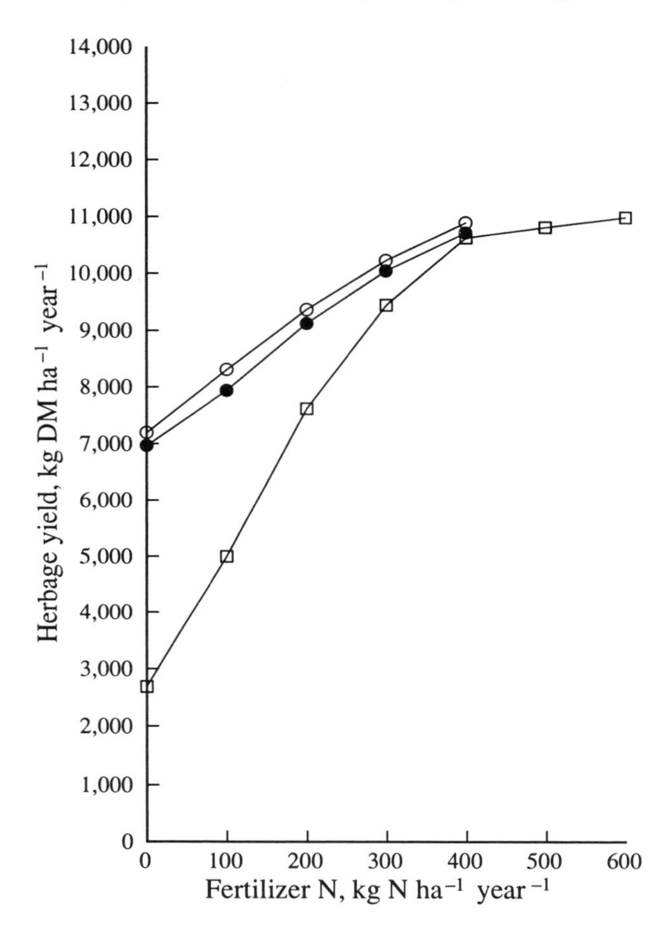

Fig. 11.4. Response of grass–clover (Blanca, ○; S100, ●) to fertilizer N in comparison with response of grass alone (□): means of data for 22 sites in the UK during 1978–1981. (Morrison *et al.*, unpublished data.)

per kg N. Grass–clover swards are more variable in response because the growth of clover varies more widely from site to site and from year to year. When the proportion of clover is small, or conditions are unfavourable for clover growth, the herbage yield and response to fertilizer N will be similar to those of an all-grass sward under similar conditions. On the other hand, when there is a large proportion of clover in the sward, the herbage yield without fertilizer N is relatively high (sometimes > 10,000 kg DM ha^{-1} year^{-1}) and the response to fertilizer N is small. The response of a grass–clover sward may occasionally be negative, at least for some individual harvests during a season (Cowling, 1961a; Laidlaw, 1980).

In conditions that are favourable for white clover, it may account for more than half of the total herbage yield of a grass–clover sward receiving no fertilizer N and, in this situation under lowland UK conditions, grass alone would require at least 150 kg N ha^{-1} year^{-1} to produce a similar yield. Data from the two sets of experiments described above indicated that, on average, the yield from a well-managed grass–clover sward was equivalent to the yield from a pure grass sward receiving about 160–180 kg N ha^{-1} (Figs 11.3 and 11.4). In each set of experiments, there was considerable variation in the 'fertilizer equivalent' of grass–clover swards from site to site, and from year to year at any one site, reflecting the vigour of the grass–clover swards not receiving fertilizer N.

As indicated in Chapter 3, there are wide regional differences in the vigour of clover growth and these tend to be reflected in differences between countries in the average use of fertilizer N on grassland. Thus, in the Netherlands, clover growth is often poor and the average use of fertilizer N is high. On the other hand, in New Zealand, clover usually grows well and grass–clover swards produce herbage yields of 13,000–17,000 kg DM ha^{-1} year^{-1}, equal to those from all-grass swards receiving 300–400 kg fertilizer N ha^{-1} year^{-1} (Ball and Field, 1985). However although the general use of fertilizer N is uneconomic in New Zealand, small amounts of fertilizer N applied during winter and early spring to grass–clover swards do produce an economic response in parts of the country (O'Connor, 1990).

In many countries, there is continued discussion on the relative merits of clover and fertilizer N, and on the extent to which it is feasible to combine the two sources in the same sward. The choice between the two sources of N, or the balance between them on an individual farm, is influenced by the climate of the particular location and by the type of livestock involved. Clover tends to be favoured in areas with high rainfall, where most of the grassland is long-term and where sheep are the main type of livestock, whereas fertilizer N is usually the dominant source of N in somewhat drier areas, where the grassland is often re-seeded at intervals of a few years and dairy cattle are the most important livestock. Whether clover and the use of fertilizer N can be combined effectively in the same sward is more difficult to resolve. The data shown in Figs 11.3 and 11.4 indicate that, on average in the UK, higher yields are obtained from grass–clover than from all-grass swards at rates of fertilizer N up to at least 300 kg ha^{-1} year^{-1}, and this suggests that it may be economic to include clover in swards intended to receive moderate rates of fertilizer N. The use of grass–clover swards in combination with some fertilizer N has been recommended by several authors (e.g. Brockman and Wolton, 1963; Frame, 1987; Laidlaw, 1980; Mackenzie and Daly, 1982; Morrison *et al.*, 1983; Reid, 1983), who also recommend that most of the fertilizer should be applied early in the season when the difference in growth rate between N-fertilized grass and unfertilized grass–clover is greatest. However, the longer term effectiveness of combining clover and fertilizer N depends on establishing and maintaining an adequate proportion of clover in the sward, and this is difficult to achieve when fertilizer N is used (see p. 48).

In mixtures of grass and white clover, the choice of variety of white clover (and probably the species and variety of grass) may have an influence on the response of the mixed sward to fertilizer N. For example, mixtures of white clover with perennial ryegrass produced a greater response to fertilizer N when the clover variety was the large-leaved long-petioled Blanca than when it was the smaller-leaved short-petioled S100 (Reid, 1983). However, the data in Fig. 11.4 show that, although cut herbage yields were slightly higher from Blanca than from S100, the responses to fertilizer N were similar. Other investigations have shown little difference between large-leaved and small-leaved varieties of white clover in the effect of fertilizer N applied in spring and, with both types, the yield of the clover component was reduced by fertilizer N (Frame and Boyd, 1987a; Laidlaw, 1984, 1988). However, differences between cultivars of white clover in their tolerance to fertilizer N applied in spring were found in a study in New Zealand (Caradus *et al.*, 1993).

As there is often a trend for the vigour of clover in grass–clover swards to decline after the second or third year, one possible strategy for a farmer wishing to maximize the contribution of clover is to rely mainly on clover in the early years after sowing, and to increase the rate of fertilizer N as the vigour of the clover declines (Castle, 1965). An alternative strategy is to divide the available grassland area into two parts, with one dependent entirely on clover and the other dependent entirely on fertilizer N (Frame and Boyd, 1987b; Linehan and Lowe, 1961; Walker, 1962).

The Influence of Age of Sward

The species composition of grassland is very much influenced by the age of the sward, with the largest number of species usually occurring in grassland that has remained unploughed for many decades. In sown swards, and in swards developed on land where arable cultivation has ceased, the number of species tends to increase with increasing age. In the UK, about half the total area of enclosed grassland consists of long-term grassland that was either sown more than 20 years ago or was never deliberately sown. This long-term grassland often contains some perennial ryegrass and other cultivated species, but the main constituents are indigenous species e.g. *Agrostis* spp., *Poa* spp., *Festuca rubra* and *Holcus lanatus* (Hopkins, 1987). About one-quarter of the enclosed grassland consists of swards of 5–20 years old, while the remaining quarter consists of swards sown within the previous 5 years. In both of these age groups, perennial ryegrass is the main sown species (Hopkins, 1987) but, with increasing age of sward, there is a tendency for the proportions of perennial ryegrass and other cultivated species to decline, and for the indigenous

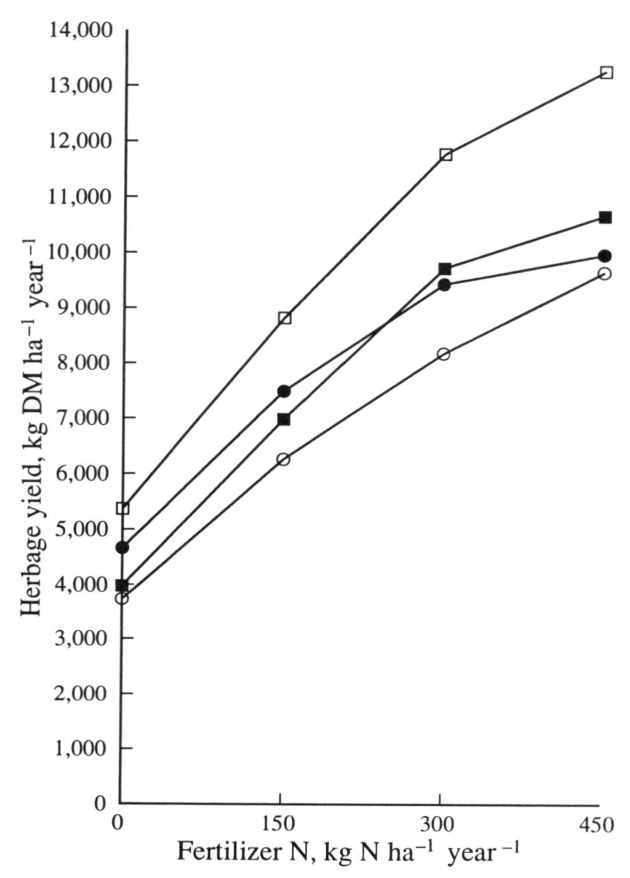

Fig. 11.5. Response of long-term grassland to fertilizer N in comparison with response of re-seeded perennial ryegrass: long-term grassland, year 1 (○); long-term grassland, years 2 and 3 (●); re-seeded ryegrass, year 1 (□); re-seeded ryegrass, years 2 and 3 (■); mean values from 16 sites in England and Wales. (Data from Hopkins *et al.*, 1990.)

species to increase. As noted on p. 224, these indigenous species generally show less response to high rates of fertilizer N.

When old grassland swards and swards re-seeded to ryegrass were compared at 16 sites in England and Wales, higher yields were obtained from the re-seeded swards in the first year after re-seeding, at all sites and at all rates of fertilizer N (Hopkins *et al.*, 1990). Presumably, the cultivation for re-seeding caused a flush of mineralization of organic N in the following year. However, in subsequent years, yields from the re-seeded swards were no more than those from the old grassland, unless high rates of fertilizer N were applied (Fig. 11.5). Even at high rates of fertilizer N, the benefits of re-seeding were offset by the inevitable loss in production during the year of re-seeding. Although the proportion of perennial ryegrass in the old grassland varied considerably among the 16 sites, the proportion did not appear to be related to the response to fertilizer N (Hopkins *et al.*, 1990). In a similar investigation, a re-seeded sward outyielded long-term grassland (> 25 years) only when the rate of fertilizer N was more than 360 kg N ha^{-1} $year^{-1}$ (Daly and Mackenzie, 1988).

In farming practice, the yield from sown grassland has sometimes been observed to decline after about 3 years, with the trend sometimes being reversed from about the seventh year (Hoogerkamp, 1979). The decline in yield is most likely to occur with grass–clover swards, and to be due partly to reduced clover growth. However, it has also been noted with all-grass swards and is then due to factors such as reduced mineralization of soil N, progressive compaction of the soil and, sometimes, the use of short-lived grass varieties (Hoogerkamp, 1979). The yield of herbage does not always show this type of decline, and in an investigation in which a moderate rate of fertilizer N (about 220 kg N ha^{-1} $year^{-1}$) was applied to individual swards of four grasses, including perennial ryegrass, there was no fall in production throughout a 6-year period (Hunt, 1966b). In a comparison of ageing leys (5–12 years since sowing) with re-seeded swards at eight sites in England and Wales, herbage yields (assessed for 3 years after re-seeding) from plots receiving fertilizer N at 250 kg N ha^{-1} $year^{-1}$ generally showed little effect of re-seeding (Hopkins *et al.*, 1995). With no fertilizer N, yields were usually higher from the ageing leys than from the re-seeded swards, probably due to the presence of some clover in the ageing leys. With rates of fertilizer N of 375 and 500 kg N ha^{-1} $year^{-1}$, yields at most (though not all) sites were higher from the re-seeded swards (Hopkins *et al.*, 1995).

This evidence that the re-seeding of both old grassland and ageing leys often has little economic benefit contrasts with the policy, widely recommended from about 1940 to the mid-1960s, of ploughing and re-seeding long-term grassland in order to increase productivity (Davies, 1960). Although the output from grassland was undoubtedly increased by this policy, the direct benefits of ploughing and re-seeding appear to have been over-emphasized, while the benefits of other factors that usually accompanied re-seeding, such as increased fertilizer application, improved drainage and higher stocking rate, were under-emphasized.

The Influence of the Pattern of Fertilizer Application

Fertilizer N is normally applied to grassland on a number of successive occasions during the growing season, a practice that enables the N to be taken up and utilized as effectively as possible. It also makes it possible to modify the seasonal pattern of growth so that herbage production can be matched a little more closely to the needs of livestock. The first application of fertilizer N is generally made in early spring and, with cut or rotationally grazed swards, subsequent applications are made after each defoliation except the last. With continuously grazed swards, applications are generally made at intervals of between 4 and 8 weeks from early spring to late summer.

In experimental studies of fertilizer rates involving, say, six applications during the growing season, the annual amount of fertilizer N is often divided equally among the six applications, but this pattern of application is not necessarily the most effective. The effect of six equal applications was compared with two additional patterns in an investigation at 21 sites in England and Wales (Morrison *et al.*, 1980). The additional patterns involved apportioning the annual amount of fertilizer N as follows: (i) 1/4 + 1/4 + 1/8 + 1/8 + 1/8 + 1/8 and (ii) 1/8 + 1/8 + 1/4 + 1/4 + 1/8 + 1/8. In this comparison, there were no consistent differences between the patterns in the total yield of herbage for the year as a whole, but the pattern of application did influence the distribution of herbage yield over the growing season. The effect on the distribution of yield was greater at a rate of 300 kg N ha^{-1} $year^{-1}$, than at rates of 150 kg N ha^{-1} and 450 kg N ha^{-1}. When half the fertilizer N was applied for the first two cuts, production at these cuts was increased but production during mid- and late-season was reduced. On the other hand, when half the fertilizer N was applied for cuts 3 and 4, less herbage was produced at cuts 1 and 2 than with the uniform distribution of fertilizer, but more was produced in mid and late season. A typical example of the effect of the pattern of distribution is shown in Fig. 11.6. Other studies have also shown that changing the pattern of fertilizer application through the season can modify the seasonal distribution of herbage growth while having little effect on total annual yield (Castle *et al.*, 1965). In farming practice, it is possible to be flexible in deciding on the pattern of application and in modifying the rates as appropriate to particular circumstances.

For both cut and rotationally grazed swards, the most efficient use of fertilizer N is achieved when the application for each growth period (except the first) is made immediately after cutting or grazing, and when the rate of application is matched to the requirement for the planned period of growth. A delay in application results in some loss of yield at the next harvest. In an investigation of various timings, with swards harvested at intervals of 28 days, a delay of 8 days in application produced 94% of the maximum yield, and a delay of 15 days produced only 84% of the maximum (Brockman, 1974).

The Influence of Intensity of Defoliation

The intensity of defoliation is determined partly by frequency, and partly by the height to which the sward is cut or grazed. In general, under a cutting

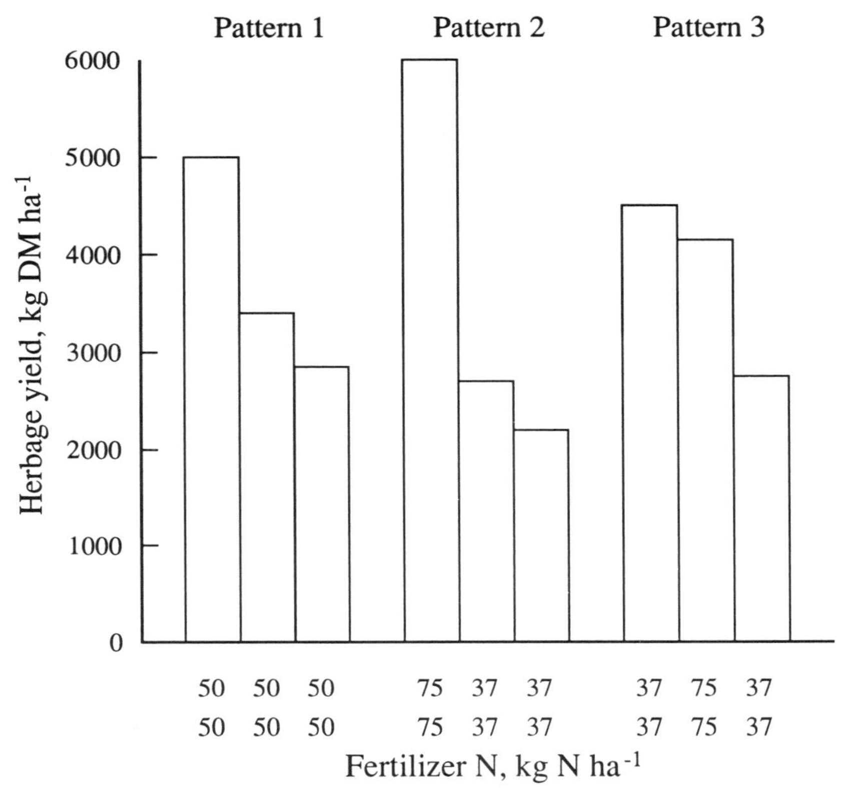

Fig. 11.6. Herbage yield from cuts 1 plus 2, 3 plus 4, and 5 plus 6, from perennial ryegrass receiving a total of 300 kg N ha^{-1} $year^{-1}$ as ammonium nitrate in three different patterns: (1) 50 kg N for each cut, (2) 75 kg N for cuts 1 and 2 and 37 kg N for subsequent cuts and (3) 75 kg N for cuts 3 and 4, and 37 kg N for the other cuts. (Data for a site near Oxford, UK, from Morrison *et al.*, 1980.)

regime, the yield of herbage produced during the year is greatest when the interval between defoliations is at least 6 weeks and when the cutting height is no more than 4–5 cm (Binnie *et al.*, 1974; Davies, 1988; MacLusky and Morris, 1964). However, a higher quality of herbage, in terms of digestibility, is obtained when defoliation is more frequent than every 6 weeks (Frame *et al.*, 1989). The optimum timing of defoliation is therefore a compromise between quantity and quality.

In the UK, the effects of the frequency of defoliation on yield are greater in the period before early July than subsequently (Frame *et al.*, 1989). With cut swards, the optimum date for maximizing the yield of high digestibility herbage at the first harvest is when the ears have emerged on about 50% of the flowering stems. This date, which varies from variety to variety, from site to site and from year to year, is towards the end of the period of maximum growth rate but, with most varieties, it precedes the decrease in digestibility that occurs if flowers and seeds are allowed to develop (Holmes, 1989).

In general, when fertilizer N is applied at rates up to about 400 kg N ha^{-1} year^{-1}, a regime of four or five cuts per year produces a greater response than a regime of eight or ten cuts though, at high rates of fertilizer N, the incremental response to additional N is greater with the more frequent cutting (Bartholomew and Chestnutt, 1977; Frame *et al.*, 1989; Reid, 1978; Sibma and Alberda, 1980). In a comparison of three cutting frequencies at four rates of fertilizer N on a grass–clover sward, yields with two cuts were always greater than with six, which in turn were greater than with ten; but the largest response to the highest rate of fertilizer N (470 kg N ha^{-1} year^{-1}) was obtained with 10 cuts per year (Fig. 11.7).

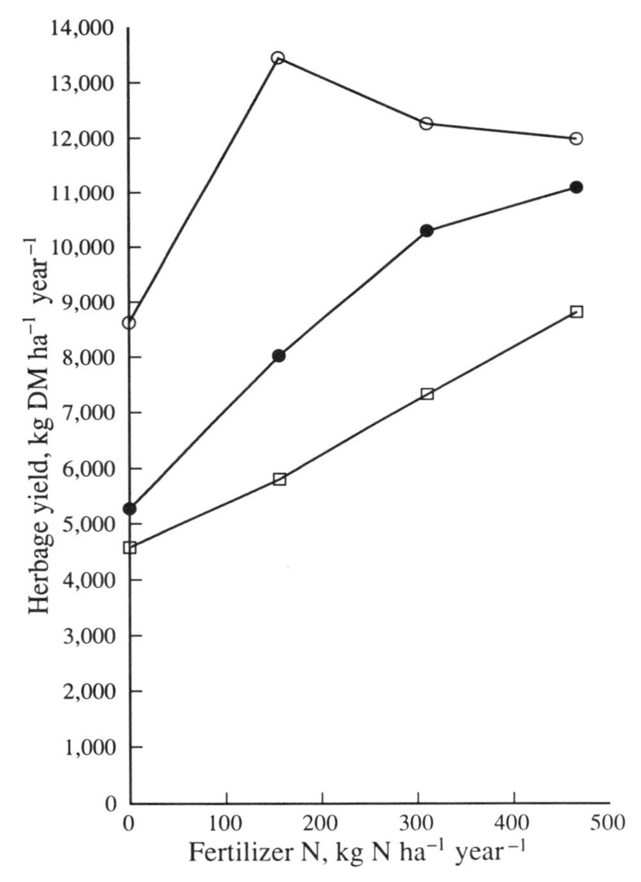

Fig. 11.7. Response of a grass–clover sward to fertilizer N as influenced by the frequency of defoliation: 2 cuts per year (○); 6 cuts per year (●); 10 cuts per year (□). (Data from Holliday and Wilman, 1965.)

Assuming that fertilizer is applied at the beginning of the season and after each defoliation, then frequent defoliation implies a relatively short time interval between the application of fertilizer and harvest. The effect of this interval on the yield response to fertilizer N is illustrated by a study in which separate plots of Italian ryegrass were harvested at weekly intervals following the application of three rates of fertilizer N (Fig. 11.8). Herbage yield increased slowly during the first 2 weeks and the rate of fertilizer N had little effect but, during the next 4 weeks, the yield increased more rapidly and fertilizer N had a considerable effect (Wilman, 1965). After 6 weeks' growth, the difference between rates of application of 28 and 84 kg N ha^{-1} represented a response of 29 kg DM per kilogram extra N, but that between the 84 and 140 kg rates was only 7 kg DM per kilogram extra N. In a similar investigation, the rate of fertilizer N applied in mid-April had little effect during the first 16 days but, by the 23rd day, there was a response to 116 kg N, compared with 58 kg, of

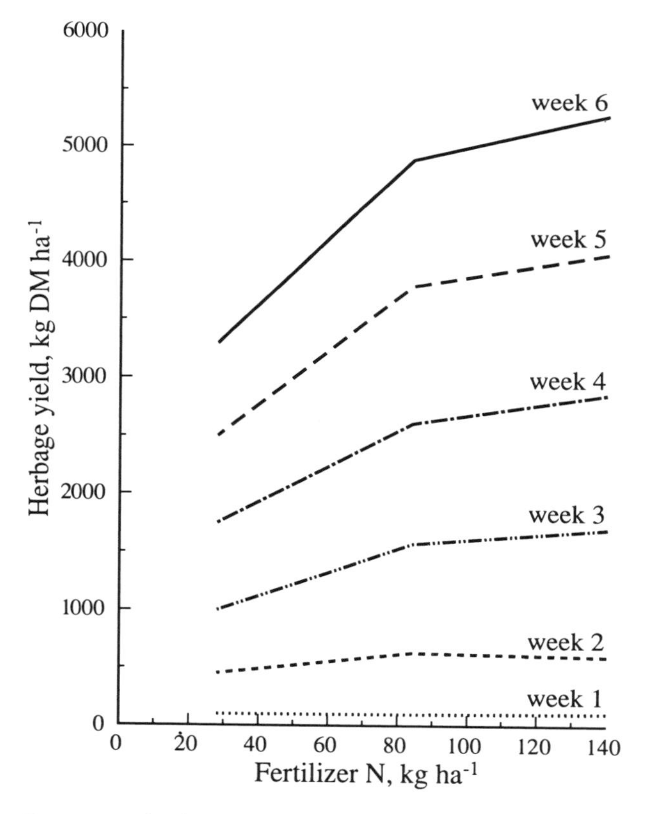

Fig. 11.8. Response of Italian ryegrass to a single application of fertilizer N with different plots harvested at intervals of 1–6 weeks after the application. (Data from Wilman, 1965.)

10 kg DM per kilogram of additional N: by the 57th day the response had increased to 24.4 kg DM per kilogram of additional N (Hunt, 1966a). Changes in the DM yield and N yield of the herbage from plots receiving 116 kg N ha^{-1} also illustrate the way in which the length of the growth period influences the response to fertilizer N (Fig. 11.9). When the plots were cut for the first time at intervals between 8 and 45 days and were cut again at the end of the 57day period, the time of the first cut had a much greater effect on the total DM yield for the 57-day period than on the herbage N yield. When the first cut was at 23 days after the application of N, only 27% of the DM yield obtained after 57 days had been produced, but 82% of the 57day N uptake had been attained (Fig. 11.9). The subsequent regrowth to day 57 provided 22% of the uninterrupted 57-day uptake of N but only 17% of the corresponding herbage yield.

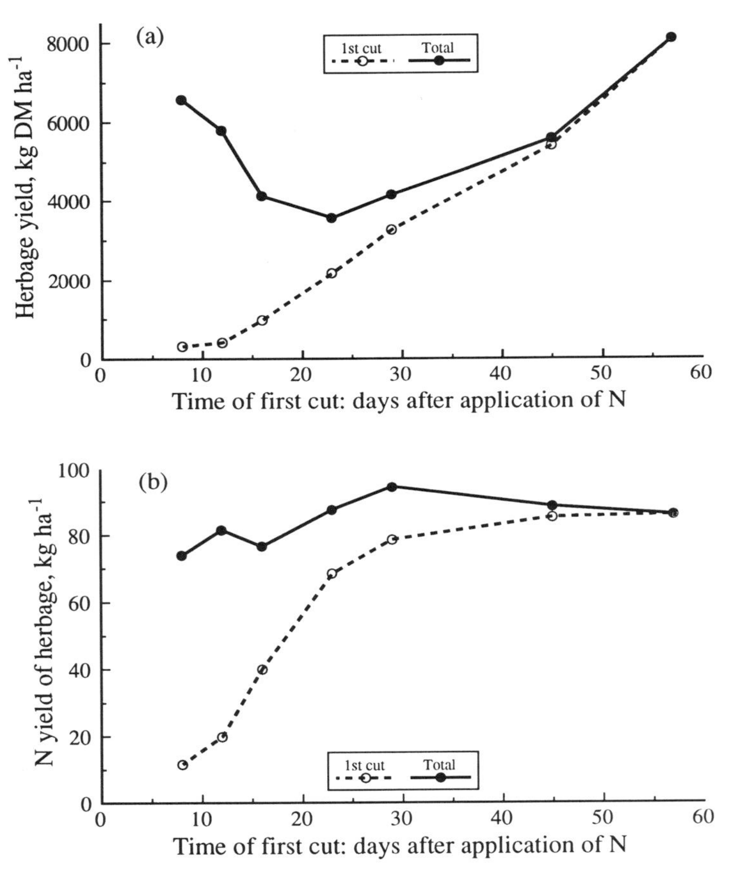

Fig. 11.9. Accumulation of (**a**) herbage DM and (**b**) herbage N following the application of fertilizer N (116 kg N ha^{-1}) to ryegrass, with different plots cut after 8–45 days and all allowed to re-grow to day 57. (Data from Hunt, 1966b.)

The effect of defoliation height has received less attention than the effect of defoliation frequency. With a ryegrass and white clover sward cut at either 4 cm or 8 cm, the lower cutting height increased the annual yield of herbage but had little effect on the response to fertilizer N (Frame and Boyd, 1987a), confirming earlier observations (Reid, 1962). Also, with cocksfoot at each of three rates of fertilizer N, the annual yield was greater when the cutting height was 5 cm rather than 10 cm, though the response to 225 kg N ha^{-1} was the same at both cutting heights (Wilson, 1964). However, when Italian ryegrass was given fertilizer N at rates between 170 and 390 kg N ha^{-1} $year^{-1}$, the response to the additional 220 kg N declined substantially with an increase in cutting height (at 6-week intervals) from 2.5 cm to 7.6 cm and again to 12.7 cm (Binnie *et al.*, 1974). In general, in order to maximize herbage yield per year, it is best either to combine severe defoliation with a long regrowth period or more lenient defoliation with a shorter period of regrowth (Robson *et al.*, 1989).

The Response of Grazed Compared with Cut Swards

When swards are grazed, a large proportion of the N consumed in the herbage is returned in excreta, enabling some of the N derived from fertilizer to be utilized more than once during a growing season. Taking this into account, grazed swards would be expected to show a greater response than cut swards to fertilizer N at rates below the optimum. However, the return of excreta often entails large losses of N (see Chapters 7–9) and, in addition, grazing animals have adverse effects on the sward through treading, selective grazing and the fouling of herbage by dung. These adverse effects vary in their impact and are influenced by soil characteristics, weather and grazing management. For example, the damage caused by treading is generally greatest on clay soils, and when the soil is saturated or almost saturated with water (Patto *et al.*, 1978). The pressure exerted per unit area of hoof is about 50% greater for cattle than for sheep, and cattle therefore tend to cause more damage in wet conditions (Davies, 1988). Dung pats, as well as reducing consumption by grazing animals, cause the death of plants by excluding light (see MacDiarmid and Watkin, 1971) and growth may remain poor for up to 2 years (Weeda, 1967). Again, the adverse effects are greater with cattle than with sheep.

The various adverse effects of grazing animals offset the positive effect of the return of nutrients, and the net effect of grazing animals on the response to fertilizer N is often small (Armitage and Templeman, 1964; Baker, 1986; Brockman and Wolton, 1963; Green and Cowling, 1961; Richards, 1977). However, it is difficult to measure herbage yield in grazed swards and it is therefore difficult to compare the responses of cut and grazed swards. This problem was highlighted in a comparison, carried out at six centres in England

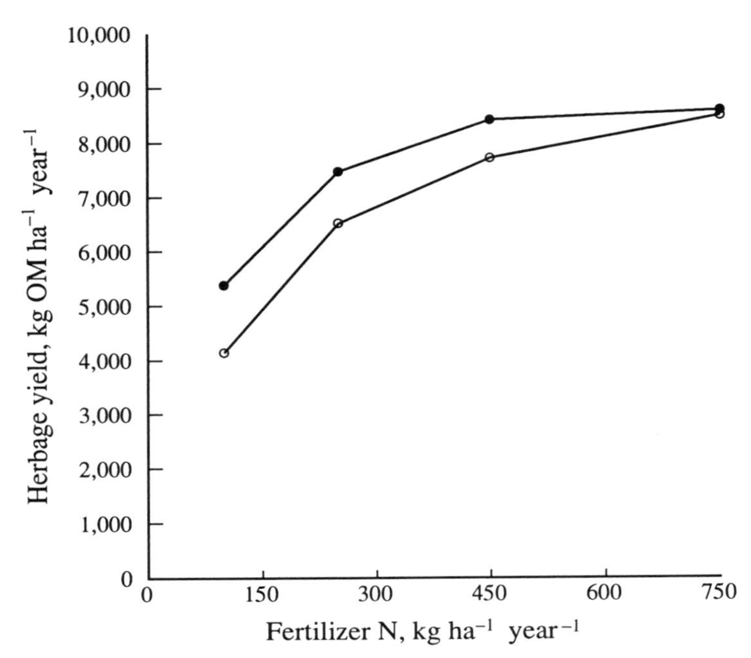

Fig. 11.10. Response of cut (○) and grazed (●) plots of perennial ryegrass to fertilizer N; mean values from five sites during 2 years, expressed in terms of organic matter (OM) to avoid the effects of differences in soil contamination. (Data from Baker, 1986.)

and Wales, in which the effects of grazing were found to be variable (Jackson and Williams, 1979). Nevertheless, an appreciable positive response was obtained in a subsequent trial involving five sites, in which the grazed plots were grazed continuously by beef cattle (Fig. 11.10). In a trial in the Netherlands, it was concluded that the return of excreta had a small positive effect on response at rates of fertilizer N of 250 and 400 kg N ha^{-1} $year^{-1}$ but, that at higher rates, there was an increasingly negative effect due to treading and poaching (Deenen and Lantinga, 1993). When a positive response to the return of excreta does occur it may sometimes be due to K rather than N (Wheeler, 1958).

In practice, with both continuous and rotational grazing, the frequency of defoliation is usually greater than with cut swards, and this is an additional factor that influences the response to fertilizer N. In an investigation in which both all-grass and grass–clover plots were defoliated by both cutting and grazing at the same times, the response to fertilizer N, at rates up to 200 kg fertilizer N ha^{-1}, was greater on the grazed plots and the effect was greater on the all-grass than on the grass–clover plots (Shaw *et al.*, 1966). But if grazed plots are defoliated more frequently than the corresponding cut plots, there may well be lower yields and lower responses to N on the grazed plots.

12

Response to Fertilizer Nitrogen: Influence of Weather, Seasonal Factors and Soil Type

Introduction

The yield of grass herbage responds most to fertilizer N when the grass is growing actively, and therefore when conditions of light, temperature and water supply are all favourable. Differences in these factors are largely responsible for differences in herbage yield and response to fertilizer N, at different times of year, and from one year to another. And differences in light, temperature and water supply, as well as soil factors, contribute to the differences in yield and response among a range of sites. An illustration of the range of variation in response that can occur from year to year, and from site to site, is provided by data from a trial in which perennial ryegrass was grown at 21 sites in England and Wales over a period of 4 years (Morrison *et al.*, 1980). For example, during the 4 years, the response to fertilizer N, applied at a rate of 300 kg N ha^{-1} $year^{-1}$, at a site with a light soil and relatively low rainfall in the east midlands of England, varied from 10 to 19 kg DM kg^{-1} N (Fig. 12.1). The extent of site-to-site variation in response, averaged over the 4 years, is illustrated in Fig. 12.2. The lowest average response occurred at a dry site in eastern England while the greatest response of the 21 sites occurred at a site with a much higher rainfall in south Wales (Fig. 12.2). Year-to-year variation during the 4-year period was much greater at some sites than at others and, although the reasons for this were not clear, they were thought to involve interactions between soil and weather factors during the course of each growing season.

Although the main reason for applying fertilizer N to grass is often to increase the total annual yield of herbage, applications are sometimes intended primarily to reduce the seasonal variation in production or to extend the

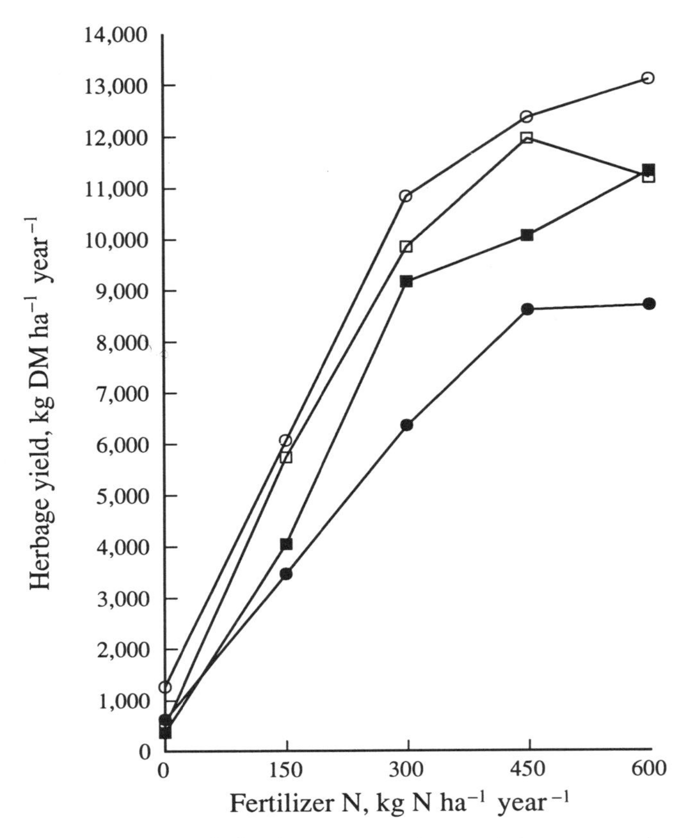

Fig. 12.1. Year-to-year variation in response of perennial ryegrass to fertilizer N at Gleadthorpe, Nottinghamshire, UK: 1971 (○); 1972 (●); 1973 (□); 1974 (■). (Data from Morrison *et al.*, 1980.)

growing season. In these situations, the effects of light, temperature and water supply are particularly important.

The Influence of Light

Grass growth and response to fertilizer N are influenced both by variation in light intensity and by variation in daylength. The effect of light intensity on response was clearly shown in an experiment in the Netherlands, in which three controlled lighting regimes were applied to plots of perennial ryegrass (Deinum, 1966). The response to fertilizer N, applied at rates of 25 and 125 kg N ha^{-1} was greatest at the highest light intensity, equivalent to that of midsummer, and was least with a light intensity equivalent to a dull day in midwinter

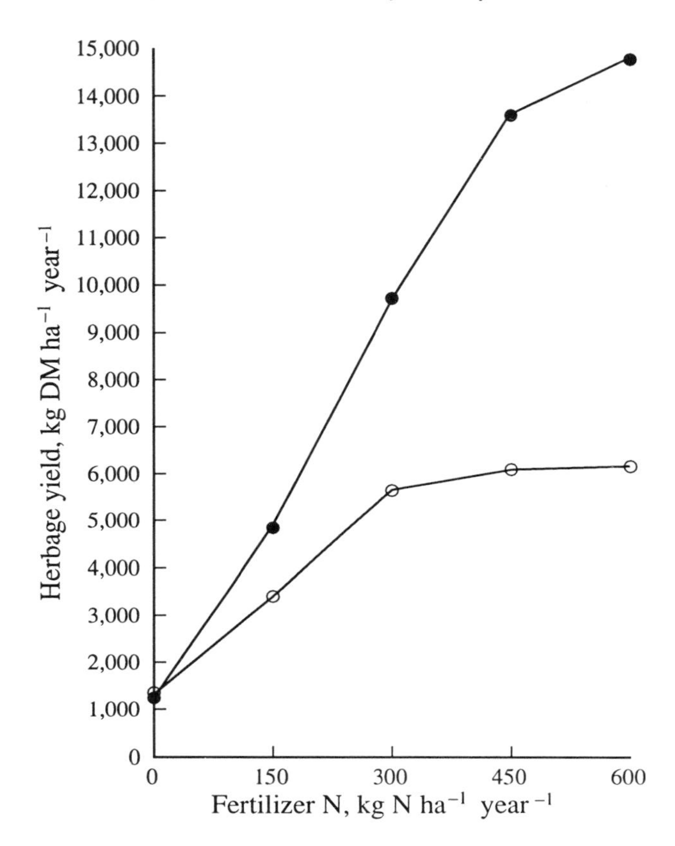

Fig. 12.2. Response of perennial ryegrass to fertilizer N at two contrasting low-land sites in the UK: Cambridge (○); Wenvoe, south Wales (●); mean values for 4 years. (Data from Morrison *et al.*, 1980.)

(Fig. 12.3). Similar results were obtained in a pot experiment with Italian ryegrass, in which the response to fertilizer N was inversely related to the degree of shading of the summer light intensity (Cunningham and Nielsen, 1965).

There is less information on the influence of daylength but it is likely that, in temperate areas with mild winters such as southern Australia, the short daylength in winter limits the response to fertilizer N.

The Influence of Temperature

Although temperature has a large effect on grass growth (Keatinge *et al.*, 1979), there is little information on its interaction with fertilizer N, especially at

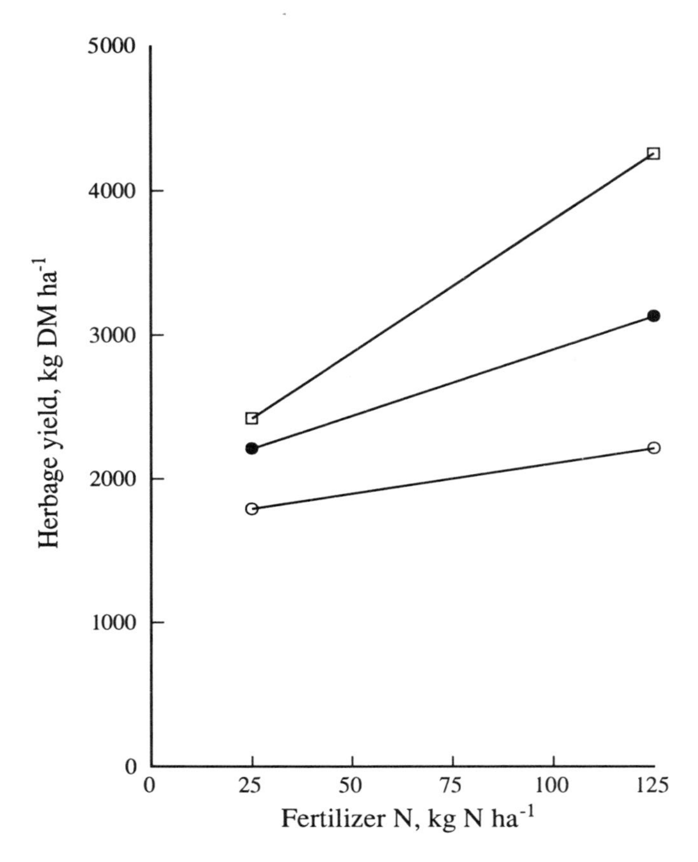

Fig. 12.3. Response of perennial ryegrass to fertilizer N applied in autumn, assessed at three average light intensities obtained with supplementary lighting and shading: ○, 0.20 kJ cm^{-2} day^{-1}; ●, 1.00 kJ cm^{-2} day^{-1}; □, 2.27 kJ cm^{-2} day^{-1}. (Data from Deinum, 1966.)

temperatures of less than about 10°C. However, studies with long-term grassland receiving fertilizer N in early spring indicated that no growth occurred when the soil temperature at 10 cm was < 5°C: at temperatures between 5°C and 8°C, growth was increased by fertilizer N and, at temperatures > 8°C, the growth rate of grass receiving fertilizer N was not appreciably greater than that of unfertilized grass (Blackman, 1936). The results suggested that, at this location in southern England, the mineralization of soil N at temperatures > 8°C was equivalent to the application of about 75 kg N ha^{-1}. The benefit of the fertilizer N was due to growth starting earlier in the season.

In glasshouse experiments with ryegrass, increasing temperature over the range 10–23°C generally increased yield but had no effect on the response to fertilizer N (Deinum, 1966). However, in a pot experiment in which Italian ryegrass was grown at soil temperatures of 10°C, 20°C and 30°C, a slightly higher response to fertilizer N was obtained at 20°C than at 10°C and 30°C (Parks and Fisher, 1958). In a rather similar experiment, also with Italian ryegrass, in which the temperatures were 11°C, 19.5°C and 28°C, the response to fertilizer N, applied at the rate of 100 mg N per kg soil, was almost identical at 11°C and 19.5°C (though yields were higher at 19.5°C) but substantially lower at 28°C (Nielsen and Cunningham, 1964). However, in farming practice, the effect of temperature on the growth of grass is most important in early spring, when the temperature is less than 10°C and when temperature is often more important than light in restricting grass growth (Wareing and Allen, 1977).

The Influence of Water Supply

The amount of water available to a grass sward during the growing season is a major factor in year-to-year variation in yield. It is also responsible for much of the variation between sites in any one year. In general, the amount of plant-available water reflects the water holding capacity of the soil plus the rainfall (and/or irrigation) during the growing season. Soils are normally at field capacity at the beginning of the growing season, but the amount of plant-available water held at field capacity may be two or three times greater in clay/loam soils than in sandy soils. Although irrigation is used to supplement rainfall on some intensively-managed farms, supplies of irrigation water are limited and, on a world scale, little grassland is irrigated.

When fertilizer N is applied during dry conditions, it remains on the sward surface until dissolved by subsequent rainfall or irrigation. Once in the soil, further movement of the fertilizer N is also dependent on the moisture content of the soil. Both grass roots and nutrients are usually concentrated in the top few centimetres of soil, and the movement of ions in this zone by mass flow and diffusion requires adequate moisture. When the topsoil is dry, grass growth may be curtailed by the restricted movement of nutrient ions, even though the grass is able to obtain sufficient water from lower depths in the soil (Garwood and Tyson, 1973; Garwood and Williams, 1967; Lemaire and Denoix, 1987). In dry conditions, the more frequently the surface is re-wetted by light rainfall, the better the availability of N (Lemaire and Denoix, 1987).

Although the supply of water has a large effect on the yield of grass, its effect on the response to fertilizer N is variable. In several experiments in the UK, fertilizer N and irrigation have been shown to act independently in their effects on the annual yield of herbage (Tayler, 1965). However, some experiments have shown a positive interaction (i.e. the response to N plus irrigation has exceeded the sum of the responses to each applied separately) and others have shown a negative interaction (i.e. irrigation, although increasing herbage yield, has decreased the response to N). Factors that favour a positive interaction between water supply and fertilizer N include (i) low rainfall during irrigation experiments, or during some of the years of year-to-year comparisons, (ii) high rates of application of fertilizer N and (iii) the absence of clover. Differences in rainfall between southeast England and southwest Scotland explain why a positive interaction between irrigation and fertilizer N is often obtained with grassland in southeast England, where the average annual rainfall is 600–700 mm (e.g. D'Aoust and Tayler, 1968) but not in southwest Scotland where the average annual rainfall is about 900 mm (Reid and Castle, 1965). Studies in Ohio, USA, showed that the response of Bermudagrass, to rates of fertilizer N up to 900 kg N ha^{-1} $year^{-1}$, was much greater in a wet year than in a dry year (Prine and Burton, 1956).

Irrigation was shown to have a greater effect at a high rate of fertilizer N in a study with Italian ryegrass: it had no significant effect on the annual response to fertilizer N when the rate was increased from 125 to 250 kg N ha^{-1} $year^{-1}$, but it did increase the response when the rate of fertilizer N was increased to 500 kg ha^{-1} $year^{-1}$ (D'Aoust and Tayler, 1968). Both positive and negative interactions occurred during the six consecutive 4-week growth periods: they were approximately equal over the whole season for the comparison between fertilizer rates of 125 and 250 kg N, but the positive interactions dominated for the comparison between the rates of 250 and 500 kg N ha^{-1}. When examined in relation to weather data, the results of this trial were consistent with the view that the benefit of irrigation was due mainly to an effect on the top few centimetres of soil, and hence on the accessibility of N and other nutrients to the roots.

A negative interaction between fertilizer N and water supply is most likely to occur with grass–clover swards because clover often shows a greater response than grass to irrigation. The positive effect of irrigation is therefore greatest on the unfertilized plots which generally have the highest contents of clover. The irrigation of grass–clover swards is likely to increase the response to fertilizer N only if the weather is particularly dry (Stiles and Williams, 1965) or if the proportion of clover is low (Munro, 1958). Otherwise irrigation is likely either to produce no effect on the response to fertilizer N (Low and Armitage, 1959) or to cause a reduction in response, despite increasing yield (Stiles and Williams, 1965).

When the supply of water is less than optimum but not severely deficient, irrigation and fertilizer N are to some extent interchangeable, though fertilizer N is usually much cheaper than irrigation for each unit of extra grass produced. For example, with cocksfoot, doubling the rate of fertilizer N over the range 19–75 kg N ha^{-1} per cut had an effect equivalent to adding 13 mm of readily available water (Penman, 1962). With only 19 kg N per cut, growth was checked at a soil moisture deficit of 25 mm whereas, with 38 kg N per cut, growth was not checked until the deficit reached 38 mm. The interchangeability of fertilizer N and irrigation, and the extent to which this is influenced by

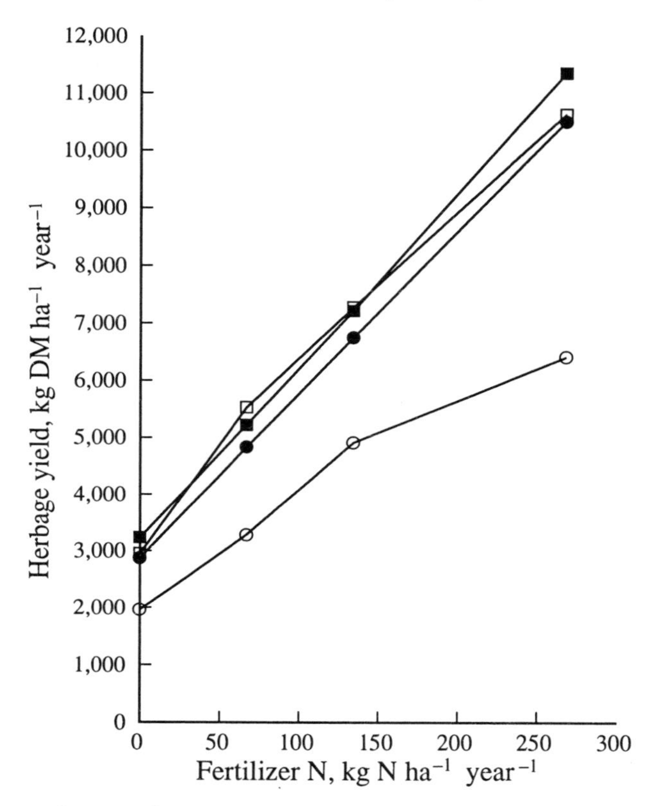

Fig. 12.4. Influence of irrigation on response of grass to fertilizer N in a dry year and a normal year in Ohio, USA (without and with irrigation in a dry year, ○, ●; without and with irrigation in a normal year, □, ■; mean values for cocksfoot and bromegrass). (Data from Prine *et al.*, 1963.)

weather conditions, is also shown by data from the USA (Fig. 12.4). Although irrigation often enables grass to respond to a higher rate of fertilizer N, it may sometimes result in an increased loss of N by leaching, with a possible reduction in yield in the year following the irrigation (Prine *et al.*, 1963). Conversely, irrigation applied during dry periods within the growing season normally increases the uptake of fertilizer N and reduces the amount leached during the subsequent winter (Garwood *et al.*, 1980).

The Influence of Season of the Year

In general in temperate regions, the growth of grass follows a characteristic seasonal pattern. Growth begins slowly in late winter or early spring, accelerates to a peak in early summer, declines in mid-summer, often increases again in late summer and then declines slowly until it stops in late autumn (see p. 59). The maximum rate of growth occurs shortly before the average date of ear emergence of the grass and, with cut swards, the response to fertilizer N is normally greatest during the period immediately before this date (see p. 233). The yield and response to fertilizer N at subsequent harvests during summer and autumn are influenced by variations in water supply, light intensity and temperature, and also by the density of tillers in the sward and the severity of defoliation at the previous harvest. A high yield at any one harvest tends to restrict the development of new tillers, and is therefore often followed by a relatively low yield at the next harvest.

In warm temperate areas, fertilizer N may stimulate the growth of grass in winter, if light is adequate. For example, in parts of Australia, winter temperatures are generally too low for the effective mineralization of soil N, but grass will grow if fertilizer N is provided (Newman *et al.*, 1962). Similarly in cooler areas, it is possible to lengthen the active growing season to some extent by the use of fertilizer N, particularly in early spring.

When grass is growing actively, most of the nitrate-N or ammonium-N applied as fertilizer is taken up during the 3–4 weeks after its application though, when growth is restricted by cold or drought, the uptake of N occurs more slowly. Whatever the rate of uptake, herbage dry matter is produced more slowly than N is taken up (Fig. 11.9) but, because the digestibility of the herbage declines with the increasing length of the regrowth period, the choice of harvest date is a compromise between yield and digestibility. Grass is usually defoliated, and given an additional application of fertilizer N, at intervals of between 4 and 8 weeks during the growing season (see p. 232). The optimum timing of defoliation, and the optimum rate of application, differ between cut and grazed swards. With cut swards, the interval between defoliations, and therefore between fertilizer applications, is usually 6–8 weeks, and the highest annual yields may well be obtained if the amount of fertilizer N applied in successive applications is reduced progressively during the course of the season. However, with grazed swards, the amount of grass available in late summer often decides the overall stocking rate, and the application of equal amounts of fertilizer at intervals of 4–6 weeks during the season is often satisfactory (Brockman, 1974). Alternatively, the best match with the requirements of grazing animals may be achieved by making the largest application in mid-season (Morrison, 1987).

In field experiments, the differences in the response to fertilizer N in successive growth periods are influenced by the way in which the plots are managed. Thus, if the same plots receive fertilizer in successive periods, there may be a residual effect of fertilizer from one application to the next. However, if different and previously unfertilized plots are used for successive applications, the plots used for the later applications will differ in sward morphology from those that have been fertilized. The residual effects are often positive (e.g.

when fertilizer N is not completely taken up during the intended growth period), but are sometimes negative (e.g. when a large crop of herbage suppresses developing tillers and curtails yield at the next harvest). An example of consistently positive residual effects was obtained in a field experiment in which perennial ryegrass was cut five times during the season, and the residual effect often amounted to 30–35% of the direct effect of the fertilizer applied for the second of the two periods being considered (Reid, 1984).

In a number of investigations, residual effects have been avoided by using separate plots in which the morphology of the sward has been maintained as uniform as possible by cutting or grazing all plots on the same dates. In a study of the response of an irrigated sward of perennial ryegrass during four successive growth periods, using this procedure, the response to fertilizer N was greatest in the first period, but there was little difference between the other three periods in response to rates of up to 150 kg N ha^{-1} (Fig. 12.5). In experiments on permanent pasture in the Netherlands, in which separate plots grazed by sheep were used for each application date, the responses to fertilizer N applied in April and June were approximately the same when the growth period was 4–5 weeks, but were greater from fertilizer applied in April when the growth period was 7 weeks: the response to fertilizer N applied in August was smaller than to applications in April or June (van Burg, 1961). In a subsequent experiment, fertilizer N at 240 kg N ha^{-1}, in comparison with none, was applied to irrigated swards on one of 17 successive dates, and the herbage harvested 31 days later (van Burg, 1961). The response to fertilizer N increased as the date of application progressed from early March to late April, declined during midsummer, increased again for N applied during July and August, and then dropped sharply to virtually nil for N applied in late September.

In general, in these and other experiments, the response to fertilizer N was greatest for grass harvested during May and/or early June: the response then declined. The experiments differed, however, in whether or not the response increased again during the period mid-July to early September.

In some areas (e.g. Lusignan, France) the application of fertilizer N in autumn may stimulate growth in the following spring, an effect thought to be due to increased tiller production during the winter (Culleton and Lemaire, 1988).

Timing of the First Fertilizer Application in Spring

The timing of successive applications of fertilizer N influences the pattern of herbage growth through the season. Timing for the first growth period in spring is particularly important, and the optimum time is influenced by weather conditions, soil type and the form of fertilizer N. If the application is

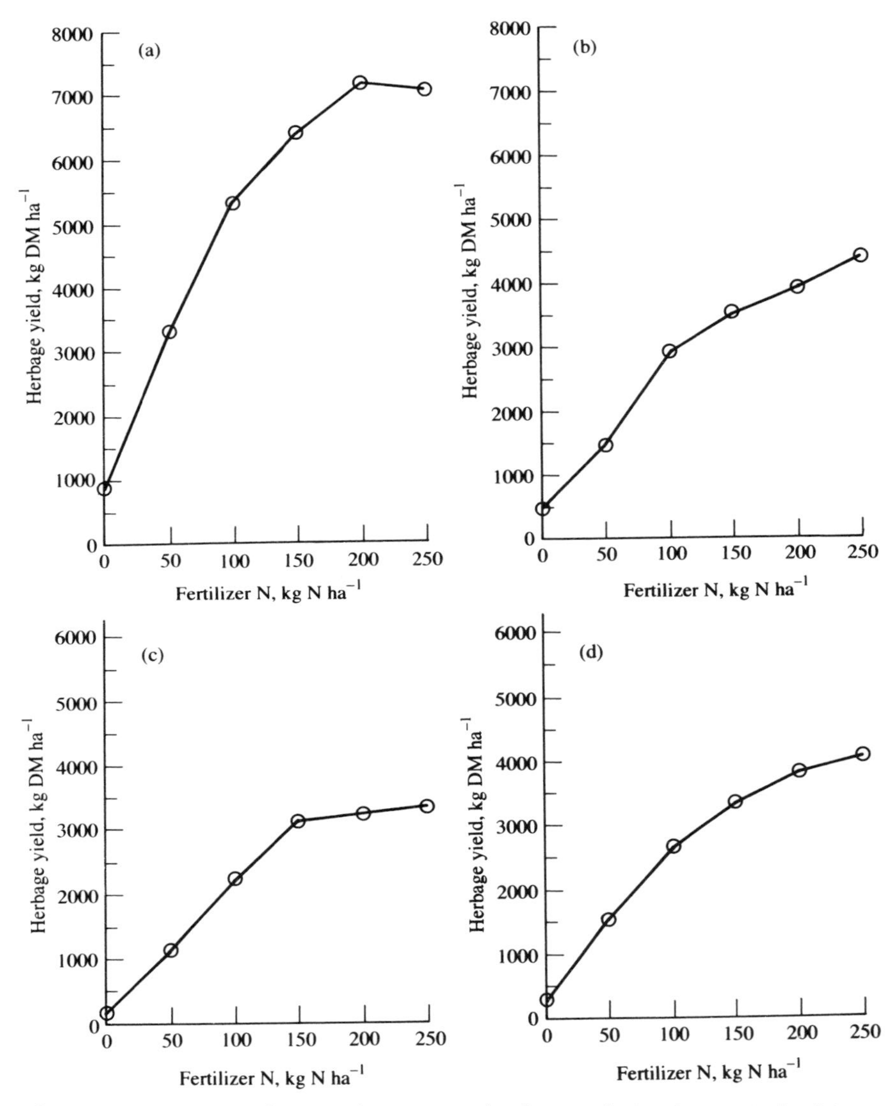

Fig. 12.5. Response of irrigated ryegrass to fertilizer N during four periods of the year: (**a**) start of growth – 6 June, (**b**) first regrowth, 7 June – 11 July, (**c**) second regrowth, 12 July – 17 August, (**d**) third regrowth, 18 August – 7 October. (Data from Cowling and Lockyer, 1970.)

made too early in the season, there is a greater risk of leaching and more opportunity for N to be lost by ammonia volatilization and denitrification. However, if the fertilizer is applied too late, the potential response is restricted. Assuming that there is no loss of N, the earlier the application, the greater the herbage yield attained by a specific date. Expressing the effect in another way, the earlier the date of fertilizer application, the earlier the date by which a herbage yield suitable for grazing will be reached.

Several attempts to provide general guidelines for the timing of the first application have been proposed and some take into account, so far as possible, year-to-year and site-to-site variations in weather conditions. This has been achieved by measuring soil and/or air temperature at intervals, and sometimes rainfall in addition. An early recommendation from the Netherlands was that fertilizer N should not be applied before 20 February; that it should be applied only if the mean minimum temperature had remained above 0°C for 10 days, and that it should be delayed until the first 10 days of April if the February rainfall exceeded 100 mm, which is more than twice the normal rainfall. In six out of seven years, these criteria provided good results in field trials (van Burg, 1968). Subsequent work in the Netherlands suggested that, in general, the best date for applying fertilizer N to grassland was that on which the accumulated mean daily air temperature since 1 January reached 250°C (Jagtenberg, 1970), subsequently modified to 200°C (Postmus and Schepers, 1980; van Burg *et al.*, 1981). To obtain this date, the means of the minimum and maximum daily temperature are summated daily until the total reaches 200, with negative values being ignored. This guideline has been quite widely adopted in Europe, including the UK, though evaluations of its effectiveness have produced mixed results (Baker, 1986; Stevens *et al.*, 1989a; van Burg *et al.*, 1981). In the UK, it appears that normally there is a period of 2–5 weeks during which the application of fertilizer N would produce a yield within 10% of the maximum, and that usually the date of T-sum 200 falls within this period (Baker, 1986). Studies carried out during 3 years at four sites in Northern Ireland, indicated that the T-200 system, and also the date on which the soil temperature at 10 cm first exceeded 5.5°C, were no more successful in predicting the optimum date for the first application than was a simple date range of 4 weeks (Stevens *et al.*, 1989a). Average data for the T-sum in the UK are published weekly for each of 97 grid squares but it should be appreciated that individual fields within a grid square may vary considerably in aspect and moisture status, and hence in the T-sum of soil temperature which, in practice, is more important than air temperature. A well-drained south-facing field may reach T-200 for soil temperature as much as 4 weeks earlier than one that is poorly drained and facing north. For this reason, the on-farm recording of soil temperature may provide a better prediction of the optimum date of application (Harkess and Frame, 1986).

In countries such as Australia, seasonal patterns of response to fertilizer N differ from those in cooler and more humid areas, and the greatest responses may occur during the late autumn or early spring. At several sites in southern Victoria, fertilizer N applied to perennial ryegrass with clover swards in late autumn produced a response of about 28 kg DM kg^{-1} for rates of application up to 100 kg N ha^{-1} (Newman *et al.*, 1962).

Site-to-site Differences in Response to Fertilizer N

Site-to-site differences in grass yield and response to fertilizer N, as illustrated in Fig. 12.2, are due partly to differences in weather and partly to differences in soil type. The available water capacity of the soil was shown to be important in a study involving eight grassland sites in the Netherlands: there was a positive relationship between the response to high rates of application (> 300 kg N ha^{-1} $year^{-1}$) and soil moisture status, assessed from soil texture and the depth of soil to the water table (Steenbergen, 1977). As explained on p. 207, a response curve for an individual site enables an estimate to be made of the optimum rate of application for that site, if it assumed, for example, that the optimum rate is the rate that produces a marginal response of 10 kg DM per kg fertilizer N. Some data, including the optimum rate of application, obtained from the 4-year investigation carried out at 21 sites in England and Wales are shown in Table 12.1. The herbage yield at the optimum rate of fertilizer application for each site was positively related ($r = 0.75$) to the amount of water available during the growing season, as assessed from the available water holding capacity of the soil plus rainfall during the growing season (Morrison, 1980; Morrison *et al.*, 1980). However, where sites differ more widely in latitude or altitude than in this investigation, it is likely that differences in temperature will have a greater impact, relative to the available water supply.

The potential effect of soil type, and its interaction with weather, is illustrated by a study in which four rates of fertilizer N were applied to lysimeters containing monoliths of long-term grassland on both a sandy and a clay soil (Dowdell and Webster, 1984). The lysimeters were covered, and four patterns of application of simulated rainfall were examined. While the distribution of simulated rainfall, within a uniform total amount, had little effect on herbage yield and N uptake on the clay soil it did influence yield and N uptake on the sandy soil. In particular, when heavy rainfall followed cutting and the application of fertilizer N on the sandy soil, there was a substantial reduction in yield and N uptake.

In the UK, soil texture and depth, the main determinants of available water capacity, together with rainfall between April and October can be used to allocate any field to one of five 'grass growth classes' as illustrated in Table 12.2. The classes differ in their responses to fertilizer N and in the optimum rate of fertilizer N required (Baker *et al.*, 1991). In general, deep loam soils in areas where the average summer rainfall is more than 400 mm provide the best conditions for grass growth and are designated Class 1: an example is shown by the upper curve of Fig. 12.2. Shallow gravelly or chalk soils in areas with a summer rainfall of less than 300 mm are the least productive and are designated Class 5: an example is shown by the lower curve of Fig. 12.2. Differences in yield and response between the site classes tend to increase as the growing

Table 12.1. Response of perennial grass to fertilizer N at 21 sites in England and Wales. (Data from Morrison *et al.*, 1980.)

	Herbage yield with no N (kg DM ha^{-1} $year^{-1}$)	Optimum rate of fertilizer N (kg ha^{-1} $year^{-1}$)	Herbage yield with optimum fertilizer N (kg DM ha^{-1} $year^{-1}$)	Overall response to optimum fertilizer N (kg DM/kg N)	Herbage yield with 300 kg fertilizer N (kg DM ha^{-1})	Response to 300 kg fertilizer N (kg DM/kg N)
Cambo, Northumberland	6,080	374	12,340	17	11,760	18
Harewood, Yorks	4,420	338	12,190	23	11,880	25
High Mowthorpe, Yorks	1,640	292	7,350	20	7,500	19
Leeds, Yorks	640	423	11,190	25	9,570	29
Bangor, Gwynedd	5,650	337	13,160	22	12,100	23
Aberystwyth, Dyfed	2,960	372	11,890	24	11,240	27
Wenvoe, Glamorgan	1,240	530	14,350	25	9,710	29
Rocester, Staffs	4,400	317	11,210	21	10,930	22
Rosemaund, Hereford	2,900	358	11,330	24	10,410	26
Drayton, Warwicks	1,530	422	9,790	20	8,140	22
Morley, Derby	4,100	336	11,940	23	11,920	25
Gleadthorpe, Notts	690	415	10,410	23	9,050	27
Cambridge	1,350	260	5,240	15	5,640	14
Seale Hayne, Devon	2,850	485	13,640	22	11,280	27
Cannington, Somerset	1,610	358	9,100	21	8,670	23
Bridgets, Hants	1,980	355	9,290	21	8,550	22
Oxford	3,120	442	12,510	21	10,680	25
Rowsham, Bucks	3,940	412	11,690	19	10,120	21
Hurley, Berks	1,820	450	9,880	18	7,620	20
Wye, Kent	2,750	357	9,490	19	8,590	20
Pluckley, Kent	820	475	10,800	21	7,890	25

Table 12.2. Allocation of fields to UK grass growth classes 1–5 (class 1, excellent; class 5, poor) on the basis of soil texture and depth, and rainfall during the months April–September. (From Baker *et al.*, 1991.)

	April–September rainfall (mm)			
Soil texture/depth	> 500	425–500	350–425	< 350
Clays, loams, sandy loam; depth > *c.* 20 cm	1	2	2	3
Shallow soils over chalk or rock; gravelly and coarse sandy soils	2	3	4	5

season advances and the soil moisture deficit increases (Lazenby, 1988). The allocation of a particular field to a 'grass growth class' provides a means (alternative to the MAFF guidelines described on p. 212) of making a recommendation of the optimum rate of fertilizer N (Baker *et al.*, 1991), though no account is taken of N supply from the soil.

13 Response to Fertilizer Nitrogen: Influence of Type of Fertilizer and Supplies of Other Nutrients

Yield Responses to Ammonium and Nitrate Salts

Most fertilizer N, in whatever form, is taken up by grass as nitrate, though some may be taken up as ammonium when this form, or urea, is applied. Although urea can be absorbed by plant roots, it is normally hydrolysed rapidly to ammonium in the soil. The differences in yield response to the various forms of fertilizer N are due mainly to differences in the loss of N from the soil rather than to differences due to the form of uptake. In general, ammonium and nitrate produce similar responses in grass (and other crops) and these are at least equal to the responses produced by other forms of fertilizer N. Occasionally, the response to ammonium is significantly greater than that to nitrate or vice versa and, when these differences do occur, they reflect the soil and weather factors that influence ammonia volatilization, denitrification and leaching. Ammonium-N is susceptible to ammonia volatilization (particularly on calcareous soils) but is not susceptible to denitrification or leaching until nitrified, whereas nitrate-N is not susceptible to ammonia volatilization but is susceptible to denitrification and leaching.

In many grassland experiments, comparisons between ammonium and nitrate have shown only small differences in yield and response (e.g. Heddle, 1968; van Burg *et al.*, 1982) but in some, particularly those on calcareous soils, nitrate has produced larger yields than ammonium. For example, in 89 experiments on grassland, ammonium nitrate and ammonium sulphate usually produced similar responses but, on soils containing more than 10% calcium carbonate, the response to ammonium sulphate was only about 80% of that to ammonium nitrate (Devine and Holmes, 1964). On the other hand, under cool wet conditions, yields are sometimes larger from ammonium than from nitrate (Watson, 1986; Watson and Adams, 1986). The timing of the application can influence the result of such a comparison: ammonium sulphate produced a greater response than did calcium ammonium nitrate when both were applied in late February, though they produced similar responses when applied in late March (McAllister *et al.*, 1965). Leaching and/or denitrification of the nitrate probably occurred from the application in February, and nitrification of the ammonium-N would have been slow at that time. Similar results were obtained from other experiments (Devine and Holmes 1965; Stevens and Laughlin, 1988).

Yield Responses to Urea

The response to urea is often lower than that to ammonium nitrate, due largely to the volatilization of ammonia (see p. 168). Other factors that may possibly curtail the response to urea are the occurrence of chemo-denitrification (Watson *et al.*, 1990a) and, occasionally, the leaching of the urea itself. Although urea is only weakly adsorbed by soil and therefore leaches readily (Broadbent *et al.*, 1958; Gould *et al.*, 1986), it is usually hydrolysed rapidly to ammonium which is retained in the soil by cation exchange, until taken up by roots or nitrified to nitrate. Even at 5°C, 90% of urea applied to a moist soil was found to be converted to ammonium in 2 days, though only about 14% was nitrified (Low and Piper, 1961). Some of the urea manufactured during the 1950s and 1960s contained biuret (the dimer of urea) at a concentration of about 1% or more and, as a result, was toxic to plants (Cooke, 1967; Gould *et al.*, 1986). However, this problem is unlikely to occur in the urea currently manufactured. When urea is applied to a seedbed, the ammonia resulting from hydrolysis may sometimes damage germinating seedlings (Low and Piper, 1961), but this type of damage is unlikely with established grassland. Although urea causes an initial rise in soil pH following its hydrolysis to ammonium, it tends to acidify the soil in the longer term as a result of nitrification.

A number of field experiments on grassland have shown that urea is often as effective as calcium ammonium nitrate when applied early in the growing season, but is less effective in summer (Lloyd, 1992b; Watson *et al.*, 1990a). However, urea is sometimes less effective than ammonium nitrate even in spring (e.g. Chaney and Paulson, 1988; Swift *et al.*, 1988). In 20 comparisons of urea with calcium ammonium nitrate carried out in spring, the relative herbage yield ranged rather widely from 91 to 121% whereas, in 17 comparisons carried out in summer, the range was 72 to 102%: in spring, the relative yield from urea was more than 95% in 17 of 20 studies whereas, in summer, this relative yield was achieved in only six of the 17 studies (Watson *et al.*, 1990a). The actual yield variation in spring appeared to be no greater with urea than with ammonium nitrate, and losses of N undoubtedly occurred from both forms. Losses of N from urea are likely to have been due mainly to ammonia volatilization, whereas losses from ammonium nitrate are likely to have been due to leaching and denitrification (Watson *et al.*, 1990a). Lower responses to urea than to ammonium nitrate have also been reported in Kentucky, USA, with tall fescue (Murdock and Frye, 1985). The influence of soil properties, and weather conditions at the time of application, on the response of grassland to urea is shown by data from Ireland and the Netherlands (O'Toole and Morgan, 1988; van Burg *et al.*, 1982). The highest efficiency of urea, compared with ammonium nitrate, occurs on well-buffered soils (especially peat soils) and, in general, the efficiency of urea is increased if about 5 mm or more of rain falls within 2 days of the application (van Burg *et al.*, 1982).

Various possible means of improving the efficiency of urea have been examined, including slow release forms, the addition of urease inhibitors, other chemical additives and the alteration of granule size. Of these possibilities, the urease inhibitor, NBPT (*N*-(*n*-butyl) thiophosphoric triamide) is considered to have the greatest potential for improving response in the field: when incorporated into urea at 0.5% by weight, it improved the response of ryegrass by about 10% (Watson *et al.*, 1990b). Studies of the effects of nitrification inhibitors have shown variable results, but yield increases appear most likely when the urea plus inhibitor is injected into the sward (Rodgers *et al.*, 1984, 1987). The granule size of the urea has been shown to have little effect (Watson and Kilpatrick, 1991).

Yield Responses to Anhydrous and Aqueous Ammonia

Both anhydrous and aqueous ammonia are applied to grassland by injection into the soil, usually to a depth of 10 cm or more, and with a spacing between the tines of 15–30 cm (van Burg, 1969). The resultant ammonium ions are largely retained by cation exchange, and are progressively nitrified to nitrate. The effectiveness of injected ammonia is influenced by the stoniness and texture of the soil. With stony soils, some loss of gaseous ammonia is almost inevitable and, with clay soils that are either too wet or too dry, the slit behind the injector may fail to close, allowing ammonia to volatilize (Nelson, 1982). Some ammonia may also volatilize from sandy soils due to poor retention by cation exchange, particularly if the soil is dry and low in organic matter.

When ammonia is injected, high concentrations occur in a narrow band of soil which is partially sterilized, and this partial sterilization delays nitrification and therefore tends to reduce denitrification and leaching. It has been suggested that a single injection of ammonia in winter or early spring might provide a steady supply of available N through the season, thus avoiding the need for fertilizer to be applied on several occasions. This suggestion is based on three assumptions, first that the resulting ammonium-N will be adsorbed by the soil within about 15 cm of the line of injection; second that the ammonia will kill roots and delay nitrification close to the line of injection, and third that the surviving roots will take up N progressively from the outer edge to the centre of the adsorption zone. Another potential advantage is that ammonia might be applied when the pressure of other farm work is less, provided that temperatures are low enough at that time to restrict nitrification and thus minimize denitrification and leaching. Comparisons of ammonia with solid fertilizers, applied as large single applications in spring, have often shown little difference in annual herbage yield (Burton and Jackson, 1962; van Burg *et al.*, 1967; Widdowson *et al.*, 1972). The ammonia generally produced a lower yield at the first cut and rather higher yields at later cuts. However, in an experiment in which ammonia and calcium ammonium nitrate were applied for each cut, and in which the sward receiving calcium ammonium nitrate was damaged by the passage of injection tines to the same extent as that receiving ammonia, considerably higher yields were obtained from the calcium ammonium nitrate than from the ammonia (van Burg *et al.*, 1967). Some comparisons of single applications of ammonia injected in spring with split applications of solid fertilizer have shown annual yields of herbage to be similar (Hodgson and Draycott, 1968; Widdowson *et al.*, 1972) though, at high rates of application, single applications of ammonia have often produced yields 10–40% lower than those obtained from split applications of solid fertilizer (Burton and Jackson, 1962; Jeater, 1967; Swift *et al.*, 1988; van Burg *et al.*, 1967; van Burg, 1969). A single injection of ammonia usually produces a seasonal pattern of herbage growth different from that obtained with split applications of solid fertilizer. If the injection is made during the winter or early spring, growth in the early part of the season may be equal to, or better than, that from the normal application of solid fertilizer but growth in the later part of the season is likely to be poor. On the other hand, if the injection is made in mid to late spring, growth is likely to be relatively poor in the early part of the season but may well improve, relative to solid fertilizer, later in the season (Swift *et al.*, 1988).

Possible reasons why ammonia is often less effective than solid fertilizers include:

1. damage to the sward caused by injection equipment,
2. the loss of gaseous ammonia,
3. the release of adsorbed ammonium not matching sward requirements.

Damage by injection equipment may reduce herbage yield by 6–10%, or even more in dry conditions (van Burg *et al.*, 1982). Damage was found to increase with increasing depth of injection from 5 to 15 cm, and was considerably greater on a clay soil than on sandy or peat soils (van Burg and van Brakel, 1965).

The extent to which ammonia is lost to the atmosphere depends on the depth of injection, the efficiency of the equipment, the soil type and soil moisture status. On sand and peat soils, ammonia produced slightly higher yields when injected at a depth of 10 cm than at 5 cm, and a higher yield at 5 than at 15 cm, but on a clay soil the yield obtained with injection at a depth of 5 cm was 13% higher than from injection at 10 cm (van Burg and van Brakel, 1965). The lower yields at shallow injection depths, on the sand and peat soils, were apparently due to the volatilization of ammonia. The extent to which ammonia is lost depends partly on the design of the injection equipment (van Burg *et al.*, 1967) as well as on the depth and timing of injection, and on weather factors (Swift *et al.*, 1988).

In an investigation to examine whether the release of available N from adsorbed ammonia matched the requirements of the sward, aqueous ammonia was injected into the soil (at rates equivalent to 200 and 400 kg N ha^{-1}) using a hypodermic syringe at a depth of 10 cm and at points 15 cm apart, thus avoiding sward damage and the loss of gaseous ammonia (Cowling, 1968). Even with this technique, ammonia injected in late winter on a loam soil was no more effective than a single application of ammonium nitrate in providing a continuous supply of N throughout the season, and was considerably less effective than split applications of ammonium nitrate. In another experiment in which ammonia was injected as early as mid-November, responses appeared to be considerably influenced by winter rainfall (Jeater, 1967). With a rainfall of 330 mm between the November injection and the first cut, the yield response to ammonia, at 336 kg N ha^{-1}, was 64% of that to split applications of ammonium nitrate whereas, at a second site, with a rainfall of 590 mm during the same period, the comparable figure was only 42%, probably due to increased leaching of nitrate after nitrification. With a typical rate of application of about 250 kg N ha^{-1} and a typical spacing between injection tines of 30 cm, the concentration of ammonia in the injection zone is probably insufficient to retard nitrification throughout the season (Ashworth and Flint, 1974). The injection of a nitrification inhibitor with aqueous ammonia might be expected to improve the total response by prolonging its period of availability for uptake, but tests of nitrapyrin, carbon disulphide and trithiocarbonates over 3 years at two sites produced inconsistent results (Ashworth *et al.*, 1980).

Aqueous ammonia is not suitable for surface application to grassland, as even solutions diluted to 5% N cause damage to the herbage (Nowakowski, 1961).

Yield Responses to Fertilizers Applied in Solution

When various forms of N applied in solution have been compared with the same compounds applied in solid form, the yield from the solution has generally been similar to, or lower than, the yield from the corresponding solid. For example, in two field experiments in which ammonium nitrate, ammonium sulphate, calcium nitrate and urea were applied, both in solid form and as a 5% N solution sprayed on to the sward, the method of application had no appreciable effect on yield (Nowakowski, 1961). However, the grass absorbed more N, and concentrations of N in the herbage were higher, with the solid forms than with the solutions. When solutions of ammonium nitrate and urea, and a mixture of the two, were sprayed on to grass swards, yields were less than those from solid calcium ammonium nitrate (van Burg, 1964). Fertilizer solutions injected into the soil also generally gave lower yields than solid top-dressings, partly as a result of mechanical damage caused by the injection equipment (Jameson, 1959). However, solutions of ammonium sulphate and ammonium nitrate + urea, when injected into grassland at a depth of 7.5–10 cm and at 30 cm spacing, at rates of more than 224 kg N ha^{-1}, sometimes produced higher annual yields than the same amount of N supplied as split applications of solid fertilizer (Draycott *et al.*, 1967). The advantage of injected solutions was greater during dry periods.

Yield Responses to Slow-release Nitrogen Fertilizers

A means of supplying fertilizer N continuously but slowly over a period of months would be particularly valuable for grass because it has a long season of potential growth. However, as noted above, the injection of ammonia, either aqueous or anhydrous, is not generally successful for this purpose. A second approach to developing a slow-release form of fertilizer is to coat a soluble form of N (such as ammonium nitrate or urea) with a membrane which is semipermeable, perforated, or degradable by soil microorganisms, but the additional cost of such a coating makes the fertilizer uneconomic for agricultural use (Parr, 1967). A third approach is to use a more complex N compound which has a low solubility and which is decomposed by soil microorganisms (Parr, 1967). One of the most effective compounds is urea-formaldehyde, a condensation product which consists of polymer molecules ranging in size from methylenediurea to tetramethylenepentaurea. The rate at which it releases soluble N is influenced partly by the distribution of molecular size, and partly by soil microbial activity and hence by temperature and soil moisture (Alexander and Helm, 1990). In the year of application, the response of Italian ryegrass to urea-formaldehyde was little more than half that to Nitro-chalk, when both forms of N were given as single applications in spring, and it was even less when the Nitro-chalk was applied in split dressings (Widdowson *et al.*, 1962). Field trials with Kentucky bluegrass showed that, with a single application of 157 kg N ha^{-1}, a mixture containing about half the N as urea-formaldehyde and half as ammonium nitrate produced annual yields as high as those from ammonium nitrate alone and gave more uniform growth through the season (Kilian *et al.*, 1966). The urea-formaldehyde had an appreciable residual effect for at least 2 years after application. Formalized casein (a waste product of the plastics industry containing about 12.2% N) has also been investigated as a slow-release source of N. In studies with Italian ryegrass, little N was mineralized during the first few weeks following application in early spring but considerable amounts became available later in the season, and some in the following year (Widdowson and Shaw, 1960). Aggregate yields over 1 or 2 years were, however, lower from formalized casein than from split applications of ammonium sulphate, calcium nitrate or urea.

In general, urea-formaldehyde and other slow-release N fertilizers are useful for sports turf areas and for horticulture (Alexander and Helm, 1990) but are too expensive for use on most agricultural grassland.

Influence of Supplies of Other Nutrients on Response to Fertilizer N

The yield response of grass to fertilizer N is sometimes limited by deficiencies of other nutrient elements such as P, K and S. Such deficiencies are most likely to occur in fields where high rates of fertilizer N have been applied over many years, and especially where most of the grass has been cut and removed. The harvesting of silage or hay entails the removal of substantial amounts of P, K and S from the soil; for example, an annual yield of 10,000 kg of herbage DM ha^{-1} with average contents of P, K and S will remove about 35 kg P, 200 kg K and 30 kg S ha^{-1}. Such removals will eventually deplete the soil supplies of these elements unless they are restored through fertilizers or manures.

It is unusual for nutrient elements other than P, K and S to limit the response of grassland to fertilizer N, though soil acidity may do so. One effect of soil acidity is to encourage the less productive grass species and, in order to obtain the maximum productivity from grassland, soils with a pH of less than 4.5 should be limed. In some regions, particularly parts of Australia and New Zealand, deficiencies of trace elements such as Mo or Co may restrict plant growth and the yield of grassland. However, clovers are generally more susceptible than grass to such deficiencies.

Influence of Phosphorus on Response to Fertilizer N

Phosphorus deficiency is generally more common in the wetter upland areas than in drier lowland areas, partly because the availability of P is low in acid soils, and acid soils are more widespread in upland areas. Another factor is that very little fertilizer P has been applied to upland areas in past decades, whereas

Table 13.1. Herbage yield (kg DM ha^{-1} $year^{-1}$) of a mixed grass–clover sward at four rates of fertilizer N, with and without fertilizer P. (Data from Gething, 1963.)

	Fertilizer N (kg ha^{-1} $year^{-1}$)			
	0	98	196	392
With 52 kg P ha^{-1}	7,310	8,780	12,240	19,160
Without additional P	6,780	7,714	10,173	14,920

much of the arable land in lowland areas, especially in Europe, has received substantial amounts of fertilizer P.

Some trials in UK have shown little or no effect of P on the response of grass to fertilizer N (Castle and Holmes, 1960; Holmes and MacLusky, 1954; Reith *et al.*, 1961) but deficiencies have been reported in lowland areas and are most likely when large crops of herbage are regularly cut and removed. A deficiency that restricted the response to fertilizer N was observed in moist hay meadows in southwest England (Kirkham and Wilkins, 1994). Another example of fertilizer P increasing the response to N occurred with a cut grass–clover sward on a clay soil in Warwickshire, UK (Table 13.1). When P is only marginally deficient, the deficiency is likely to become more marked from year to year if no fertilizer P is supplied, as shown in a 3-year investigation (Wolton *et al.*, 1968). A gradual depletion of soil P was also shown in a 12-year experiment at the North of Scotland College of Agriculture (1965), in which a mixed sward gave an appreciable yield response to phosphate only in the last 3 years and then only when the rate of fertilizer N was as high as 310 kg ha^{-1} $year^{-1}$.

Influence of Potassium on Response to Fertilizer N

Potassium deficiency in grassland would be widespread if fertilizer K were not applied, especially on sandy soils which have little capacity to retain K by cation exchange. However, many clay soils are inherently well supplied with K and, on many loam soils in agriculturally productive areas, regular applications of fertilizer K have resulted in some accumulation of soil K. An important factor affecting the incidence of K deficiency in grassland is whether the sward is cut or grazed: when swards are cut for silage or hay, large amounts of K are removed in the herbage whereas, when swards are grazed, much of the herbage K is returned to the soil in urine. However, the return of K in urine results in a patchy distribution and, within a field, areas of deficiency may co-exist with areas of surplus.

In the UK, several grassland experiments have shown no effect of fertilizer K on the response to fertilizer N, but a major effect did occur in a study in

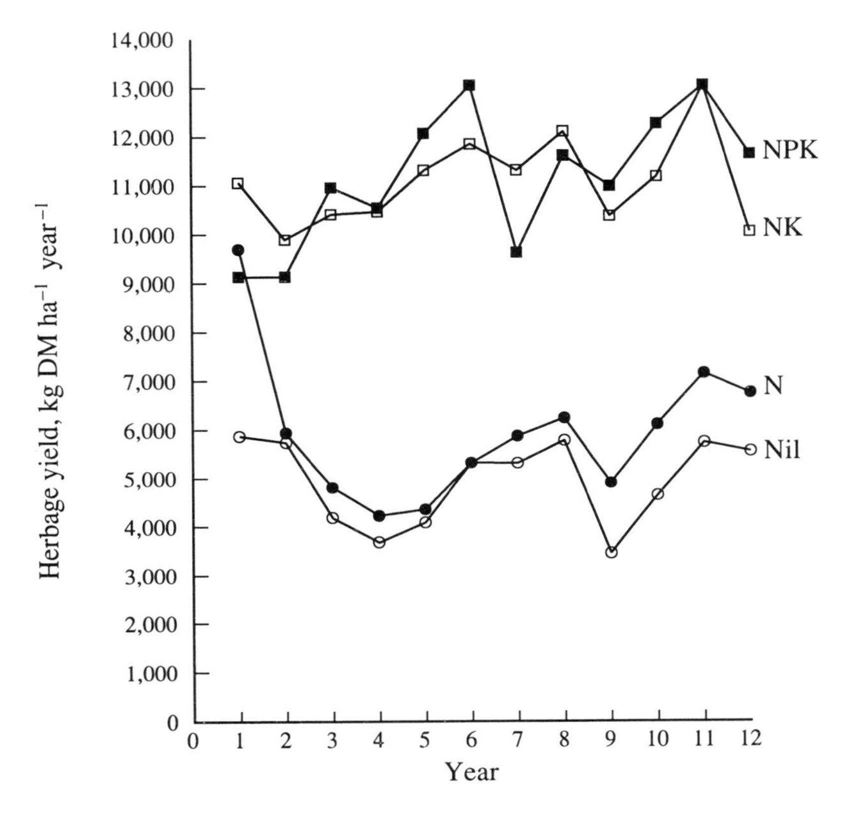

Fig. 13.1. The influence of K (500 kg K ha^{-1} $year^{-1}$) and P (89 kg P ha^{-1} $year^{-1}$) on response to N (580 kg N ha^{-1} $year^{-1}$) during a 12-year period. (Data from Holmes and MacLusky, 1954, and Castle and Holmes, 1960.)

southwest Scotland (Castle and Holmes, 1960; Holmes and MacLusky, 1954). Although the application of fertilizer K had no effect on the response to fertilizer N in the first year of this experiment, in the second and subsequent years, a response to fertilizer N was dependent on the application of K (Fig. 13.1). In another investigation, the application of fertilizer K increased the response of cut swards to fertilizer N at five out of six sites in Scotland (Reith *et al.*, 1961). The effect of K was greatest at the highest rate of fertilizer N (390 kg N ha^{-1} $year^{-1}$) and increased during the second and third years of the trial, indicating that soil K was progressively depleted.

Several other examples of the application of K increasing the response of grassland to fertilizer N have also been reported (Chestnutt, 1965; Griffith and Teel, 1965; Hemingway, 1963; Kirkham and Wilkins, 1994; Lowe, 1967; Widdowson *et al.*, 1965a,b; Wolton *et al.*, 1968).

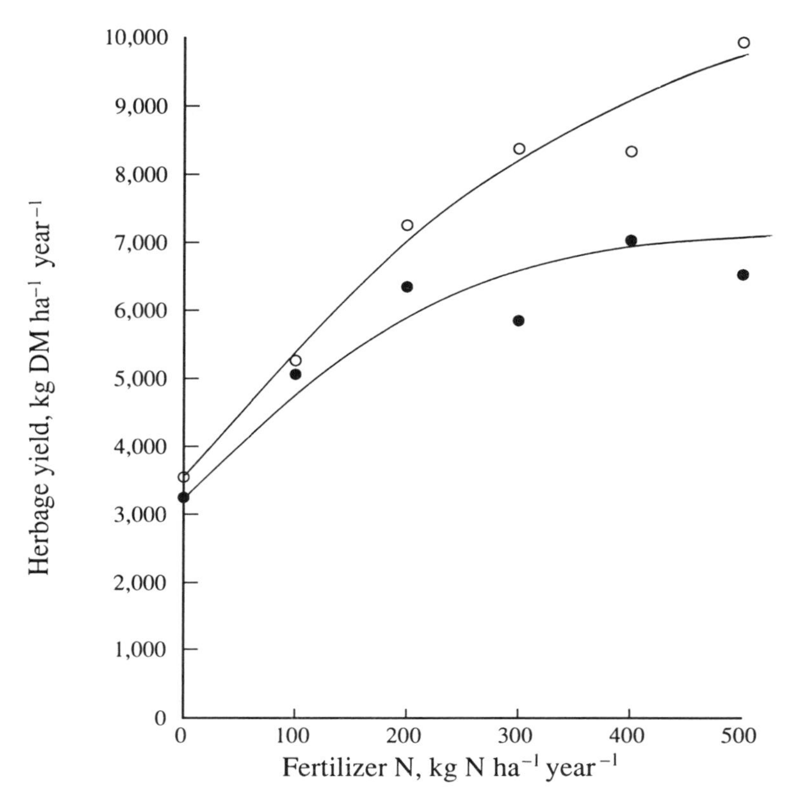

Fig. 13.2. Response of perennial ryegrass to fertilizer N in Ireland, with (○) and without (●) the addition of sulphur (as $CaSO_4$) at 25 kg S ha^{-1}. (Data from Murphy and O'Donnell, 1989.)

Influence of Sulphur on Response to Fertilizer N

A number of field trials in Ireland and in the UK have shown grassland to respond to the application of fertilizer S, increasingly so since the early 1970s (Syers *et al.*, 1987). The incidence of S deficiency has become more widespread as concentrations of SO_2, H_2SO_4 and sulphate aerosols in the atmosphere have declined. In trials at a number of sites in England and Wales, the response to S was greatest at high rates of fertilizer N (Skinner, 1987) and responses occurred at several sites in Northern Ireland at a high rate of fertilizer N (Stevens and Watson, 1986). Responses to S occur most commonly on light-textured soils (Syers *et al.*, 1987) which are low in adsorbed sulphate and organic S. A clear interaction between applications of fertilizer N and S was shown in a trial on grassland in Ireland (Fig. 13.2).

Table 13.2. Amounts of lime (expressed as kg CaO ha^{-1}) required to neutralize the acidity induced by various forms of fertilizer N applied to grassland at 100 kg N ha^{-1}. (From van Burg *et al.*, 1982.)

Fertilizer	% N	Chemical form of N	kg CaO ha^{-1} required
Ammonium nitrate	34	NH_4-N 17%; NO_3-N 17%	80
Calcium ammonium nitrate	23	NH_4-N 11.5%; NO_3-N 11.5%	0
Calcium ammonium nitrate	26	NH_4-N 13%; NO_3-N 13%	30
Urea	46	$CO(NH_2)_2$-N 46%	80
Anhydrous ammonia	82	NH_3-N 82%	80
Ammonium sulphate	21	NH_4-N 21%	270
Calcium nitrate	16	NO_3-N 16%	0

Influence of Fertilizer N on Soil pH, and Influence of Liming on Response to Fertilizer N

All fertilizers that supply ammonium (or a precursor of ammonium) tend to acidify the soil, an effect that is due mainly to nitrification. Nitrate salts, other than ammonium nitrate, do not cause acidification. The acidifying effect of fertilizer N is greatest with ammonium sulphate (Table 13.2) because, in addition to acidity resulting from the nitrification of ammonium, there is an effect from the sulphate, much of which is leached from the soil accompanied by calcium ions. With calcium ammonium nitrate, the calcium carbonate component tends to neutralize the acidifying effect of the ammonium nitrate, and this neutralization is approximately complete with the form of fertilizer containing 23% N though not with the form containing 26% N (Table 13.2).

Soil pH has little effect on the response of grassland to fertilizer N if it is above about 4.5, but the response is curtailed at lower pH values (van Burg *et al.*, 1982). Soil acidity is one of the factors that tend to restrict the response to fertilizer N in upland areas. Liming had little if any effect in four trials in Northern Ireland, when the initial pH of the soil (in water) was in the range 4.9–5.7 but, when the lime was applied just a few days before the fertilizer N, and when this was in the form of ammonium sulphate or urea, the lime actually caused some reduction in the response to the fertilizer N, presumably due to an increase in ammonia volatilization (Adams, 1986).

14

Influence of Fertilizer Nitrogen on the Composition and Nutritional Quality of Grassland Herbage

Influence of Fertilizer N on the Botanical Composition of Swards

The application of fertilizer N to grass–clover swards generally reduces the proportion of clover (see Chapter 3 and especially Table 3.3). With swards that contain a variety of grass species, moderate to high rates of fertilizer N usually encourage the more productive species, such as perennial ryegrass, relative to the less productive species such as *Agrostis*, *Holcus* and *Poa* spp. (Garstang, 1981; Holmes, 1949; Hopkins, 1986; Reith *et al.*, 1964; de Vries and Kruijne, 1960). However, in some instances, fertilizer N has been found to increase *Poa* spp. (Browne, 1966; Kreil *et al.*, 1966; McAllister and McConaghy, 1960) or (when K was deficient) red fescue (Heddle, 1967). Observations on permanent pasture in the Netherlands indicated that increasing the rate of fertilizer N increased both perennial ryegrass and *Poa* spp. at the expense of cocksfoot and *Agrostis* (t'Hart, 1957). The influence of fertilizer N on the balance of grass species is modified by factors such as the frequency of defoliation and the supply of other nutrients. For example, the proportions of perennial ryegrass, meadow fescue and cocksfoot were found to be higher in plots that had received both N and K over a 12-year period, than in plots that had received N but not K (Castle and Holmes, 1960).

The application of fertilizer N to swards containing non-legume dicotyledonous species has usually reduced the proportion of these species (Garstang, 1981; de Vries and Kruijne, 1960; Jones and Haggar, 1993; van Strien *et al.*, 1988; Wilkins *et al.*, 1989) though occasionally there has been little effect (Kreil *et al.*, 1966) or even an increase (Templeton and Taylor, 1966). On moist hay meadows in southwest England, increasing the rate of ammonium nitrate over the range from nil to 200 kg N ha^{-1} $year^{-1}$ reduced botanical diversity, while increasing both perennial ryegrass and *Holcus lanatus* (Mountford *et al.*, 1993). As little as 25 kg N ha^{-1} $year^{-1}$ increased the proportion of the more productive grasses within a period of 2 years; and 50 kg N ha^{-1} $year^{-1}$ significantly reduced botanical diversity within 3 years. In some instances, the loss of species diversity has been due to the soil acidity induced by the repeated application of ammonium fertilizers. Such an effect was observed over a period of more than 80 years in some of the Park Grass plots at Rothamsted, especially when the soil pH dropped to less than 4.5 (Silvertown, 1980).

Changes in the botanical composition of a sward may influence the chemical composition of the herbage: in particular, a lower proportion of clover as associated with a higher concentration of cellulose and lower concentrations of calcium and magnesium (see below).

Influence of Fertilizer N on the Dry Matter Content of Herbage

The content of dry matter in grassland herbage is generally reduced by fertilizer N (Cowling and Lockyer, 1970; Sprague and Taylor, 1970). This is due partly to an increase in the amount of internal water in the leaves, and partly to the fact that a larger crop, resulting from the application of fertilizer N, retains a greater quantity of superficial water from dew or rainfall. An example of the overall effect of fertilizer N is shown in Fig. 14.1.

Influence of Fertilizer N on Total and Organic N in Herbage

When fertilizer N is applied at a low rate to grass that is seriously deficient in N, there is usually an increase in growth but little or no change in N concentration. As the rate of fertilizer N is increased, both the yield of herbage and the concentration of N increase until, eventually, the yield reaches a maximum (Fig. 14.2). With still higher rates of application, the concentration of N continues to increase but the yield shows little change. Sometimes, when a small amount of fertilizer N (equivalent to 50 or even 100 kg N ha^{-1} $year^{-1}$) is applied to a deficient sward, the increase in yield is associated with a small decrease in the concentration of N in the herbage (e.g. Cowling and Lockyer, 1970; ICI, 1966). However at higher rates, and at all rates with many soils, increasing the rate of fertilizer N up to the economic optimum generally results in a progressive increase in the concentration of N in the herbage (Cowling and Lockyer, 1967, 1970; ICI, 1966; Morrison *et al.*, 1980; Reid, 1966). The results of 32 sets of data from experiments in various countries indicated that,

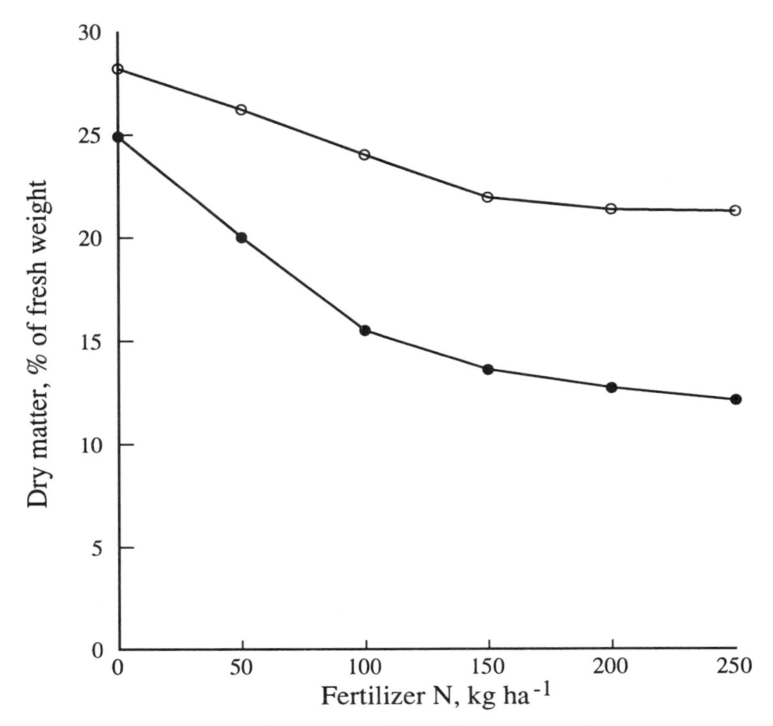

Fig. 14.1. Influence of fertilizer N (single application) on dry matter content (% of fresh weight) of perennial ryegrass harvested after 2 growth periods, to 6 June (○), and 7 June – 11 July (●). (Data from Cowling and Lockyer, 1970.)

on average, the application of 100 kg N ha^{-1} $year^{-1}$ increased the concentration of N in the herbage from a mean of 1.97% to 2.13%, an increase of 0.16 percentage units: the average increase for the application of 200 kg N compared with 100 kg N was 0.24 percentage units and the response then declined steadily with increasing rate of application to 0.17 percentage units for the increment between 400 and 500 kg N ha^{-1} $year^{-1}$ (Wilman and Wright, 1983). When herbage is cut (and fertilizer applied) on several successive occasions during a growing season, the concentration of N at each rate of fertilizer N often decreases from the first to the second cut, and then increases progressively during the latter part of the season (Fig. 14.3).

The effect of fertilizer N on the concentration of N in grass herbage is influenced by the time interval between the application and sampling. With good growing conditions, the effect is often greatest 2–3 weeks after the application, as shown in trials with Italian ryegrass (Fig. 14.4).

In some situations there may be a long-term effect on herbage N concentration. For example with a prairie grassland, the effect on herbage N concentration of a single application of 1000 kg N ha^{-1} was apparent in each of the

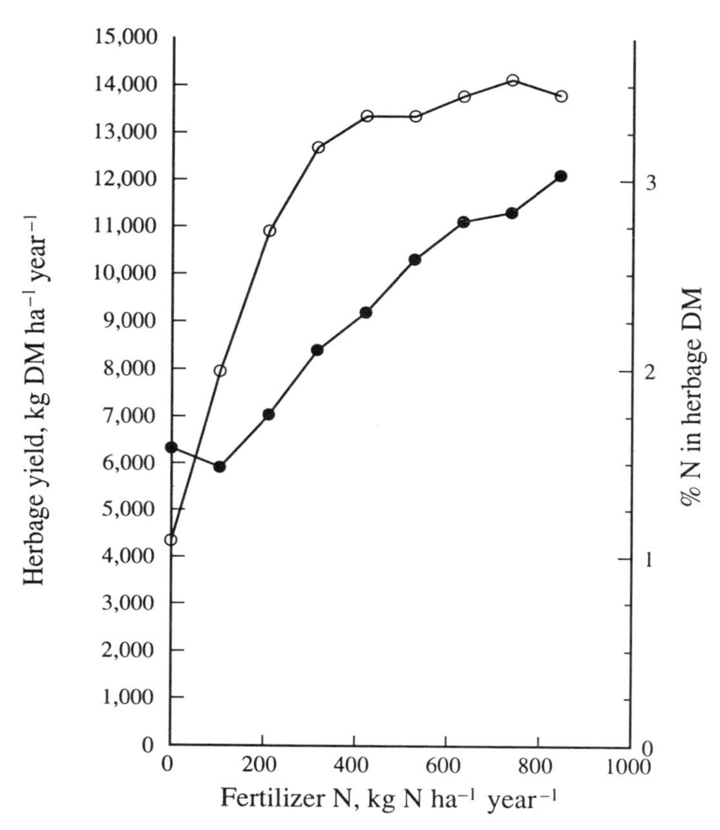

Fig. 14.2. Herbage yield (o) and N concentration in the herbage (•) of Italian ryegrass (cut four times per year) receiving a wide range of rates of fertilizer N as calcium ammonium nitrate. (Data from ICI, 1966.)

next 7 years, the average N concentration after this time being increased from 0.93% to 1.22% (Black and Wight, 1979). Grass species grown under uniform conditions vary slightly in herbage N concentration. For example, timothy (cv. S51) and meadow fescue (cv. S53) had higher concentrations than did perennial ryegrass (cv. S24) at each of four rates of fertilizer N (Chestnutt, 1972).

As well as increasing the total N concentration, fertilizer N has an effect on the forms in which the N is present in herbage. The main effect is often on the concentration of nitrate (see below) but there is also some effect on the organic N. Although there appears to be little effect on the proportion of the herbage protein that is water-soluble (Gillet, 1982), the amount of free amino acid N tends to be increased (Goswani and Willcox, 1969). In studies of Italian ryegrass grown under controlled conditions with various amounts of ammonium and nitrate, the proportion of herbage N present in soluble

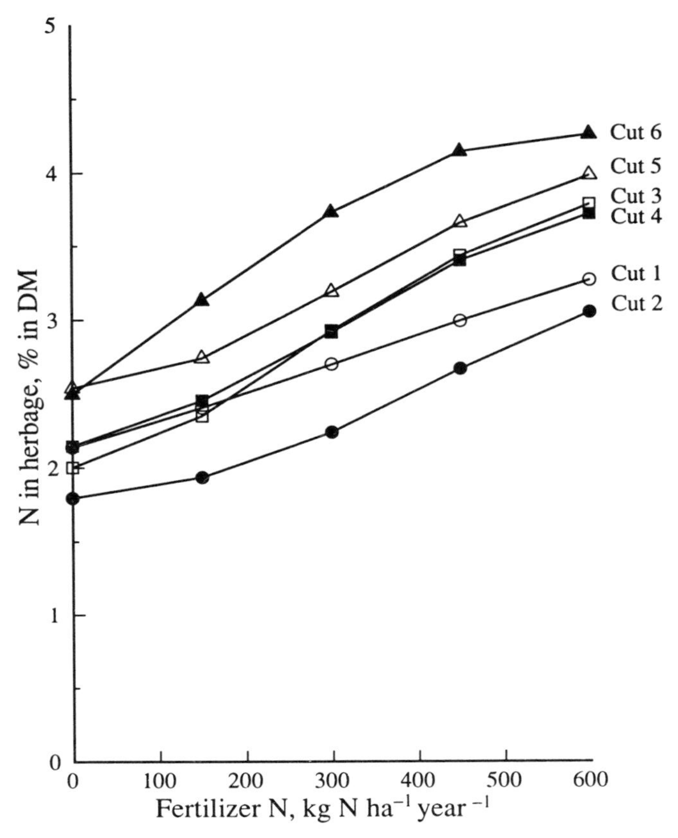

Fig. 14.3. The influence of the rate of fertilizer N on the concentration of total N in six successive harvests of perennial ryegrass (mean values for 21 sites). (Data from Morrison *et al.*, 1980.)

organic non-protein form was greater when fertilizer N was applied as ammonium than when it was applied as nitrate (Nowakowski and Cunningham, 1966; Nowakowski *et al.*, 1965). The proportion of the total N present in soluble organic forms, such as asparagine, tends to increase when light intensity is low and when there is a deficiency of another nutrient such as K (Nowakowski, 1964; Griffith *et al.*, 1964) or S (Millard *et al.*, 1985). In general, fertilizer N has little effect on the amino acid composition of the herbage protein (Goswani and Willcox, 1969; Reid and Strachan, 1974; Syrjala-Qvist *et al.*, 1984) though, in an investigation with *Poa pratensis*, the proportion of aspartic acid was found to increase somewhat with increasing rate of fertilizer N (Nelson and Sosulski, 1984).

Herbage that is high in protein is difficult to ensile, partly because more lactic acid is required to produce a given change in pH, and partly because herbage with a high concentration of N is often low in available carbohydrate

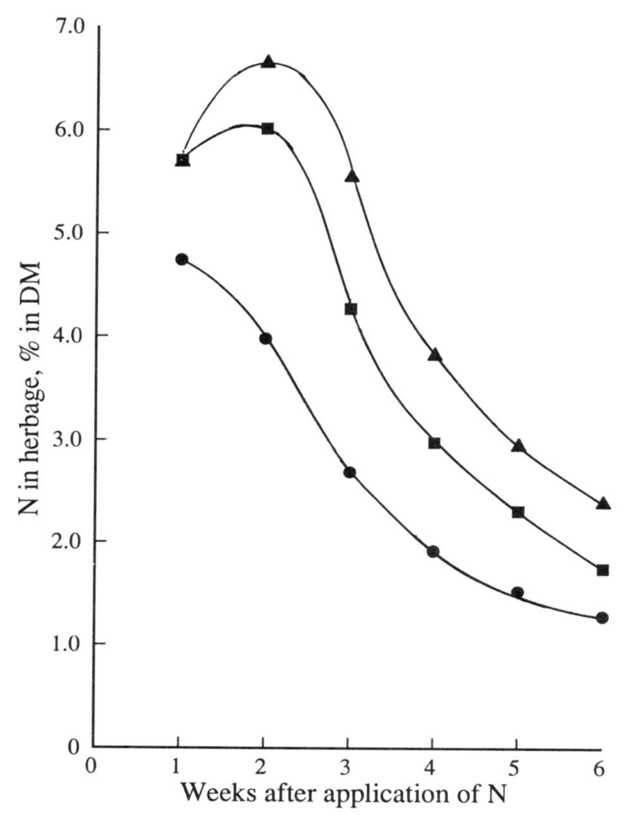

Fig. 14.4. The influence of three rates of calcium ammonium nitrate on the concentration of total N in Italian ryegrass sampled at weekly intervals for 6 weeks after the application (28 kg N ha^{-1}, ●; 84 kg N ha^{-1}, ■; 140 kg N ha^{-1}, ▲). (Data from Wilman, 1965.)

which is the main precursor of lactic acid. When herbage is ensiled, changes occur in the forms of N present and, in particular, the true protein fraction decreases and water-soluble forms, including ammonium, increase (Noller and Rhykerd, 1974). The true protein may be only 40–65% of the total N in silage (Thomas *et al.*, 1980).

The application of fertilizer N normally has little if any effect on the concentration of N in the herbage of white clover (Chestnutt, 1972; Cowling, 1961b; Simpson *et al.*, 1988). When grass and clover are grown in a mixed sward without fertilizer N, the concentration of N in the herbage is normally greater in the clover than in the grass component. For example, at two sites in Wales, average concentrations of N in perennial ryegrass were 3.15% and 3.00%, whereas the corresponding concentrations in white clover were 4.42% and 4.57% (Wilman and Hollington, 1985). At one site, the concentration of N in the ryegrass grown with clover was greater than the concentration in grass grown alone with 200 kg fertilizer N ha^{-1} $year^{-1}$ (3.15% cf. 2.62%) whereas, at the other site, this was not so (3.00% cf. 3.14%). In a study in southeast England, the herbage of white clover generally contained > 4% N while that of associated grasses (except *Agrostis*) contained < 3% N. The concentration in the clover generally exceeded that in grass alone receiving fertilizer N at a rate of 390 kg N ha^{-1} $year^{-1}$ (Cowling and Lochyer, 1967). When white clover was grown in association with various grass species, the concentration of N in the clover was found to be slightly higher with timothy or meadow fescue than with perennial ryegrass (Chestnutt, 1972).

Influence of Fertilizer N on Nitrate-N in Herbage

Nitrate is the only inorganic form of N that accumulates in herbage when the supply of N exceeds requirements for growth: ammonium does not accumulate. A small amount of inorganic N in plant tissue is essential for growth and, unless the supply of N is entirely as ammonium, some will be present as nitrate. The concentration of nitrate required in young leaves for the maximum yield of grass is about 0.05–0.15% nitrate-N on a dry-matter basis (see p. 25). Higher concentrations do not impair grass growth but, at concentrations greater than about 0.4%, the nitrate may be harmful to livestock consuming the herbage (see p. 282).

Nitrate accumulates in grass herbage when the rate of uptake by the roots exceeds the rate of conversion to organic N. Usually, the rate of uptake depends mainly on supply, whereas the rate at which nitrate is converted to organic N depends on carbohydrate and therefore on photosynthesis. Factors that restrict photosynthesis without restricting nitrate uptake to an equivalent extent, therefore tend to increase the concentration of nitrate. Low light intensity is the most important factor (Alberda, 1968; Bathurst and Mitchell, 1958; Deinum, 1966; Deinum and Sibma, 1980; Nowakowski and Cunningham, 1966). In contrast, low temperature tends to curtail nitrate uptake more than growth, and the concentration of nitrate in Italian ryegrass was lower at 11°C than at 28°C at each of six rates of nitrate application (Nowakowski *et al.*, 1965). Other experiments have also shown small increases in the concentration of nitrate with increasing temperature (Deinum and Sibma, 1980). Defoliation greatly reduces photosynthesis and, when defoliation is followed by the application of fertilizer N, high concentrations of nitrate are likely to occur temporarily in the regrowth (Deinum and Sibma, 1980).

The concentration of nitrate in grass herbage following the application of fertilizer N depends on the rate applied and on the time interval between the application and sampling. The data in Fig. 14.5 suggest that, with a single application of 100 kg N ha^{-1}, the maximum concentration of nitrate in the

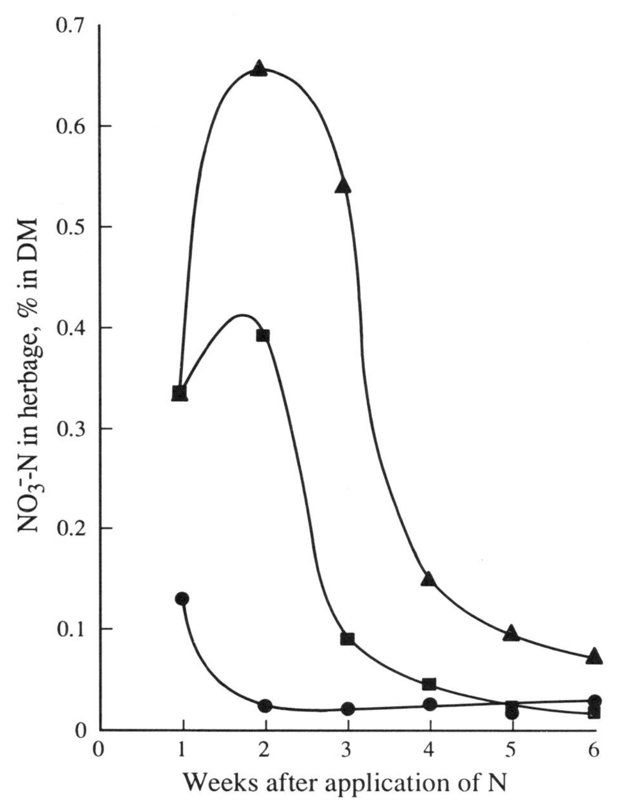

Fig. 14.5. The influence of three rates of calcium ammonium nitrate on the concentration of nitrate-N in Italian ryegrass sampled at weekly intervals for 6 weeks after application (28 kg N ha^{-1}, ●; 84 kg N ha^{-1}, ■; 140 kg N ha^{-1}, ▲). (Data from Wilman, 1965.)

herbage occurred about 2 weeks later. However, this time interval would clearly be influenced by weather conditions and the rate of growth of the grass. Exceptionally, nitrate may constitute more than half the total herbage N. For example, Italian ryegrass grown from seed for 9 weeks in soil supplemented with 500 mg nitrate-N kg^{-1} contained 2.8% nitrate-N in the herbage, representing 51% of the total herbage N (Nowakowski *et al.*, 1965).

The relationship between the rate of fertilizer application and the concentration of nitrate in the herbage is illustrated in Table 14.1. Similar results were obtained in another experiment with a ryegrass–timothy sward, in which there was little accumulation of nitrate when the rate of fertilizer N was less than about 300 kg N ha^{-1} $year^{-1}$ but, when the rate of fertilizer application was between 450 and 900 kg N ha^{-1} $year^{-1}$, more than 60% of the additional herbage N was present as nitrate (Reid, 1966). Also with cut grass swards in

Table 14.1. Total N and nitrate-N in ryegrass–timothy herbage with increasing rate of fertilizer N, applied as calcium nitrate in equal quantities in early spring and after each cut except the last; means of five cuts. (From Reid, 1966.)

	Fertilizer N (kg ha^{-1} year^{-1})								
	0	112	224	336	448	560	672	784	896
Total N, % in DM	2.27	2.40	2.59	2.85	3.30	3.20	3.54	3.54	3.74
Nitrate-N, % in DM	0.011	0.019	0.035	0.084	0.254	0.313	0.408	0.491	0.531
Nitrate-N as % of total N	0.48	0.79	1.35	2.95	7.70	9.78	11.5	13.9	14.2

the Netherlands, the application of 360 kg N ha^{-1} year^{-1} as calcium nitrate had little effect on the concentration of nitrate but, with the application of 600 kg N ha^{-1} year^{-1} (120 kg N per cut), 61% of the samples (cut at a yield of 2.0–2.5 t DM ha^{-1}) exceeded 0.17% nitrate-N (Prins, 1983). High concentrations of nitrate occurred most commonly in herbage harvested in the summer, occasionally in the autumn but never in the spring (Prins, 1983).

Average results from 12 sets of data for grass in the first half of the growing season indicated that the nitrate-N content increased gradually from 0.019% of dry matter with no fertilizer N to 0.027% with 200 kg N ha^{-1} year^{-1} as calcium nitrate, and to 0.060% with 500 kg N (Wilman and Wright, 1983). In the second half of the growing season there was a larger effect with nitrate-N increasing from 0.024% with no fertilizer N to 0.052% with 200 kg N ha^{-1} year^{-1} and to 0.164% with 500 kg N ha^{-1} year^{-1}. The proportion of total N present as nitrate is higher in stem than in green leaf material, though the concentration of total N is usually lower in the stem (Wilman and Wright, 1986).

Ammonium sulphate and calcium nitrate, at rates up to 440 kg N ha^{-1} year^{-1}, produced similar concentrations of nitrate in the herbage of Italian ryegrass (Kershaw, 1963), presumably as a result of nitrification of the ammonium before uptake. However, when applied with a nitrification inhibitor, high rates of ammonium N had little effect on the nitrate content of Italian ryegrass grown in pots (Nowakowski and Cunningham, 1966) or in the herbage of long-term grassland in the field (Nowakowski and Gasser, 1967).

Grass species differ in their tendency to accumulate nitrate, and concentrations tend to be higher in Italian ryegrass than in perennial ryegrass (Wilman and Wright, 1986).

Influence of Fertilizer N on Fibre and Carbohydrates in Herbage

Fertilizer N generally has little effect on the amount of fibre (cell wall material) in grass herbage (Morrison *et al.*, 1980; Ramage *et al.*, 1958; Reid *et al.*, 1966;

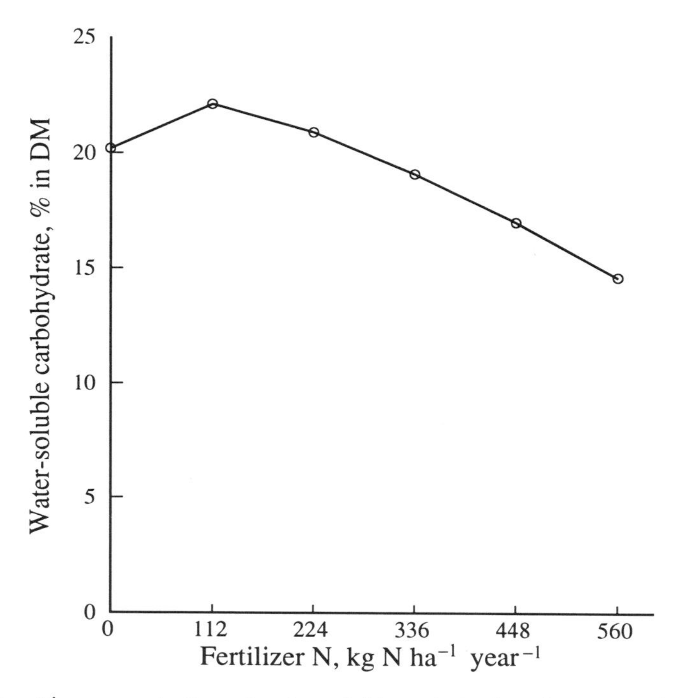

Fig. 14.6. The concentration of water-soluble carbohydrate in perennial ryegrass as influenced by the rate of fertilizer N; means of five harvests in each of 3 years. (Data from Reid and Strachan, 1974.)

Woelfel and Poulton, 1960) when the herbage is sampled at uniform time intervals. Similarly, there is usually little effect on the constituent materials of the fibre, cellulose (Bailey, 1973; Blaser, 1964; Bryant and Ulyatt, 1965) and lignin (Blaser, 1964; Reid *et al.*, 1967; Waite, 1970). However, high rates of fertilizer N (> *c.* 500 kg N ha^{-1} year^{-1}) may cause small reductions in cellulose and lignin (Reid *et al.*, 1967). Also, because fertilizer N enables grass to be harvested at a shorter growth interval, the herbage tends to have a lower content of fibre at the time of harvesting (Waite, 1970). However, with grass–clover swards, this effect may be offset by the tendency of fertilizer N to increase the proportion of grass and thus increase the overall content of fibre.

The content of water-soluble carbohydrate in grass herbage is often substantially reduced by the application of fertilizer N (Jones *et al.*, 1965; Reid and Strachan, 1974, Smith, 1973), especially at rates of application of more than about 200 kg N ha^{-1} year^{-1} (Fig. 14.6). In experiments carried out in Wales, the average concentration of water-soluble carbohydrate was reduced by 0.9 percentage units, from 17.6% in the dry matter, by the application of 100 kg fertilizer N ha^{-1} $year^{-1}$, by 1.5 percentage units per 100 kg N over the range 100–400 kg N, and by 1.1 percentage units from 400 to 500 kg N (Wilman and Wright, 1983). The change in water-soluble carbohydrate content is proportionately much greater in the polysaccharide fraction than in the monosaccharides and disaccharides (Bryant and Ulyatt, 1965; Nowakowski, 1962).

When grass is consumed directly by livestock, the reduction in soluble carbohydrate caused by fertilizer N has little effect on overall digestibility, though it may influence digestion processes in the rumen and the utilization of the dietary N (see p. 67). However, with grass intended for silage, a reduction in soluble carbohydrate may result in less lactic acid being formed and hence a poorer quality silage (Woolford, 1984).

Influence of Fertilizer N on Concentrations of Mineral Elements in Herbage

The application of fertilizer N is one of many factors that influence the concentrations of mineral elements in grassland herbage. Other factors include the grass:clover ratio, the species of grass, whether dicotyledenous species are present, the stage of growth at harvesting, recent weather conditions and soil characteristics. Grasses and clovers differ substantially in their concentration of some nutrient elements as indicated in Tables 14.2 and 14.3. The concentration of calcium, in particular, is higher in clovers than in grasses, and concentrations of Mg, Cu and Co are also generally higher in clovers. With Mg, the

Table 14.2. Typical concentrations of major nutrient elements in perennial ryegrass and white clover harvested at a leafy stage of growth, e.g. in spring or after 4 weeks regrowth (% in dry matter).

	N	P	K	S	Ca	Mg	Na
Perennial ryegrass	2.0–3.5	0.3–0.6	1.5–2.8	0.2–0.5	0.4–0.8	0.10–0.30	0.05–0.40
White clover	3.5–5.2	0.3–0.6	1.6–3.0	0.2–0.5	1.0–2.0	0.15–0.40	0.05–0.40

Table 14.3. Typical concentrations of some trace elements in perennial ryegrass and white clover harvested at a leafy stage of growth (mg kg^{-1} dry matter).

	Fe	Mn	Zn	Cu	Co	Se	I
Perennial ryegrass	50–200	30–300	15–60	3–15	0.03–0.20	0.02–0.04	0.10–0.30
White clover	100–300	30–200	20–40	5–12	0.06–0.40	0.04–0.05	0.10–0.30

difference is greatest in spring. With Cu and Co, the difference between grass and clover is proportionately greater when the supply of these elements is low. Dicotyledenous herb species, such as chicory and plantain, also tend to have higher concentrations of mineral elements than do the more widely cultivated grasses (Thomas and Thompson, 1948; Thomas *et al.*, 1952; Wilman and Derrick, 1994) and, as noted above, the proportion of these species tends to be reduced by fertilizer N.

With all-grass swards, the effects of fertilizer N on mineral composition depend partly on the supply of individual elements. When an element is in limited supply, the increased growth resulting from the fertilizer N tends to dilute its concentration in the herbage. However, when there is a plentiful supply of the element, the effect of dilution may be more than offset by enhanced uptake. An enhancement of uptake is likely to occur if root activity is increased (see p. 34), and/or if there is a synergistic effect of fertilizer N on the element in question. During the uptake of nutrient ions by plant roots, there is generally a synergism between ions of opposite charge and an antagonism between ions of the same charge. Thus, there is a synergism between nitrate and cations such as calcium and magnesium, and an antagonism between ammonium and these cations. However, with some combinations of ions of the same charge (e.g. NH_4^+ and Na^+; NO_3^- and $H_2PO_4^-$), the antagonism appears to be negligible.

The mineral composition of herbage may also be influenced when the fertilizer contains a nutrient element, additional to N, as an incidental constituent. Examples include calcium ammonium nitrate and ammonium sulphate. There may also be an effect if the fertilizer changes the pH of the soil. The major forms of fertilizer N currently in use, ammonium nitrate and urea, do not contain any nutrient elements other than N, but both tend to acidify the soil as a result of nitrification. With calcium ammonium nitrate, the $CaCO_3$ component of the fertilizer supplies Ca, and also tends to counteract the acidity caused by the ammonium nitrate. The effect of fertilizers on soil pH is particularly important for those elements (Fe, Mn, Co) whose uptake is most influenced by pH.

Phosphorus

The effects of fertilizer N on the concentration of P in herbage have been inconsistent: decreases in P concentration have been reported in some trials, presumably reflecting a dilution effect (Adams, 1973b; Heddle and Crooks, 1967; Lambert and Toussaint, 1978; Macleod, 1965; Whitehead, 1966a,b); increases have been reported in others (e.g. Kershaw and Banton, 1965) while in others there has been little or no effect (Reith *et al.*, 1964; Rinne *et al.*, 1974a; Sillanpaa and Rinne, 1975; Stewart and Holmes, 1953; Whitehead *et al.*, 1978; Hopkins *et al.*, 1994). In a study at 16 UK sites, fertilizer N tended to increase herbage P at the first harvest in May but to decrease it at a later harvest in August (Hopkins *et al.*, 1994). There is some evidence that the concentration of P in herbage is reduced by fertilizer N on soils that are low in P, and increased on soils that are well supplied (Walker *et al.*, 1952). Since the uptake of P is normally greatest at around pH 6.5 and declines under both acid and alkaline conditions, it is possible that fertilizer N might affect herbage P concentration through changing soil pH. However, no appreciable difference in herbage P concentration was found in comparisons of ammonium sulphate and calcium nitrate (Kershaw and Banton, 1965) or of sodium nitrate, ammonium nitrate, ammonium sulphate and urea (Reid *et al.*, 1966).

Potassium

The effect of fertilizer N on herbage K is influenced by the supply of K. Thus, fertilizer N has been found to decrease herbage K when the concentration of K in the control plots was below about 2%, but to increase herbage K, or to have no effect, when the concentration of K was above 2% (Heddle and Crooks, 1967; Kemp, 1960). However in some experiments, decreases have been reported when the concentration of K in the herbage from control plots was more than 2%, as when calcium ammonium nitrate at 390 kg N ha^{-1} $year^{-1}$ reduced herbage K concentration from 2.25% to 1.65% (McConaghy *et al.*, 1962), and ammonium nitrate at 560 kg N ha^{-1} caused a reduction from 3.2% to 2.7% K (Mortensen *et al.*, 1964). In other experiments, fertilizer N has had little effect on the concentration of K in the herbage of both ryegrass and ryegrass with white clover swards (Adams, 1973b; Hopkins *et al.*, 1994; Reid and Strachan, 1974; Whitehead *et al.*, 1983).

There was no significant difference between calcium nitrate and ammonium sulphate in their effect on the K content of Italian ryegrass (Kershaw and Banton, 1965) but anhydrous ammonia produced herbage with lower K contents than did the same amount of N applied as calcium ammonium nitrate (van Burg and van Brakel, 1965). There is also evidence that herbage K is lower when fertilizer N is applied as urea than as nitrate or ammonium salts (Reid, 1966).

Calcium

The herbage concentration of Ca is generally higher with grass–clover than with all-grass swards due to the higher concentration in clovers (see Table 14.3). With all-grass swards, the influence of fertilizer N on herbage Ca depends on the form of fertilizer applied. In general, nitrate has a synergistic effect on the uptake of Ca and tends to increase the concentration, as shown with Italian ryegrass grown in pots (Nielsen and Cunningham, 1964). However, ammonium, if taken up as such, has an antagonistic effect and tends to decrease the concentration of Ca. Decreases resulting from the application of anhydrous ammonia (van Burg and van Brakel, 1965) and from urea, in comparison either with no fertilizer N or with other forms of N, have been reported (Adams, 1984; Reid, 1966). When ammonium nitrate is applied, there may be a small decrease in herbage Ca (Hopkins *et al.*, 1994) or little or no effect (Reid *et al.*, 1967). The effect of calcium ammonium nitrate is sometimes to decrease the Ca content of herbage at moderate rates of application and increase it at high rates (Rinne *et al.*, 1974a; Whitehead, 1966a,b); and sometimes there may be a decrease in herbage Ca at the first harvest in the season and an increase subsequently (Whitehead *et al.*, 1978).

Magnesium

Although the concentration of Mg is usually higher in clover than in grass, the difference is smaller than with Ca (Table 14.3). The influence of fertilizer N on all-grass herbage depends on the form of the fertilizer, with nitrate having a synergistic effect on Mg, and ammonium an antagonistic effect. Thus, in pot experiments, nitrate but not ammonium increased the Mg content of Italian ryegrass (Gardner *et al.*, 1960; Mortensen *et al.*, 1964), and herbage concentrations of Mg were higher with calcium nitrate than with ammonium sulphate (Follett *et al.*, 1977; Kershaw and Banton, 1965). In several field experiments, the application of ammonium nitrate and calcium ammonium nitrate increased the concentration of Mg in the herbage of all-grass and grass–clover swards (Adams 1973b; Black and Richards, 1965; Hemingway, 1961; Hopkins *et al.*, 1994; Stewart and Holmes, 1953; Whitehead *et al.*, 1978). However, herbage from a sward treated with anhydrous ammonia contained about 15% less Mg than one treated with an equal quantity of N as calcium ammonium nitrate (van Burg and van Brakel, 1965). Urea was found to increase the Mg concentration of herbage from long-term pastures in Northern Ireland, presumably because much of the N was taken up as nitrate (Adams, 1984).

Sulphur

Fertilizer N often has little effect on the concentration of total S in herbage (Conroy, 1961; Rahman *et al.*, 1960; Whitehead *et al.*, 1978) which is usually in the range 0.2–0.5% S. However, in some instances a decrease, apparently due to a dilution effect, has been reported (e.g. Panditharatne *et al.*, 1986; Whitehead *et al.*, 1986a). A decrease in S due to a dilution effect is most likely to occur in the latter part of the growing season (Hopkins *et al.*, 1994), a time when supplies of S from the soil and the atmosphere may well be low. Even when fertilizer N has no effect on the concentration of total S, it does reduce the proportion of the herbage S present as sulphate (Whitehead *et al.*, 1983).

When fertilizer N is applied in the form of ammonium sulphate, there is an increase in herbage S concentration (Conroy, 1961; Jones, 1960).

Sodium

The effect of fertilizer N on the concentration of Na in herbage varies markedly with grass species. There is little effect with timothy and meadow fescue, which are inherently low in Na, but substantial increases in perennial ryegrass, Italian ryegrass and cocksfoot, which have the potential to take up large amounts (Rinne *et al.*, 1974b; Sillanpaa and Rinne, 1975; Whitehead, 1966a,b; Whitehead *et al.*, 1978). The application of ammonium nitrate at 300 kg N ha^{-1} $year^{-1}$ to ryegrass almost doubled the concentration of Na, from an average of 1.6% to 2.9% Na (Hopkins *et al.*, 1994).

With mixed grass–clover swards, the mean concentration of herbage Na from four sites was increased from 0.39% to 0.64% by the application of 195 kg N ha^{-1} $year^{-1}$ as calcium ammonium nitrate and to 0.69% by the application of 390 kg N ha^{-1} (Reith *et al.*, 1964). Increases in Na in long-term pastures following the application of urea have also been reported (Adams, 1984).

Chlorine

Reported effects of ammonium nitrate and calcium ammonium nitrate on herbage Cl have been inconsistent. In a pot experiment, ammonium nitrate had little effect on the Cl content of ryegrass, though it did cause a reduction when KCl was also added (Dijkshoorn, 1958). In field trials, ammonium nitrate applied at 377 kg N ha^{-1} $year^{-1}$ reduced the Cl content of perennial ryegrass herbage from an average of 0.53% to 0.46% and that of perennial ryegrass with white clover from 0.57% to 0.42% (Rahman *et al.*, 1960); and ammonium nitrate applied at 200 kg N ha^{-1} $year^{-1}$ had no consistent effect (Whitehead *et al.*, 1983). However, herbage Cl was increased by calcium ammonium nitrate when KCl was also applied (Heddle, 1967).

Iron

The uptake of Fe by plants increases with increasing acidity, and fertilizers that reduce soil pH, such as ammonium sulphate, therefore tend to increase the herbage concentration of Fe. On the other hand, fertilizers that increase soil pH, such as calcium ammonium nitrate, tend to reduce the concentration of Fe (e.g. Rinne *et al.*, 1974b). However, no increase was found during a 3-year period in which ammonium sulphate was applied at the rate of 310 kg N ha^{-1} $year^{-1}$ (Hemingway, 1962).

Manganese

The uptake of Mn by plants normally increases with increasing acidity, and there is therefore a tendency for herbage Mn to be increased by ammonium sulphate (Conroy, 1961; Hemingway, 1962) and decreased by calcium ammonium nitrate (Reith and Mitchell, 1964; Rinne *et al.*, 1974b). However, in a study at 16 UK sites, ammonium nitrate generally decreased herbage Mn (Hopkins *et al.*, 1994). Also, urea was found to cause a decrease in herbage Mn despite slightly increasing soil acidity (Adams, 1984).

Zinc

Low rates of application of fertilizer N have little effect on herbage Zn (Rinne *et al.*, 1974b; Sillanpaa and Rinne, 1975) but rates of 300 kg ha^{-1} $year^{-1}$ or more have increased herbage Zn in some investigations (e.g. Hopkins *et al.*, 1994; Miller *et al.*, 1964; Whitehead, 1966a,b) though not in all (e.g. Reid *et al.*, 1966).

Copper

Reported effects of fertilizer N on herbage Cu have been inconsistent. Increases have been found in some investigations (e.g. Havre and Dishington, 1962; Rinne *et al.*, 1974b; Stewart and Holmes, 1953), decreases in others (e.g. Kreil *et al.*, 1966) and variable effects in others (Hopkins *et al.*, 1994). It is possible that the effect differs between cut and grazed swards since, on cut grass swards, the application of 314 kg N ha^{-1} slightly reduced herbage Cu whereas, on grazed swards, the application of high rates of fertilizer N (> 650 kg N ha^{-1} $year^{-1}$) increased herbage Cu (Whitehead, 1966a,b).

Cobalt

Cobalt is similar to Fe and Mn in that its uptake by plants increases with soil acidity, and the concentration in herbage therefore tends to be reduced by calcium ammonium nitrate (Reith and Mitchell, 1964). Ammonium nitrate and urea have been found to have no consistent effect on herbage Co (Hopkins *et al.*, 1994; Reid *et al.*, 1966). The effect of fertilizer N on the concentration of Co in herbage also appears to be influenced by soil type. The application of fertilizer N as calcium ammonium nitrate increased herbage Co on a peat soil but reduced it on mineral soils (unless Co was added), mainly by eliminating clover (Reith *et al.*, 1983).

Iodine

Fertilizer N is most likely to influence herbage iodine through the dilution effect of increased growth, and reported reductions in herbage iodine (Alderman and Jones, 1967; Hartmans, 1974) were probably due to this effect.

Molybdenum

In contrast to Fe, Mn and Co, the uptake of Mo is reduced by soil acidity, and a reduction in herbage Mo content following the application of ammonium sulphate has been reported (Hemingway, 1962). Ammonium nitrate at a rate of 300 kg N ha^{-1} $year^{-1}$ also reduced herbage Mo (Hopkins *et al.*, 1994). Calcium ammonium nitrate has been reported to have no consistent effect (Reith and Mitchell, 1964) and to reduce herbage Mo (Rinne *et al.*, 1974b).

Influence of Fertilizer N on the Palatability of Herbage

Herbage palatability is usually assessed with the animals having a free choice between herbage from several treatments. When the choice has been between grass grown with various rates of fertilizer N, palatability has sometimes declined with increasing fertilizer N (e.g. McAllister and McConaghy, 1960; Reid and Jung, 1965) and has sometimes increased (e.g. Burton *et al.*, 1956; Reid *et al.*, 1967). Contrasting results were obtained in two trials, in both of which cocksfoot receiving fertilizer N at rates up to 448 kg ha^{-1} $year^{-1}$ was fed to sheep, but in one trial the herbage was made into hay while in the other it was grazed (Reid *et al.*, 1966). In the trial in which grass grown in spring was made into hay, its palatability to sheep (intake of individual hay sample as percentage of total *ad lib.* intake) declined with increasing rate of fertilizer N whereas, in the trial in which herbage grown in autumn was grazed, palatability was greater at the higher rates of fertilizer N. Two factors may have contributed to this difference between the trials. First, in the grazing trial the application of fertilizer increased the quantity of herbage available and this would have encouraged consumption from the high-N plots. Second, with the herbage grown in the spring and made into hay, the fertilizer N consistently decreased soluble carbohydrate but not cell-wall material whereas, with the grazed herbage in autumn, there was little effect on soluble carbohydrate but there was some reduction in cell-wall material.

The form in which fertilizer N is applied also influences herbage palatability. Ammonium sulphate appears to produce herbage of lower palatability than does calcium ammonium nitrate (Moloney and Murphy, 1963), ammonium nitrate (Reid *et al.*, 1966) or calcium nitrate (Widdowson *et al.*, 1966). In two comparisons of five forms of fertilizer N, each at 112 kg N ha^{-1}, the palatability of hay decreased in the sequence: sodium nitrate > ammonium nitrate > ammonium sulphate > urea = ammonium phosphate, whereas in a trial involving grazing during the autumn, palatability decreased in the order: ammonium nitrate = ammonium phosphate > sodium nitrate > ammonium sulphate > urea (Reid *et al.*, 1966). It is possible that the low palatability of grass receiving ammonium sulphate is due partly to its high content of S (Conroy, 1961).

Influence of Fertilizer N on the Animal Intake of Herbage

Differences in palatability are not necessarily reflected in differences in intake by livestock when only one type of herbage is available, and the rate of fertilizer N has been found to have little or no effect on intake in a number of experiments (Blaxter *et al.*, 1971; Cameron, 1966, 1967; Holmes and Lang 1963; Mahoney and Poulton, 1962; Reid and Jung, 1965; Reid *et al.*, 1966). However, in some instances, intake by sheep has been reduced by high rates of fertilizer N (Bryant and Ulyatt, 1965; Reid *et al.*, 1974) and, in other instances, intake has been increased (e.g. Odhuba *et al.*, 1965; Reid *et al.*, 1967). A reduced intake may sometimes result from higher moisture content (including superficial water) induced by the application of fertilizer N. Conversely, an increased intake may reflect the fact that more herbage is available, a factor that is most likely to be significant with intensive grazing.

Influence of Fertilizer N on the Digestibility of Herbage

In general, the application of fertilizer N has little effect on herbage digestibility, at least with all-grass swards (Dent and Aldrich, 1968; Raymond and Spedding, 1965; Reid *et al.*, 1966; van Vuuren *et al.*, 1991; Wilman *et al.*, 1976a) though small increases (Holmes and Lang, 1963; Reid and Jung, 1965) and small decreases (Cameron, 1967; McCarrick and Wilson, 1966) have been reported. There is likely to be a decrease in digestibility when the fertilizer reduces the proportion of clover in grass–clover swards, but an increase in digestibility is likely when the fertilizer allows grass herbage to be harvested at a younger stage (Waite, 1970).

Influence of Fertilizer N on the Metabolic Utilization of Herbage

The metabolic utilization of herbage by livestock may be impaired when the herbage has a high concentration of N and there is an imbalance between available carbohydrate and N, relative to the needs of the rumen microorganisms (Bryant and Ulyatt, 1965; Gill *et al.*, 1989; Raymond and Spedding, 1965). In general, the best balance between carbohydrate and N is obtained when the dietary concentration of N is in the range 2.2–2.8%, i.e. 14–18% crude protein (Gill *et al.*, 1989).

The balance between N and S in ruminant diets may also affect utilization, and the optimum ratio is usually between 10 : 1 and 15 : 1. The precise value depends on factors such as whether the animal is producing meat, milk or wool, and the relative availabilities of the N and S in the diet (Bird *et al.*, 1978).

The metabolic utilization of the diet may also be reduced if the concentration of nitrate is so high as to affect microbial fermentation in the rumen: *in vitro* studies with cocksfoot showed less fermentation in heavily fertilized herbage containing 0.72% nitrate-N than in unfertilized herbage (Perez and Story, 1960).

Influence of Fertilizer N on Metabolic Disorders of Livestock

In several investigations in which the effects of high rates of fertilizer N on animal performance have been specifically examined, there has been little if any harmful effect (e.g. Coombe and Hood, 1980; Hodgson and Spedding, 1966; Large and Spedding, 1966; Mudd, 1970; Phipps, 1975). Also, at the Nitrogen Experimental Farms in the Netherlands, on which rates of 250–550 kg N ha^{-1} $year^{-1}$ were used regularly, the health of the animals was as good as on the average farm (De Groot, 1963). However, metabolic disorders have sometimes been associated with the use of high rates of fertilizer N, and there is still some uncertainty about the effects of the continued use of high rates over a period of years.

Nitrate toxicity

Although the nitrate ion itself has a low toxicity to animals, much of it is converted to nitrite and thence to ammonia by rumen microorganisms. Sometimes, especially with ruminant diets that are low in available carbohydrate, the conversion of nitrite to ammonia occurs slowly and nitrite, which is toxic, is then absorbed into the bloodstream. Once in the blood, nitrite converts oxyhaemoglobin into methaemoglobin, and thus reduces the capacity of the blood to transport oxygen. In addition, the blood pressure is reduced, and heart beat and respiration rate are increased. Concentrations of methaemoglobin in the blood are greatest between 1 and 5 h after the animals begin consuming the high-nitrate feed, and the higher the nitrate content, the longer the time to attain the peak value for methaemoglobin (Geurink *et al.*, 1982).

In addition to the effects of nitrite on haemoglobin, it is possible that chronic nitrate toxicity may interfere with vitamin A and iodine metabolism, and that it may decrease milk yield and/or increase the incidence of abortion (Wright and Davison, 1964). Because nitrate interacts with other factors, there is considerable variation in the dietary concentration that is critical for the production of toxic effects. A concentration of about 0.4% nitrate-N has been regarded as potentially toxic to ruminants (Wright and Davison, 1964) but studies of lambs grazing herbage containing 0.3–0.6% nitrate-N throughout most of the season (Large and Spedding, 1966) and of sheep fed grass with contents up to 0.72% nitrate-N (Bryant and Ulyatt, 1965) showed no apparent harmful effects on health. With cows grazing ryegrass containing 0.19–0.76% nitrate-N resulting from successive applications of 103 kg N ha^{-1} during a season, there was no increase in blood methaemoglobin (Phipps, 1975). There is some evidence that the dietary concentration of nitrate that is toxic for cattle varies with the nature of the feed. A concentration above 0.17% nitrate-N is regarded as potentially toxic for conserved grass, whereas the critical concentration is thought to be about 0.34% for freshly mown grass fed indoors, and 0.45% or more for grazed grass (Geurink *et al.*, 1982). The differences are due to nitrate being released at a slower rate from fresh grass than from silage or hay, and to the longer period of time required by cattle that are grazing, rather than fed indoors, to consume a certain amount of feed.

Hypomagnesaemia

Fertilizer N has several effects which may influence the incidence of hypomagnesaemia. As noted on p. 277, the effect of fertilizer N on the concentration of Mg in grass herbage depends on the form in which the N is taken up: there is a tendency for Mg to be increased by nitrate and to be decreased by ammonium. With grass–clover swards, the decline in the proportion of clover tends to reduce herbage Mg. However, although the concentration of Mg in the diet is a major factor in hypomagnesaemia, differences in the availability of Mg may also be important. Availability is likely to be low when the herbage contains a large amount of non-protein organic N and only a small amount of available carbohydrate, a combination that increases the concentration of ammonia and raises the pH in the rumen. Decreases in Mg availability with increasing N concentration in the herbage have been reported (Grunes *et al.*, 1970) and a possible mechanism is the precipitation of $MgNH_4PO_4$ (Wilcox and Hoff, 1974).

In some field investigations, the incidence of hypomagnesaemia has been increased by fertilizer N, particularly when K has also been applied (Kemp, 1960; Metson *et al.*, 1966; van der Molen, 1964; Walshe and Conway, 1960) but, in other investigations, fertilizer N has been found to have no effect on blood Mg in animals consuming the herbage (Hodgson and Spedding, 1966;

Large and Spedding, 1966; L'Estrange *et al.*, 1967). Assessments of the effect of fertilizer N on the availability of herbage Mg to animals have also produced inconclusive results. In one investigation, there was no difference in the apparent availability to sheep of the Mg in herbage receiving either 45 or 450 kg N $\approx$ ha^{-1} year^{-1} (L'Estrange *et al.*, 1967) whereas, in another, the apparent availability of Mg was reduced by the application of 560 kg N ha^{-1} year^{-1} (Stillings *et al.*, 1964). In a study with dairy cows, the concentration of Mg in the blood serum decreased as the concentration of ammonia in the rumen increased at the beginning of the spring grazing period, and there was also a decrease in serum Mg following the introduction of ammonium salts into the rumen (Head and Rook, 1955). On the other hand, with sheep, no reduction in serum Mg was caused by feeding urea, despite a marked increase in rumen ammonia, indicating that ammonia in itself did not affect Mg availability (Wilson, 1963). The consumption of herbage with a high N content liberates N compounds other than ammonia into the rumen, and it is possible that these influence serum Mg. Compounds such as amino acids and amides, which have chelating properties, might be responsible for the decreased availability of Mg, and the stability of such Mg chelates would be greater at the high pH values induced by large amounts of ammonia. Some support for this type of mechanism was provided by the finding that when both EDTA and ammonium carbonate were fed to sheep, there was a decrease in the level of serum Mg (Ashton and Sinclair, 1965). An alternative possibility is that fertilizer N may reduce the availability of Mg by increasing the lipid content of herbage (Kemp *et al.*, 1966). A linear correlation was found between the N content of herbage and the content of higher fatty acids and, in a feeding experiment with dairy cows, it was found that adding fat to the ration tended to reduce the retention of dietary Mg: it is possible that excretion of Mg in the faeces was increased by the formation of insoluble Mg soaps (Kemp *et al.*, 1966).

15

Nitrogen Balances in Contrasting Grassland Systems

The Basis of N Balance Calculation

The concept of N balance can be applied to any ecosystem ranging in size from a few square millimetres to thousands of square kilometres. However, in the context of grassland, the concept of N balance is usually applied to an individual field, a microplot or lysimeter representing a field, or a whole farm system. Basically, the concept involves summating all the inputs of N to the system and all the outputs of N from the system: the difference between the sum of the inputs and the sum of the outputs must then equal the change in the total amount of N in the system. Clearly, the amounts must be expressed in the same units and must apply to the same time period (Floate, 1987) and, when calculated on a field scale, the items in the balance are usually expressed in kg N ha^{-1} $year^{-1}$. The calculation is a useful means of organizing quantitative information on N transformations, and a comparison of balances from different locations, or under different management systems, often illustrates how a change in one transformation is reflected in others. If a measured value for one particular input or output is missing from the balance, but the change in the total amount of N in the system is known, the preparation of the balance may enable the missing value to be estimated.

With grassland, the total amount of N in the system at any one time is the amount in the soil (including the roots and soil fauna) plus the much smaller amounts in above-ground plant material and in grazing animals. In general, the soil (including roots and soil fauna) contains at least 98% of the total N in the system at any one time. If the content of soil N is not at equilibrium, the increase or decrease in soil N during the time period being considered must equal the difference between inputs and outputs. In the context of grassland, including grass–clover swards, potential inputs of N from external sources are:

1. wet and dry deposition of N from the atmosphere;
2. symbiotic biological fixation of atmospheric N_2;
3. non-symbiotic biological fixation of atmospheric N_2;
4. application of fertilizer N;
5. application of organic manures or slurries.

Potential outputs of N are:

6. removal in cut grass herbage or in animal liveweight gain, milk or wool;
7. leaching, mainly of nitrate;
8. volatilization of ammonia;
9. volatilization of N_2, N_2O and NO through denitrification/nitrification.

The difference between the totals of these inputs and outputs can be corroborated by measuring the change in total soil N. However, a period of several years is necessary before the change in total soil N can be measured accurately because the annual change is much smaller than the total amount present (see p. 83).

While a balance for total soil N includes all inputs and outputs from the field in question, some important transformations of the N cycle within the soil are not included. In particular, there is no indication of the amount of N occurring in plant-available form (inorganic N) during the year, or the amount actually taken up by the plants. In order to assess these components of the system, it is necessary to assess inputs to, and outputs from, the pool of plant-available N. These inputs include, in addition to (1) to (4) above:

10. mineralization of N from plant residues, animal excreta and manures;
11. mineralization of N from the soil organic matter.

The outputs relevant to the pool of plant-available N are (6) to (9) above plus:

12. retention in unharvested portions of the plant material;
13. immobilization in soil organic matter;
14. excretion by animals.

The removal of N in the harvested herbage, plus retention in the unharvested portion of the plant material, plus the transfer of N in unharvested material to the soil organic matter, is equivalent to total plant uptake.

Total N Balances in Grassland Systems

It is extremely difficult to obtain accurate measurements for all the inputs and outputs of N that occur in a grassland sward during a year, and the difficulty is greater for the outputs than the inputs (Ryden, 1984). Accurate values for fertilizer input, and for output in herbage and/or animal products, can be obtained from straightforward measurement and analysis and, in much of Europe, the input from atmospheric deposition, at least via wet deposition, can be estimated with reasonable accuracy from existing data. However, more information is needed on regional variations in inputs from the atmosphere, especially through dry deposition. An estimate of symbiotic fixation by a grass–clover sward can be made if herbage N is measured in both the grass–clover sward and an all-grass sward grown without fertilizer N on the same soil (see p. 40). As noted in previous chapters, methods are available for measuring leaching, ammonia volatilization and denitrification in the field, but the methods are time-consuming. Consequently, the data that are currently available for the gaseous losses generally include some interpolation for periods without actual measurements. There is, therefore, a lack of precision in many of the published accounts of N balances, especially if there is no information on the change, if any, in soil N content (Allison, 1966; Kolenbrander, 1977; Legg and Meisinger, 1982).

As indicated in Chapter 5, if constant management is maintained over many years, field soils tend to gain or lose organic N until an equilibrium content is reached. The equilibrium soil N content is influenced by various factors involving climate, soil type, vegetation and management; and changes in any of the factors will change the equilibrium towards which the existing soil N content tends to shift. In some balance studies, it has been assumed that there was no change in soil organic N (i.e. that the amount was at a steady state), but this situation rarely occurs. Usually in grassland soils, the amount of soil organic N tends to increase from year to year, but sometimes there may be a decline, at least temporarily, due to liming or an increase in the intensity of management. Evidence of declining soil organic N under intensively grazed grassland has been reported from New Zealand, the decline being attributed to an increased proportion of the herbage being consumed by livestock, and hence more of the circulating N being concentrated into urine patches from which losses are large (Ball and Field, 1987).

N Balance in Cut Grassland Fertilized with N

A number of the N balance studies reported for cut grass swards have been based on lysimeters, and these have enabled leaching, and changes in soil N content, to be measured more accurately than would have been possible from field sampling. However, it is almost as difficult to measure gaseous losses from lysimeters as it is from the field and, in some investigations in which changes in soil N have been measured, the gaseous losses have been assessed by difference. An example of this type of N balance, obtained with timothy grown

Table 15.1. N balances for timothy grown at four rates of fertilizer N in lysimeters in New York State, USA; mean values for an 8-year period. (Data of Bizzell, cited by Allison, 1955.)

	kg N ha^{-1} year^{-1}			
Inputs				
Fertilizer ($NaNO_3$)	104	139	174	238
Rainfall	7	7	7	7
Total	111	146	181	245
Outputs				
Removal in cut grass	87	110	136	172
Leaching	2	1	2	3
Retention in soil	17	21	16	19
Total N accounted for	106	132	154	194
N not accounted for (including denitrification)	5	14	27	51

at four rates of fertilizer N in the USA, is shown in Table 15.1. Some lysimeter studies with nil or low rates of fertilizer have shown an unexpected gain of N, presumably due to the absence of measurement (or under-estimation) of non-symbiotic fixation or inputs from the atmosphere.

With heavily fertilized grassland, the fertilizer N represents the dominant input to the system and, in this situation, the use of ^{15}N-labelled fertilizer enables a balance to be obtained for that specific input. When used on cut grass in lysimeters, it is possible to obtain a rather precise balance for the labelled fertilizer, though the balance may differ from that in the field situation. In one example of this approach, it was possible to recover in plant material, soil and leachate, more than 97% of the N from calcium nitrate applied at a rate of 395 kg N ha^{-1} in the first year of a 3-year period of measurement (Table 15.2).

Although the N balances of cut swards often show only small losses due to ammonia volatilization, leaching and denitrification, it should be remembered that the cut sward does not constitute the whole production system and, that when the herbage is fed to animals, much of the herbage N may be lost subsequently from the slurry or manure derived from the animal excreta.

N Balance in Intensively Grazed Grassland Systems

Lysimeters cannot be used to obtain a N balance for grazed grassland, though they can be used to obtain a balance for a specific application of urine or dung. It is therefore necessary to assess the various inputs and outputs of N under a

Table 15.2. N balance for perennial ryegrass grown in lysimeters with ^{15}N-labelled calcium nitrate fertilizer, assessed during a 3-year period after the application of the fertilizer. (Data from Dowdell *et al.*, 1980.)

	kg N ha^{-1}
Input	
Fertilizer ($CaNO_3$)	395*
Outputs	
Removal in cut herbage	234
Recovery in stubble	2
Recovery in soil + roots	127
Leaching	21
Denitrification	< 0.1
Total N accounted for	384

*Applied in first year of the investigation.

grazing regime from sampling in the field. The preparation of a N balance is more difficult for grazed than for cut grassland but it has the advantage of taking a wider range of transformations into account. Removals of N in liveweight gain, milk or wool are much smaller than removals in cut herbage, and outputs of N through leaching, ammonia volatilization and denitrification are much larger from grazed than from cut grassland, especially when there is a substantial input of N through the application of fertilizer or N_2 fixation.

Nitrogen balances for a heavily fertilized grazed ryegrass sward, a heavily fertilized cut sward and a grazed ryegrass–clover sward at Hurley, in southeast England, Berkshire, are shown in Table 15.3. In this investigation of three contrasting swards on the same soil type, the sward management treatments were continued for 8 years, and the annual increase in the amount of N in soil organic matter was assessed from the change over this time period. With the cut ryegrass sward receiving 420 kg N ha^{-1} $year^{-1}$, losses were small and most of the fertilizer N was either removed in the herbage or accumulated in roots and soil organic matter. With the grazed ryegrass sward receiving 420 kg N ha^{-1} $year^{-1}$, only 7% of the N was recovered in the liveweight gain of the beef cattle, and losses through ammonia volatilization, denitrification and leaching were equivalent to about 67% of the fertilizer input. With the grazed grass–clover sward, N_2 fixation by the clover was much less than the fertilizer input to the all-grass sward, and this lower input was reflected in a somewhat lower stocking rate and in much smaller losses from the system (Table 15.3).

In grazed grassland plots receiving different rates of fertilizer N on a clay-loam soil in Northern Ireland, losses through ammonia volatilization, denitrification and leaching all increased with increasing rate of fertilizer N, and increased to a much greater extent than did liveweight gain (Table 15.4). Comparison of the plots receiving 400 kg fertilizer N ha^{-1} $year^{-1}$ with the grazed plots at Hurley receiving 420 kg fertilizer N ha^{-1} $year^{-1}$ illustrates

Table 15.3. N balances for three contrasting sward types on a freely-drained soil at Hurley, UK, with two of the swards grazed by beef cattle. (Based on data of Ryden and Garwood, cited by Whitehead *et al.*, 1986b.)

	kg N ha^{-1} year^{-1}		
	Cut ryegrass	Grazed ryegrass	Grazed ryegrass–white clover
Inputs			
Fertilizer (ammonium nitrate)	420	420	0
N_2 fixation by clover	0	0	160
Atmosphere*	30	30	30
Total inputs	450	450	190
Outputs			
Removal in cut grass or animal liveweight gain	300	29	23
Ammonia volatilization†	0	80	10
Denitrification†	20	40	4
Leaching‡	29	160	23
Total outputs	349	309	60
Retention in soil	90	110	110
Total N accounted for	439	419	170

*Based on data of Goulding (1990).
†Based on measurements made at intervals during the year.
‡Based on measurements of nitrate concentration in soil and underlying chalk at various depths.

Table 15.4. N balances for grazed grassland plots on a clay loam soil in Northern Ireland at five rates of application of fertilizer N; means for two years. (Data from Watson *et al.*, 1992.)

	kg N ha^{-1} year^{-1}				
Inputs					
Fertilizer (calcium ammonium nitrate)	100	200	300	400	500
Atmosphere (wet deposition only)	8	8	8	8	8
Total inputs	108	208	308	408	508
Outputs					
Animal liveweight gain	17	26	27	35	29
Ammonia volatilization	2	6	14	23	35
Denitrification	9	22	56	85	119
Leaching	18	25	42	72	63
Total outputs	46	79	139	215	246
N not accounted for (including retention in soil)	62	129	169	193	262

Table 15.5. N balances for three swards of drained and undrained old grassland grazed by beef cattle in Devon, UK. (Based on data of Ryden and Garwood, cited by Whitehead *et al.*, 1986b.)

	kg N ha^{-1} $year^{-1}$		
	Drained	Drained	Undrained
Inputs			
Fertilizer (ammonium nitrate)	200	400	400
Atmosphere	20	20	20
Total inputs	220	420	420
Outputs			
Animal liveweight gain	25	30	29
Ammonia volatilization	70	85	85
Denitrification	53	78	114
Leaching	57	188	62
Total outputs	205	381	290
N not accounted for (including retention in soil)	15	39	130

differences due to soil type and climate. In particular, ammonia volatilization and the leaching of nitrate appear to be less from the clay loam soil in Northern Ireland than from the loam soil over chalk at Hurley, while the amount of N lost through denitrification was greater in Northern Ireland. The large amount of N not accounted for (Table 15.4) may well have been due, at least partly, to the gaseous losses being under-estimated.

Nitrogen balances illustrating the effects of installing drainage in a previously poorly-drained grassland soil in Devon, UK, are shown in Table 15.5. At a uniform rate of fertilizer application of 400 kg N ha^{-1} $year^{-1}$ as ammonium nitrate, the drained and undrained plots appeared to be similar in the amount of ammonia volatilized, but leaching was much greater and denitrification much less from the drained than from the undrained sward. In comparing fertilizer inputs of 400 and 200 kg N ha^{-1} $year^{-1}$ on the drained sward, the difference in nitrate leaching was much greater than the differences in ammonia volatilization and denitrification (Table 15.5). The nitrogen balance for a typical dairy farm in the Netherlands (Table 15.6) includes an input of N from supplementary feed and demonstrates again that large amounts of N are lost from intensively grazed grassland.

In New Zealand, symbiotic fixation by clover is the major source of N for grassland and, on many soils, the amount of fixation is increased by the application of phosphate and lime. N balances for both unimproved pasture and pasture receiving phosphate and lime are shown in Table 15.7. Although the amounts of N in the various items of the balance are somewhat smaller

Table 15.6. N balance for a typical dairy farm on sandy soil in the Netherlands. (Data from Aarts *et al.*, 1992.)

	kg N ha^{-1} $year^{-1}$
Inputs	
Fertilizer	330
Atmosphere	46
Supplementary feed for cattle	178
Total inputs	554
Outputs	
Milk	68
Liveweight gain	16
Ammonia volatilization	127
Denitrification } Leaching }	286
Total outputs	497
Retention in soil	30
Total N accounted for	527

Table 15.7. N balances for two types of grassland (grazed by sheep) in New Zealand. (Data from Lambert *et al.*, 1982.)

	kg N ha^{-1} $year^{-1}$	
	Hill pasture, unimproved	Hill pasture with P and lime
Inputs		
Fertilizer	0	0
N_2 fixation by clover	17	135
Non-symbiotic fixation	13	13
Atmosphere	3	3
Total inputs	33	151
Outputs		
Animal liveweight gain or milk	4	9
Ammonia volatilization } Denitrification }	4*	11*
Leaching (including runoff)	10	14
Total outputs	18	34
N not accounted for (including retention in soil)	15	117

*Estimated as 5% of excreted N.

than those for the grazed grass–clover sward at Hurley (Table 15.3), the distribution of N amongst the various items is broadly similar.

Mass balance studies carried out to determine the fate of urine applied to grassland have generally shown substantial losses of N due to ammonia volatilization and leaching, often with evidence of an additional loss by denitrification (Ball *et al.*, 1979; Carran *et al.*, 1982; Whitehead and Bristow, 1990). The relative importance of the three loss processes has been shown to differ with soil and weather conditions, as outlined above.

N Balance in Extensively Grazed Grassland Systems

In rough grassland, including that in upland areas, the amounts of total soil N are generally similar to those in more productive lowland grassland, but the annual inputs and outputs of N are much lower. The rates of mineralization of soil N and of N_2 fixation are usually low, and little if any fertilizer N is applied. The concentration of nitrate in the leachate from rough grassland is therefore also low, sometimes even lower than that in rainfall. Animal production from rough grassland, in terms of liveweight gain or wool, is much less than that from intensive lowland grassland. Nevertheless, the productivity of rough grassland shows considerable variation from site to site, in response to the properties and depth of the soil, and climatic and management factors. In some upland areas (and also in some prairie grassland) herbage is burned at the end of the growing season and this increases mineralization, and results in an additional loss of N to the atmosphere as NO_x or ammonia. In some situations, the input of inorganic N from the atmosphere exceeds the amount released by mineralization. At a site at an altitude of about 750 m in northern England, deposition from the atmosphere amounted to 36 kg N ha^{-1} $year^{-1}$ while mineralization was estimated to provide only 3 kg N ha^{-1} $year^{-1}$ (Harrison *et al.*, 1994). Agricultural improvements, such as drainage, liming and increased density of stocking, tend to increase the rate of mineralization of soil N and therefore the amounts susceptible to plant uptake and loss from the soil (Edwards *et al.*, 1985).

The N balance of a typical upland grassland (unenclosed moorland) in the UK, shown in Table 15.8, indicates that the amounts of N in each of the inputs and outputs (except possibly the input from atmospheric deposition) are much smaller than those in productive lowland grassland (e.g. Table 15.3). Assuming that the total soil N in the upland grassland has accumulated since the last glaciation, it can be calculated that the average net input would have amounted to 1–3 kg N ha^{-1} $year^{-1}$ (Batey, 1982). Inputs from the atmosphere would have increased in recent years but the increases may well have been offset in some areas by greater losses due to changes in management and land use. Prairie grassland also usually has low inputs and low outputs of N as illustrated for

Table 15.8. N balance for a typical upland pasture in the UK. (Data from Batey, 1982.)

	kg N ha^{-1} year^{-1}
Inputs	
Fertilizer	0
N_2 fixation by clover	8
Atmosphere	14
Total inputs	22
Outputs	
Animal liveweight gain (sheep)	2
Ammonia volatilization	3
Denitrification	1
Leaching	5
Gaseous loss during burning	5
Total outputs	16
N not accounted for (including retention in soil)	6

Table 15.9. Estimated N balances for three lightly grazed prairie sites in the western USA. (Data from Woodmansee, 1979.)

	kg N ha^{-1} year^{-1}		
	Shortgrass prairie, Colorado	Mixed prairie, S. Dakota	Tallgrass prairie, Oklahoma
Inputs			
Fertilizer	0	0	0
N_2 fixation by legumes	< 0.5	< 0.5	< 0.5
Non-symbiotic fixation	< 0.5	< 0.5	< 0.5
Atmosphere	6.0	9.0	10.0
Total inputs	6.0	9.0	10.0
Outputs			
Animal liveweight gain	1.0	3.0	4.0
Ammonia volatilization	2.0	7.0	12.0
Denitrification	0	0	0
Leaching	0	0	< 1.0
Total outputs	3.0	10.0	16.0

three lightly grazed prairie grassland sites in the western USA in Table 15.9. In the upland areas of the UK, and also in the prairie grasslands of the USA, the total amounts of N being added to, and removed from, the soil each year are often only about 5% of those in heavily fertilized lowland grassland.

N Balance in Organically Farmed Grassland Systems

In organic farming, no use is made of synthetic fertilizers, pesticides or feed additives, and reliance is placed on legumes, crop rotations, manures and natural rock products to maintain the nutrient supply of the soil. Many organic farms base their cropping systems on ley/arable rotations in order to obtain the greatest benefits from symbiotic N_2 fixation and from the N in livestock excreta. Legumes, especially clovers, are used to supply N to the ley phase of the rotation and hence to livestock, and the subsequent ploughing of such leys also contributes N to the arable phase of the rotation. These sources of N, together with animal manures, enable many organic farms to achieve yields of crop and livestock products that are comparable to those achieved with a moderate use of fertilizer N. Studies on a sandy loam soil indicated that the optimum period of the ley, and of the arable phase, was no more than 3 years (Johnston *et al.*, 1994) but different periods might be optimal in different situations.

The management of organic grassland is generally similar to that of conventional grassland, though promoting the growth of clover is a more important feature. Lime and either basic slag or rock phosphate, which are acceptable in organic farming, are often used for this purpose. The application of lime and phosphate to grassland also promotes the decomposition and mineralization of dead organic matter by increasing the proportion of clover, and hence the average concentration of N in plant residues, and by encouraging the activity of earthworms and soil bacteria. On many organic livestock farms, particularly dairy farms, soil fertility is further enhanced through the provision of purchased high-protein feeding stuffs. Often organic farmers adopt a policy of importing either feeding stuffs or manures for a number of years in order to build up soil fertility, through increasing the content of soil organic matter and microbial activity, and subsequently adopting a comparatively low-input policy (Vine and Bateman, 1981). The N balance for the ley phase of a ley/arable rotation in an organic farming system in the UK would be similar to that of the grass–clover sward shown in Table 15.3, though the various outputs of N would be greater if additional livestock feeds were provided. When manures are imported on to the farm, it is difficult to assess a N balance because the amounts of manure and their concentrations of N are often uncertain.

The N Balance Concept as a Contribution to Modelling the N Cycle

The N balance concept has contributed to the development of a number of models of N cycling, including some that relate specifically to grassland. At

Table 15.10. Balance of plant-available N for well-drained and poorly-drained grassland grazed by beef cattle, on a clay loam soil in Devon, UK. (Based on model of Scholefield *et al.*, 1991.)

	kg N ha^{-1} $year^{-1}$	
	Well-drained	Poorly-drained
Inputs		
Fertilizer	400	400
Atmosphere	25	25
Mineralization of organic N	283	168
Return in urine	333	303
Total inputs	1041	896
Outputs		
Animal liveweight gain	39	36
Ammonia volatilization	52	47
Denitrification	79	141
Leaching	185	47
Retention in unharvested plant material	275	251
Excreted by animals	411	374
Total outputs	1041	896

least two have been produced with the aim of promoting a quantitative understanding of N transformations in grazed grassland systems. Both are based on assessing inputs to and outputs from the plant-available N pool. One is based on intensively-utilized dairy farming on grass–clover pastures in New Zealand (Field and Ball, 1982) and the other on grass grazed by beef cattle and now also dairy cattle in the UK (Scholefield *et al.*, 1991).

The UK model takes into account differences in soil texture and drainage status, sward age, previous cropping history and climatic conditions and then predicts the amounts of N in the various outputs of N at any selected rate of fertilizer N (Scholefield *et al.*, 1991). Differences in soil texture and drainage, sward age, cropping history and climatic conditions influence the amount of N mineralized from the soil organic matter, and therefore influence the amount of inorganic N susceptible to plant uptake, leaching, ammonia volatilization and denitrification. Differences in soil texture and drainage also influence the distribution of N between leaching and denitrification. However, the model currently makes no allowance for symbiotic N_2 fixation by clover. The N balances, as predicted by the model, for grass swards receiving 400 kg N ha^{-1} $year^{-1}$ on two sites, with contrasting rates of mineralization due to good or poor drainage, are shown in Table 15.10. Modelled balances of this type, if they accurately represent the situation in the field, predict the effects of different rates of fertilizer N on the output of livestock product and on the various losses of N, and are therefore useful in deciding the optimum rate of fertilizer N for a given situation.

A simulated mass-balance approach was adopted in assessing the amount of N_2 fixation required in a grass–clover sward to sustain a specific output of herbage for grazing, while maintaining soil N (Thomas, 1992). The calculation involved a number of assumptions: that 50% of the herbage N not consumed by grazing animals was re-mobilized from senescing leaves and stems to growing tissues, that 90% of the N consumed was excreted, that 70% of the excreted N was either lost by the volatilization of ammonia, leaching or denitrification, or was immobilized in the soil organic matter for longer than 5 years, and that a maximum of 40% of the N in dead plant material became plant-available during 5 years. With these assumptions, it was calculated that the production of 6000–15,000 kg of herbage DM ha^{-1} $year^{-1}$, containing an average of 3.5% N, from a grass–clover sward would require N_2 fixation of 120–350 kg N ha^{-1} $year^{-1}$. These figures contrast with another estimate that less than 50 kg N ha^{-1} $year^{-1}$ of N_2 fixation would be needed to meet the needs of a moderately productive grazed grass–clover pasture (Sheehy, 1989). However, the latter estimate appears to make inadequate allowance for the localization of the N returned in excreta by grazing animals, and is probably unrealistic in practice.

16 Conclusions

Transformations of N in Grassland Systems: Research Developments

Research during the past 25 years has led to a much better understanding of the quantities of N involved in the various transformations that occur in grassland, and how these quantities are influenced by management, and by weather and soil factors. The improvement in understanding is especially marked for the transformations that result in a loss of N from the soil, viz. ammonia volatilization, denitrification and leaching. Often, the advances have followed from the development of new or improved techniques that have enabled more accurate, more detailed or longer-term measurements to be made. One feature that has become increasingly clear is that large gaseous and leaching losses are liable to occur from the livestock excreta returned to the soil by grazing animals. Urine patches in particular contain extremely high but localized concentrations of plant-available N. These concentrations greatly exceed the uptake capacity of the grass, and urine patches are therefore especially susceptible to ammonia volatilization, denitrification and leaching. The greater the degree of aggregation of the plant-available N into urine patches, the lower the proportion of the total urine N that can be taken up by the grass and, as a result, the utilization of excreted N is generally lower with cattle than with sheep.

Another aspect of N cycling, which recent research has shown to be more important than previously thought, is the movement of N through unharvested plant material. In both cut and grazed swards, substantial amounts of roots, stubble and herbage material undergo decomposition *in situ*, and substantial amounts of N are thus transferred to the soil organic matter. Part of this N is then mineralized and part accumulates in the soil. It has been estimated that in a productive grazed sward the amount of N undergoing senescence and decomposition in the soil (including that from roots) is about 1.5 times the amount in the herbage consumed by sheep or cattle (Ball and Field, 1987). Differences in the proportion mineralized influence the amount of plant-available N, and the amount of N susceptible to leaching etc. The death and decomposition of plant material are also an important pathway for the transfer of N from clover to grass, a transfer that is crucial for the optimum growth and productivity of grass–clover swards.

A concept that has provided a better insight into the transformations of N in soil is that of mineralization/immobilization turnover. Work with ^{15}N-labelled fertilizers has shown that there is often a rapid and continuing interchange between the N in inorganic forms and the N in the soil organic matter, and that this turnover results from the activity of the soil micro-organisms. Much of the organic matter involved in the turnover appears to be microbial biomass. Consequently, when the size of the microbial biomass is increasing, there is a temporary increase in the immobilization of N whereas, at times when the microbial biomass is declining, there is an increase in mineralization. In general, the greater the rate of turnover, the larger the amount of plant-available N during any particular time period. The concept of mineralization/immobilization turnover also explains why the recovery of fertilizer N in the herbage is usually less when measured by ^{15}N-labelling than by analysis for total N (see p. 211). Now that this concept is widely appreciated, less importance than formerly is attached to the priming effect that fertilizer N may have on the mineralization of organic matter (see p. 120).

The factors involved in the response of grass to fertilizer N have also become more clearly understood during the past 25 years. Results from multi-centre trials, carried out in the UK during the 1970s and 1980s, show clearly that the response depends to a large extent on soil and weather factors at an individual site. The supply of available water during the growing season is a particularly important factor, and is determined mainly by the texture and depth of the soil, and by summer rainfall. However it is possible that, in other countries, soil or weather factors other than the available water supply might be relatively more important than in the UK.

Studies involving the storage and application of slurry have shown how variations in response can be explained, at least partly, in terms of the extent to which N is lost from the slurry both before and after application. Losses of N occur by ammonia volatilization during storage, and by a combination of ammonia volatilization, denitrification and leaching after application. There is now a considerable amount of information on how management and environmental factors influence the extent of these losses, and how measures such as separating the slurry into predominantly liquid and solid fractions, or the injection of the slurry into the soil, can increase the utilization of the N.

Despite continued work on the growth of grass with white clover swards and on N_2 fixation by clover, there has been limited progress in this area. The results of multi-centre trials with grass–clover swards have confirmed that large site-to-site and year-to-year differences occur in the vigour of clover growth, but have not provided explanations for the variation. Other studies have indicated that the death of clover stolon material during the winter is often important, and may contribute to site-to-site and year-to-year variation, but the factors influencing the extent of death during the winter require further investigation. Despite the introduction of new varieties of white clover, there has not yet been an appreciable increase in the use of grass–clover swards by farmers. However, with increasing pressures to promote 'sustainable' and low input agricultural systems, it is important that work aimed at improving the persistence of clover growth and N_2 fixation in grass–clover swards should continue.

During the past few years, increasing attention has been given to establishing N budgets for whole farm systems, and to developing management practices that provide the best overall efficiency of N from whatever source. Because many processes and complex interactions are involved in whole farm systems, intensive monitoring is necessary and progress is slow. Sometimes the approach is supplemented by modelling to indicate where emphasis should be placed in the field investigations. Modelling itself is becoming increasingly important in research on N transformations, in relation to both individual processes and whole systems. As stated in an overview of models of N in soil–plant systems (Myers, 1987), there is a need to acquire data sets that enable models to be tested and therefore a need for the greater integration of experimentation and modelling. A major difficulty with models that include the process of mineralization is that the organic soil N pool is very much larger than the other pools in the soil–plant system; and therefore a small error in predicting the rate of mineralization introduces a large error in the amount of potentially plant-available N (Stevenson, 1986). The rates of the other microbiological processes in the soil are also difficult to predict though progress is being made (De Willigen and Neeteson, 1985).

Impacts of Grassland Management on the Wider Environment

Areas of natural grassland are generally stable in terms of their vegetation, soil characteristics and nutrient transformations. However, areas of grassland that are intensively managed inevitably undergo changes that may have both beneficial and harmful effects on the wider environment. The beneficial and harmful effects may occur at the same time. As an example of a beneficial effect, a period of several years or more under grass usually increases the content of soil organic matter compared with continued arable cropping, and this not only enhances the long-term stability and fertility of the soil but also has the environmental benefit of tending to reduce the amount of carbon dioxide in the atmosphere. Grassland, because it provides a cover of vegetation throughout the year and is not tilled, also provides a much greater resistance to soil erosion, both by wind and water and, for this reason, is important in sustaining the long-term productivity of many agricultural regions.

Adverse environmental effects arise mainly when grassland is managed intensively for livestock production, and most are due to the large amounts of plant-available N in circulation. In addition to the possibility of nitrate being leached into groundwater and thus into supplies of drinking water, both nitrate and organic N may contaminate surface water, and encourage the growth of weeds or algae in rivers, lakes and estuaries, if the supply of phosphate is not limiting. There are also undesirable environmental effects from the volatilization of ammonia and the gaseous products of nitrification and denitrification. Volatilized ammonia, after its deposition on the land surface as either gaseous ammonia or ammonium aerosols, contributes to soil acidification (see p. 178). It therefore accentuates the effects of acid rain and, partly by increasing the leaching of nutrient cations, tends to change the nutrient balance of semi-natural ecosystems. Although the effects are greatest with semi-natural vegetation, the increase in soil acidification may limit crop yields unless lime is applied and this, of course, incurs additional costs. Where the deposition of ammoniacal N is substantial, or continues for many years, the change in nutrient balance may impair the growth of trees, and it may also reduce the botanical diversity of areas such as heathlands. In particular, those plant species that prefer a plentiful supply of N compete more strongly with species that tolerate a poor supply of N. The effects are greatest where the soil has low cation exchange and buffering capacities.

With regard to the gaseous products of nitrification and denitrification, the impact on the wider environment depends mainly on the amounts of N_2O and NO produced. Gaseous N_2 has no environmental impact, but N_2O and NO have adverse effects. N_2O in the troposphere acts as a greenhouse gas and therefore tends to increase global warming while, in the stratosphere, it damages the ozone layer thus allowing more UV radiation to reach the earth's surface. The amounts of N_2O volatilized from nitrification/denitrification have been shown to increase with increasing use of fertilizer N, though soil and weather factors are of major importance. Although NO is quantitatively less important than N_2O, it does contribute to soil acidification and the greenhouse effect, and also to the destruction of ozone in the stratosphere.

Grassland N Now and in the Future

Both crop and livestock products are currently in surplus in western Europe and North America, and these surpluses reflect the changes that occurred in agriculture during the period from 1940 onwards. Basically, there was a progressive intensification of management inputs during the period to about 1980. Although this intensification overcame the food shortages of the 1940s and early 1950s, it was accompanied, particularly in later years, by environmental damage in terms of pollution and a decline in the visual attractiveness of the countryside and its capacity to support wildlife. In the context of grassland, pollution has been associated mainly with dairy farming and only to a small extent with beef cattle and sheep. More recently, as food surpluses increased during the 1980s, governments introduced measures to minimize the over-production of the main agricultural products including both milk and meat. For example, the EC countries and Canada introduced a quota system for milk, and the USA reduced milk support prices. Subsequently, in both the EC and USA, farmers have been paid to stop milk production or to set-aside land (van der Meer and Wedin, 1989). While these measures have resulted in smaller land areas being devoted to dairy farming, they have had little effect on the intensity of management, and their environmental impact has therefore been limited. However concerns about pollution, especially of drinking water by nitrate, have led to other measures, introduced particularly by the EC, that are designed mainly to limit nitrate leaching.

The environmental problems resulting from the intensification of livestock production have been, and still are, most severe in the Netherlands. Many dairy farmers there use high rates of fertilizer N, together with substantial amounts of cereal-based and high-protein feeds. In comparison with the smaller number of farms that use grass–clover swards and limited amounts of supplementary feeds, the intensive farms produce slightly more than twice the output of milk and beef per ha, but release about six times as much N to the wider environment (van der Meer and Wedin, 1989). Losses through leaching, ammonia volatilization and denitrification together are estimated to increase from about 75 kg N ha^{-1} $year^{-1}$ with moderately intensive farms to about 450 kg N ha^{-1} $year^{-1}$ with the highly intensive farms. In addition to measures that comply with the EC Directive to restrict nitrate in water supplies, it is the policy of the Netherlands government to reduce the emission of ammonia by at least 50%, and if possible by 70%, by the year 2000, compared with the emission in 1980 ('t Mannetje, 1994). To achieve this reduction, slurry stores constructed after 1987 have to be covered and, after 1995, slurry applied to grassland has to be applied only during the growing season and by low emission techniques such as injection. Another procedure that has been introduced in the Netherlands, but not to any appreciable extent in other countries, is the transport of manures from areas with a surplus to areas with a shortage.

In the UK, the EC Directive on nitrate in water supplies led to the government introducing restrictions on agricultural management in parts of the country from the autumn of 1990. Nineteen areas, which overlie groundwater sources or contain springs or boreholes, were designated as either Nitrate Sensitive Areas or Nitrate Advisory Areas. Within the ten Nitrate Sensitive Areas, the restrictions are of two types. One type, comprising the Basic Scheme, is designed to reduce nitrate leaching without changing the existing pattern of agriculture while the other, the Premium Scheme, involves more fundamental changes. Measures in the Basic Scheme that relate to grassland include:

1. limiting the rates of application of fertilizers and manures to no more than the economic optimum, and restricting the timing of applications, and
2. limiting the ploughing of grassland to leys in rotation with arable cropping.

In the Premium Scheme there are four options, all involving the conversion of arable land to grassland, and with restrictions on the application of fertilizer N and on grazing. Whenever possible, farmers in Nitrate Sensitive Areas are persuaded to comply with the restrictions voluntarily, with financial compensations, but the legislation allows compulsory measures to be imposed if and when necessary. Within the nine Nitrate Advisory Areas, farmers are actively advised on good farming practices that will reduce the amount of nitrate leached, but no compensation is paid for complying with the advice (Ball, 1993).

In 1991, the EC issued a further Nitrate Directive under which member countries are responsible for designating 'nitrate vulnerable zones', and enforcing within them various measures specified as 'good agricultural practice'. Nitrate-vulnerable zones are defined by the Directive as areas of land which drain, either directly or indirectly into either: (i) surface freshwater or groundwater that is intended for drinking water, and which would contain more than 11.3 mg nitrate-N l^{-1} if no action were taken; or (ii) natural freshwater, estuarine or coastal water that is either already eutrophic or may become so if no action is taken (see Ball, 1993; Heathwaite *et al.*, 1993). The measures to be enforced in these nitrate vulnerable zones include:

1. restrictions on the timing, rate and other conditions for the application of fertilizers and manures;
2. closed periods for slurry spreading and minimum storage capacities for slurry; and
3. limits on the overall quantity of N per ha which may be supplied as animal manures, including that deposited while the animals are grazing (the basic annual rate is 170 kg N ha^{-1}, equivalent to two dairy cows per hectare, though for the first 4 years after the introduction of the Directive, a less stringent limit of 210 kg N ha^{-1} is acceptable) (Smith and Chambers, 1993).

However, despite the restrictions being imposed on fertilizer use and other aspects of grassland management in parts of many countries, it is likely that grass will continue to be by far the most important food for ruminant livestock in temperate areas. Most grassland in the future is likely to be long-term and on land that is unsuitable for arable cropping and/or is on specialized livestock farms. In Europe at least, only a small proportion of the grassland is currently in rotation with arable crops, and it is likely that the high cost of establishing livestock units will preclude a major increase in ley farming (Peel and Lloveras, 1994). Although some grassland that has declined in productivity will continue to be ploughed and re-seeded, this practice is likely to become increasingly unacceptable due to the large effect on the leaching of nitrate (see p. 147). Reliance on biological fixation by clover as a source of N may expand to some extent, as a means of curtailing the energy costs of fertilizer manufacture and distribution, but any shift towards grass–clover swards is likely to be slow. The finding that there is little difference in nitrate leaching between grass–clover swards and fertilized all-grass swards, if comparisons are made at equal stocking rates, together with the problem of bloat, are factors that discourage the shift towards grass–clover (Peel and Lloveras, 1994).

Nevertheless, concerns about pollution are likely to become greater in the future and there may well be increased public and political pressure to reduce the amounts of ammonia, nitrous oxide and nitric oxide being volatilized to the atmosphere. Such pressure might result in restrictions on the rate of application of fertilizer N being imposed over wider areas than required to prevent the nitrate pollution of groundwater. In order to minimize both leaching and gaseous losses of N, it is certain that more attention will be given to adjusting the amounts of fertilizer applied to individual fields to particular needs and circumstances (see p. 306). Although it may be argued that both food surpluses and environmental damage could be curtailed by enforcing reduced fertilizer inputs, such a policy incurs the political dilemma of choosing some combination of lower incomes for farmers and/or higher food prices for the public. It also has the problem of enforcement. Any proposal to restrict rates of fertilizer application without financial compensation is likely to be opposed by farmers. In an assessment of the economic impact of restricting fertilizer N, for individual dairy farmers in the UK, it was concluded that the rate of fertilizer N was the dominant factor in stocking density and therefore in income per hectare and in profit (Gardner, 1986). Various possible options that farmers could take to offset an imposed restriction were considered, options that included a greater reliance on clover, and the use of forage crops (such as lucerne and maize) that can be grown with less fertilizer N. However, the success of all of these options would be more dependent than fertilized grass on soil and weather factors and would require greater management skills and a much higher degree of risk (Gardner, 1986). Extensification is not an attractive option financially for most European farmers under present conditions, and there is a need for multi-disciplinary studies on how to reconcile the conflicting objectives of ensuring adequate incomes for farmers and achieving the various environmental improvements outlined above (van der Meer, 1994). Work in the Netherlands and elsewhere has suggested a number of management practices, in addition to reductions in the rate of application of fertilizer N, that will reduce pollution by N. These include a reduction in the concentration of N in urine by feeding maize silage and/or sugar beet pulp in addition to grass or grass–clover herbage, the injection and/or acidification of slurry, and the covering of slurry storage tanks ('t Mannetje, 1994). For these various practices to be adopted widely, there would be a need to refine the management skills of many farmers.

Clearly, an improvement in the current efficiency of utilization of N could be achieved by the more widespread adoption of established practices but there is a need for further research on a number of topics as outlined below.

Unresolved Questions: the Need for Research

Despite the advances in research in grassland nitrogen made during the past 25 years, there are a number of areas in which the understanding of processes is inadequate and in which it is important, from scientific, economic and environmental points of view, to make further progress. Such progress would enable the various sources of plant-available N to be utilized more effectively in livestock production. In conjunction with the aim of improving utilization, it is important to assess more accurately the various losses of N in different situations so that procedures for minimizing pollution can be refined.

More effective use of soil organic N

If there is to be a shift towards more sustainable and less polluting agricultural systems, the processes involved in the mineralization of N from organic matter, and hence the availability of soil N, will assume greater importance. Mineralization does not occur in isolation but as part of mineralization-immobilization turnover and, if the rate of turnover could be increased, more inorganic N would be available for plant uptake. There is increasing evidence that differences between soils in mineralization–immobilization turnover are related to differences in microbiological activity, and to differences in the size of the 'active' fraction of the soil organic matter. However, methods to determine the size of the potentially mineralizable 'active' fraction (other than by time-consuming incubation methods) are not well developed and further work is required (Paul, 1988). Although the amount of potentially mineralizable N is influenced by soil characteristics and cropping history, attempts to relate these factors to the actual amounts of N taken up into the herbage of unfertilized grassland have had only limited success. One problem in relation to field results is that the amount of N actually mineralized is dependent on weather factors, and consequently the seasonal pattern of mineralization may not match the pattern of requirement for grass growth. In addition to the problem of measuring the 'active' fraction of the soil organic matter, there is a need to understand more fully how its size in relation to the total organic matter, and its effectiveness in releasing N, can be increased. The importance of livestock excreta in providing 'active' organic matter, and the role of the soil fauna in promoting mineralization-immobilization turnover, also need further investigation. On the basis of present evidence, it appears to be impossible to maintain a highly productive grazed sward (with a stable soil N content) by the recycling of N within the soil–plant–animal system: substantial inputs of N combined with substantial losses appear to be necessary. A major problem in this context is the uneven distribution of the N returned to the soil in excreta. Consequently, any factors that promote a more even distribution (e.g. the production of a more dilute urine by livestock) will tend to improve the utilization of the N in grazed swards.

More effective use of fertilizer N

Fertilizer N should be regarded as a supplement to naturally-occurring sources of plant-available N and, to achieve maximum effectiveness, applications should be adjusted to match the supply from the soil and the atmosphere, and the expected requirement of the grass for herbage production. One possible approach would be to make adjustments on the basis of analysis of the soil for inorganic N, a procedure which, with recent developments in analytical equipment, it is possible to carry out rapidly in the field (Scholefield and Titchen, 1995). As an alternative, assessing the adequacy of soil N by analysis of the herbage for nitrate-N or total N might provide better reproducibility, but would have the disadvantage of requiring more expensive equipment or rapid access to laboratory facilities. A third possible approach would be to develop a computer program to provide recommendations for fertilizer N in specific situations. A recommendation could be made in response to inputs of data on such factors as soil type and depth, climatic region, recent weather conditions, past management of the field, proposed grazing period or cutting date, and any recent or planned use of slurry or manure. This type of approach might be combined with the use of soil or plant analysis to corroborate the recommendations, as suggested by studies in the Netherlands (Oenema *et al.*, 1992).

More effective use of slurry N

Any improvement in the effectiveness of N from slurries and manures has the potential to reduce the need for fertilizer N. A major requirement to improve effectiveness is to minimize the loss of N before application. More work is needed on factors influencing ammonia volatilization from livestock housings and on techniques such as the injection of slurry, and its separation into predominantly liquid and solid fractions, that minimize volatilization after application.

There is evidence that N applied as liquid manure or slurry has less effect than fertilizer N in suppressing N_2 fixation by clover, but the reasons for this difference, if it occurs consistently, require elucidation. An answer to this question might enable fertilizer N to be made more compatible with N_2 fixation.

In addition to providing immediately plant-available N, slurries and manures have the potential advantage of adding to the 'active' fraction of the soil organic matter and providing mineralizable N in subsequent years. However, the quantitative importance of these residual effects is not well established and more information is needed. The effects would be understood more clearly if methods for characterizing the forms of N in slurries and manures were improved.

More effective use of symbiotic fixation by legumes

There would be considerable environmental benefits from making greater use of symbiotic fixation, mainly through reducing the energy costs of manufacturing, transporting and applying fertilizer N. However, for symbiotic fixation to be more attractive to grassland farmers, there is a need to develop a legume that will grow and fix N_2 effectively for a number of years when grown in mixtures with grass, and that will not induce bloat in livestock. Such a legume would probably, but not necessarily, be a variety of white clover. Important characteristics, in comparison with current varieties of white clover, would be reduced susceptibility to die-back in cold wet conditions, earlier growth in spring and reduced susceptibility to pests and diseases. These characteristics, together with a reduction in bloat-inducing constituents, are included in the aims of current breeding programmes. Meanwhile, in order to maximize the use of existing varieties, it is important to define more precisely the soil and management factors that govern the effectiveness and persistence of legumes.

Increases in soil organic matter under grass

In general, grassland provides long-term environmental benefits through increasing the amount of organic matter in the soil, and thus slowly improving nutrient supply and soil physical properties. The increase in soil organic matter also acts as a mechanism for the removal of carbon dioxide from the atmosphere. The increase occurs most rapidly where the initial soil organic matter content is low, e.g. in areas of previously arable cultivation and areas where land has been restored. In order to obtain most environmental benefit from such areas, it is important to know to what extent the accumulation of soil organic matter can be enhanced by factors such as the presence of clovers in the sward, by the application of small amounts of fertilizer N and/or by the presence of grazing animals.

However, in productive grassland, these long-term benefits may conflict with the aim, outlined above, of increasing the rate of mineralization–immobilization turnover in order to increase the amount of plant-available N. It is thought that during the turnover process, part of the biomass N is converted to stable humified forms of organic matter, part is re-incorporated into new biomass and part is taken up by plant roots. There is a need for a better understanding of the factors influencing the balance between these three pathways.

Nitrification

Nitrification has an important influence on the amount of nitrate in soils and therefore on the amount of N susceptible to losses by leaching and denitrification. It inevitably influences the amounts of N_2O and NO released to the atmosphere during nitrification itself, and as a result of denitrification and chemo-denitrification. However, there is lack of information on the extent to which N_2O and NO are lost from soils during nitrification and the importance of temperate grassland in particular as a source of N_2O requires investigation (IPCC, 1992).

Where substantial amounts of ammoniacal N are applied to soils either as fertilizers or slurries or via the atmosphere, nitrification is a major factor in the development of soil acidity. In some situations, there are advantages in nitrification occurring as rapidly as possible (e.g. to minimize the amount of ammonia volatilized from urine) but, in other situations, there are advantages in nitrification occurring slowly (e.g. when fertilizer N is applied as ammonium early in the season). Soils differ considerably in their rates of nitrification, and in the extent to which the rate of nitrification changes with time in response to the addition of ammonium-N (see p. 115). However, the factors responsible for these differences are not well understood and require more investigation.

Leaching of nitrate and other forms of N

Although the processes involved in nitrate leaching are well established, there is a need for more information on the influence of macropores, such as those created by earthworms, on the amounts of nitrate leached, both in the short-term and on an annual basis. This topic is particularly important in areas where the winter rainfall is insufficient to ensure complete leaching or denitrification of nitrate.

The leaching of N in forms other than nitrate has generally been ignored but, in some situations, it may be appreciable. For example, urea may be leached when high concentrations occur in the surface soil, following the addition of fertilizer urea or urine, and there is a need to examine the transformations of urea in the subsoil. Similarly, the extent to which organic forms of N in slurry can be leached, and the fate of such N, require further investigation.

Ammonia volatilization

Although much is known about the factors influencing the volatilization of ammonia from urea applied to soil in urine or fertilizer, there is a need for more information on ammonia volatilization from other sources (Lee and Dollard, 1994). These include living plant material, and plant material that is senescing and undergoing decomposition. Although the proportions of the total plant N that volatilize as ammonia appear to be small, a slight but continuous volatilization might well contribute appreciably to the total emission of ammonia to the atmosphere. More information is also needed on ammonia volatilization from the excreta of housed animals, and on the measures that might be adopted to minimize this loss of N.

In view of the evidence that the deposition of ammoniacal N via the atmosphere may have a significant impact on both agricultural and semi-natural ecosystems, there is a need to assess these effects in more detail. One problem in assessing the environmental impact is the lack of information on the average distance over which volatilized ammonia may travel before it is absorbed by vegetation or is deposited on the land or sea surface by wet or dry deposition.

Denitrification and associated processes

As indicated in Chapter 9, estimates of the amounts of N lost from the soil by denitrification and associated reactions are less precise than estimates of nitrate leaching and ammonia volatilization. This is largely due to inadequacies in the methods available for measurement, and the further development of methods suitable for longer term measurements would be valuable.

In relation to the factors that influence denitrification, there is a need to examine in more detail the effects of adding organic sources of N such as slurry, and the extent to which the soluble organic carbon in these sources promotes the occurrence of denitrification. The possibility that denitrification might be promoted in the subsoil as well as in the topsoil requires investigation.

Modification of livestock diets to minimize excretion of N

Much of the N lost from grassland and ruminant livestock production systems through leaching, ammonia volatilization and denitrification, is derived from animal excreta and reflects an inefficient utilization of dietary N. Some loss of N from animal excreta is inevitable but, if the concentration of N in the excreta could be reduced, there would be a reduction both in the amount excreted and in the proportion of the excreted N susceptible to loss. Different classes of livestock have different requirements in terms of the optimum dietary N concentration and the optimum ratio of energy to protein (though values for these optima may vary somewhat with the nature of the diet). However, as the dietary N concentration increases above the optimum, the proportion of the N utilized for metabolism decreases, and the proportion that is excreted increases. By assessing the optimal concentrations more accurately for different classes of livestock, and in relation to different types of diet, it should be possible to modify the composition of diets to reduce the concentration of N in the excreta without impairing milk production or liveweight gain. The modification might be achieved, for example, by adding materials such as maize silage or whole crop cereals to predominantly young grass or grass-silage diets, or by adjusting the rate of fertilizer N to grassland intended for specific grazing periods.

Systems research

Most of the research described in this book, apart from that in Chapter 15, has concentrated on individual transformations of N. However, in the context of maximizing utilization and minimizing environmental impact, it is important to make simultaneous assessments of the effects of soil, weather and management factors on a range of transformations within whole-farm systems. In this way it can be shown whether or not a reduction in a particular loss of N from one part of the system is accompanied by an increase elsewhere in the system. The cost, in terms of livestock production, of management practices designed to reduce losses of N can also be assessed. This systems approach to developing optimum management strategies for N is currently being adopted, for example, in both the Netherlands (Aarts *et al.*, 1992; 't Mannetje, 1994) and the UK (Laws and Pain, 1994) and should be continued. The quantification of N transformations in various grassland systems will also add to the data available to modellers and, together with the results of research on individual transformations, will enable decisions on future needs in both research and agricultural practice to be more firmly based.

References

Aarts, H.F.M., Biewinga, E.E. and Van Keulen, H. (1992) Dairy farming systems based on efficient nutrient management. *Netherlands Journal of Agricultural Science* 40, 285–299.

Adams, J.A. and Pattison, J.M. (1985) Nitrate leaching losses under a legume-based crop rotation in central Canterbury, New Zealand. *New Zealand Journal of Agricultural Research* 28, 101–107.

Adams, S.N. (1973a) The response of pastures in Northern Ireland to N, P and K fertilizers and to animal slurries. I. Effects on dry-matter yield. *Journal of Agricultural Science, Cambridge* 81, 411–417.

Adams, S.N. (1973b) The response of pastures in Northern Ireland to N, P and K fertilizers and to animal slurries. II. Effects on mineral composition. *Journal of Agricultural Science, Cambridge* 81, 419–428.

Adams, S.N. (1984) Some effects of lime, nitrogen, and soluble and insoluble phosphate on the yield and mineral composition of established grassland. *Journal of Agricultural Science, Cambridge* 102, 219–226.

Adams, S.N. (1986) The interaction between liming and forms of nitrogen fertiliser on established grassland. *Journal of Agricultural Science, Cambridge* 106, 509–513.

Adams, T.McM. (1980) Macro organic matter content of some Northern Ireland soils. *Record of Agricultural Research (Northern Ireland)* 29, 1–11.

Adams, T.McM. and Laughlin, R.J. (1981) The effects of agronomy on the carbon and nitrogen contained in the soil biomass. *Journal of Agricultural Science, Cambridge* 97, 319–327.

Adams, W.E. and Twerksy, M. (1960) Effect of soil fertility on winter killing of Coastal Bermudagrass. *Agronomy Journal* 52, 325–326.

Addiscott, T.M., Whitmore, A.P. and Powlson, D.S. (1991) *Farming, Fertilizers and the Nitrate Problem.* CAB International, Wallingford.

Afzal, M. and Adams, W.A. (1992) Heterogeneity of soil mineral nitrogen in pasture grazing by cattle. *Soil Science Society of America Journal* 56, 1160–1166.

Agricultural Research Council (1980) *The Nutrient Requirements of Ruminant Livestock*. Commonwealth Agricultural Bureaux, Farnham Royal.

Agricultural Research Council (1984) *The Nutrient Requirements of Livestock*. Commonwealth Agricultural Bureaux, Farnham Royal.

Alberda, T. (1968) Some aspects of nitrogen in plants, more specially in grass. *Stikstof* (English edn) No. 12, 97–103.

Alderman, G. and Cottrill, B.R. (1993) *Energy and Protein Requirements of Ruminants*. CAB International, Wallingford.

Alderman, G. and Jones, D.I.H. (1967) The iodine content of pastures. *Journal of the Science of Food and Agriculture* 18, 197–199.

Alexander, A. and Helm, H-U. (1990) Ureaform as a slow release fertilizer: a review. *Zeitschrift für Pflanzenernährung und Bodenkunde* 153, 249–255.

Allen, A.G., Harrison, R.M. and Wake, M.T. (1988) A meso-scale study of the behaviour of atmospheric ammonia and ammonium. *Atmospheric Environment* 22, 1347–1353.

Allison, F.E. (1955) The enigma of soil nitrogen balance sheets. *Advances in Agronomy* 7, 213–250.

Allison, F.E. (1966) The fate of nitrogen applied to soils. *Advances in Agronomy* 18, 219–258.

Allos, H.F. and Bartholomew, W.V. (1959) Replacement of symbiotic fixation by available nitrogen. *Soil Science* 87, 61–66.

Amberger, A. (1991) Ammonia emissions during and after land spreading of slurry. In: Nielsen, V.C., Voorburg, J.H. and L'Hermite, P. (eds) *Odour and Ammonia Emissions from Livestock Farming*. Elsevier, London, pp. 126–131.

Anderson, J.M. (1988) The role of soil fauna in agricultural systems. In: Wilson, J.R. (ed.) *Advances in Nitrogen Cycling in Agricultural Ecosystems*. CAB International, Wallingford, pp. 89–112.

Anderson, J.M., Leonard, M.A., Ineson, P. and Huish, S. (1985) Faunal biomass: a key component of a general model of nitrogen mineralization. *Soil Biology & Biochemistry* 17, 735–737.

Anderson, R.V., Coleman, D.C. and Cole, C.V. (1981) Effects of saprotrophic grazing on net mineralization. In: Clark, F.E. and Rosswall, T. (eds) *Terrestrial Nitrogen Cycles*. Ecological Bulletin (Stockholm) 33, 201–216.

Anslow, R.C. and Green, J.O. (1967) The seasonal growth of pasture grasses. *Journal of Agricultural Science, Cambridge* 68, 109–122.

Anslow, R.C. and Robinson, R.P.J. (1986) Nitrogen fertiliser practices and grass production patterns: yields of dry matter and of nitrogen in ryegrass herbage. *Research and Development in Agriculture* 3, 7–12.

Antonovics, J., Lovett, J. and Bradshaw, A.D. (1967) The evolution of adaptation to nutritional factors in populations of herbage plants. In: International Atomic Energy Agency. *Isotopes in Plant Nutrition and Physiology*. IAEA, Vienna, pp. 549–567.

ApSimon, H.M., Kruse, M. and Bell, J.N.B. (1987) Ammonia emissions and their role in acid deposition. *Atmospheric Environment* 21, 1939–1946.

ApSimon, H.M. and Kruse-Plass, M. (1991) The role of ammonia as an atmospheric pollutant. In: Nielsen, V.C., Voorburg, J.H. and L'Hermite, P. (eds) *Odour and Ammonia Emissions from Livestock Farming*. Elsevier, London, pp. 17–20.

Archer, J. (1988) *Crop Nutrition and Fertilizer Use*, 2nd edn. Farming Press, Ipswich.

Armitage, E.R. and Templeman, W.G. (1964) Response of grassland to nitrogenous fertilizer in the west of England. *Journal of the British Grassland Society* 19, 291–297.

Arnott, R.A. (1984) An analysis of the uninterrupted growth of white clover swards receiving either biologically fixed nitrogen or nitrate in solution. *Grass and Forage Science* 39, 305–310.

Ashton, W.M. and Sinclair, K.B. (1965) A study of the possible role of chelation in the occurrence of hypomagnesaemia in sheep. *Journal of the British Grassland Society* 20, 118–122.

Ashworth, J. and Flint, R.C. (1974) Delayed nitrification and controlled recovery of aqueous ammonia injected under grass. *Journal of Agricultural Science, Cambridge* 83, 327–333.

Ashworth, J., Penny, A., Widdowson, F.V. and Briggs, G.G. (1980) The effects of injecting Nitrapyrin (N Serve), carbon disulphide or trithiocarbonates with aqueous ammonia on yield and % N of grass. *Journal of the Science of Food and Agriculture* 31, 229–237.

Asman, W.A.H. (1992) Ammonic emission in Europe: updated emission and emission variations. *Report No. 228471008 National Institute of Public Health and Environmental Protection*, Bilthoven, Netherlands.

Atkinson, D. (1985) Spatial and temporal aspects of root distribution as indicated by the use of a root observation laboratory. In: Fitter, A.H. (ed.) *Ecological Interactions in Soil.* Blackwell Scientific Publications, Oxford, pp. 43–65.

Auda, H., Blaser, R.E. and Brown, R.H. (1966) Tillering and carbohydrate contents of orchardgrass as influenced by environmental factors. *Crop Science* 6, 139–143.

Aulakh, M.S., Doran, J.W. and Mosier, A.R. (1992) Soil denitrification – significance, measurement and effects of management. *Advances in Soil Science* 18, 1–57.

Bailey, R.W. (1973) Structural carbohydrates. In: Butler, G.W. and Bailey, R.W. (eds) *Chemistry and Biochemistry of Herbage, Volume I.* Academic Press, London, pp. 157–211.

Baker, H.K. and David, G.L. (1963) Winter damage to grass. *Agriculture, London* 70, 380–382.

Baker, M.J. and Williams, W.M. (eds) (1987) *White Clover.* CAB International, Wallingford.

Baker, R.D. (1986) Efficient use of nitrogen fertilisers. In: Cooper, J.P. and Raymond, W.F. (eds) *Grassland Manuring.* Occasional Symposium No. 20. British Grassland Society, Hurley, pp. 15–27.

Baker, R.D., Doyle, C.J. and Lidgate, H. (1991) Grass production. In: Thomas, C., Reeve, A. and Fisher, G.E.J. (eds) *Milk from Grass* (2nd edn). ICI, Billingham, pp. 1–26.

Ball, P.R. and Crush, J.R. (1985) Prospects for increasing symbiotic nitrogen fixation in temperate grasslands. *Proceedings 15th International Grassland Congress, Kyoto,* pp. 56–62.

Ball, P.R. and Field, T.R.O. (1985) Productivity and economics of legume-based pastures and grass swards receiving fertiliser nitrogen in New Zealand. In: Barnes, R.F., Ball, P.R., Brougham, R.W., Marten, G.C. and Minson, D.J. (eds) *Forage Legumes for Energy-Efficient Animal Production.* USDA, Washington, pp. 47–55.

Ball, P.R. and Field, T.R.O. (1987) Nitrogen cycling in intensively-managed grasslands: A New Zealand viewpoint. In: Bacon, P.E., Evans, J., Storrier, R.R. and Taylor,

A.C. (eds) *Nitrogen Cycling in Temperate Agricultural Systems.* Australian Society of Soil Science, Riverina, pp. 91–112.

Ball, P.R., Keeney, D.R., Theobald, P.W. and Nes, P. (1979) Nitrogen balance in urine-affected areas of a New Zealand pasture. *Agronomy Journal* 71, 309–314.

Ball, P.R. and Ryden, J.C. (1984) Nitrogen relationships in intensively managed temperate grasslands. *Plant and Soil* 76, 23–33.

Ball, S. (1993) Nitrate and the Law. In: Burt, T.P., Heathwaite, A.L. and Trudgill, S.T. (eds) *Nitrate: Processes, Patterns and Management.* John Wiley and Sons, Chichester, pp. 387–400.

Barbarika, A., Sikora, L.J. and Colacicco, D. (1985) Factors affecting the mineralization of nitrogen in sewage sludge applied to soils. *Soil Science Society of America Journal* 49, 1403–1406.

Barber, S.A. (1984) Nutrient balance and nitrogen use. In: Hauck, R.D. (ed.) *Nitrogen in Crop Production.* American Society of Agronomy, Madison, Wisconsin, pp. 87–96.

Barley, K.P. (1955) The determination of macro-organic matter in soils. *Agronomy Journal* 47, 145–147.

Barraclough, D. (1988) Studying mineralization/immobilization turnover in field experiments: the use of nitrogen-15 and simple mathematical analyses. In: Jenkinson, D.S. and Smith, K.A. (eds) *Nitrogen Efficiency in Agricultural Soils.* Elsevier, London, pp. 409–417.

Barraclough, D. (1991) The use of mean pool abundances to interpret ^{15}N tracer experiments. 1. Theory. *Plant and Soil* 131, 89–96.

Barraclough, D., Geens, E.L. and Maggs, J.M. (1984) Fate of fertilizer nitrogen applied to grassland. II. Nitrogen-15 leaching results. *Journal of Soil Science* 35, 191–199.

Barraclough, D., Geens, E.L., Davies, G.P. and Maggs, J.M. (1985) Fate of fertilizer nitrogen. III. The use of single and double labelled ^{15}N ammonium nitrate to study nitrogen uptake by ryegrass. *Journal of Soil Science* 36, 593–603.

Barraclough, D., Hyden, M.J. and Davies, G.P. (1983) Fate of fertilizer nitrogen applied to grassland. I. Field leaching results. *Journal of Soil Science* 34, 483–497.

Barraclough, D., Jarvis, S.C., Davies, G.P. and Williams, J. (1992) The relation between fertilizer nitrogen applications and nitrate leaching from grazed grassland. *Soil Use and Management* 8, 51–56.

Barrow, N.J. (1961) Mineralization of nitrogen and sulphur from sheep faeces. *Australian Journal of Agricultural Research* 12, 644–650.

Barrow, N.J. and Lambourne, L.J. (1962) Partition of excreted nitrogen, sulphur and phosphorus between the faeces and urine of sheep being fed pasture. *Australian Journal of Agricultural Research* 13, 461–471.

Bartholomew, P.W. and Chestnutt, D.M.B. (1977) The effect of a wide range of fertilizer nitrogen application rates and defoliation intervals on the dry-matter production, seasonal response to nitrogen, persistence and aspects of chemical composition of perennial ryegrass (*Lolium perenne* cv. S24). *Journal of Agricultural Science, Cambridge* 88, 711–721.

Bastiman, B. and van Dijk, J.P.F. (1975) Muck breakdown and pasture rejection in an intensive paddock system for dairy cows. *Experimental Husbandry* 28, 7–17.

Batey, T. (1982) Nitrogen cycling in upland pastures of the UK. *Philosophical Transactions of the Royal Society, London, B* 296, 551–556.

Bathurst, N.O. and Mitchell, K.J. (1958) The effect of light and temperature on the chemical composition of pasture plants. *New Zealand Journal of Agricultural Research* 1, 540–552.

Beauchamp, E.G., Gale, C. and Yeomans, J.C. (1980) Organic matter availability for denitrification in soils of different textures and drainage classes. *Communications in Soil Science and Plant Analysis* 11, 1221–1233.

Beauchamp, E.G., Kidd, G.E. and Thurtell, G. (1978) Ammonia volatilization from sewage sludge applied in the field. *Journal of Environmental Quality* 7, 141–146.

Beauchamp, E.G., Kidd, G.E. and Thurtell, G. (1982) Ammonia volatilization from liquid dairy cattle manure in the field. *Canadian Journal of Soil Science* 62, 11–19.

Beauchamp, E.G. and Paul, J.W. (1989) A simple model to predict manure N availability to crops in the field. In: Hansen, J.A. and Henriksen, K. (eds) *Nitrogen in Organic Wastes Applied to Soils.* Academic Press, London, pp. 140–149.

Beauchamp, E.G., Trevors, J.T. and Paul, J.W. (1989) Carbon sources for bacterial denitrification. *Advances in Soil Science* 10, 113–142.

Beever, D.E., Ulyatt, M.J., Thomson, D.J. Cammell, S.B., Austin, A.R. and Spooner, M.C. (1980) Nutrient supply from fresh grass and clover fed to housed cattle. *Proceedings of the Nutrition Society* 39, 66A.

Belford, R.K. (1979) Collection and evaluation of large soil monoliths for soil and crop studies. *Journal of Soil Science* 30, 363–373.

Benke, M., Kornher, A. and Taube, F. (1992) Nitrate leaching from cut and grazed swards influenced by nitrogen fertilization. *Proceedings 14th General Meeting, European Grassland Federation,* Lahti, Finland, pp. 184–188.

Berendse, F., Oomes, M.J.M., Altena, H.J. and de Vesser, W. (1994) A comparative study of nitrogen flows in two similar meadows affected by different groundwater levels. *Journal of Applied Ecology* 31, 40–48.

Bergström, L. (1986) Distribution and temporal changes of mineral nitrogen in soils supporting annual and perennial crops. *Swedish Journal of Agricultural Research* 16, 105–112.

Bergström, L. (1987) Nitrate leaching and drainage from annual and perennial crops in tile-drained plots and lysimeters. *Journal of Environmental Quality* 16, 11–18.

Besnard, C. (1980) Balance and evolution of nitrogen compounds during the treatment of slurry. In: Gasser, J.K.R. (ed.) *Effluents from Livestock.* Applied Science Publishers, London, pp. 496–507.

Betteridge, K., Andrewes, W.G.K., and Sedcole, J.R. (1986) Intake and excretion of nitrogen, potassium and phosphorus by grazing steers. *Journal of Agricultural Science, Cambridge* 106, 393–404.

Bijay-Singh, Ryden, J.C. and Whitehead, D.C. (1988) Some relationships between denitrification potential and fractions of organic carbon in air-dried and field-moist soils. *Soil Biology and Biochemistry* 20, 737–741.

Bijay-Singh, Ryden, J.C. and Whitehead, D.C. (1989) Denitrification potential and actual rates of denitrification in soils under long-term grassland and arable cropping. *Soil Biology and Biochemistry* 21, 897–901.

Binnie, R.C. and Chestnutt, D.M.B. (1994) Effect of continuous stocking by sheep at four sward heights on herbage mass, herbage quality and tissue turnover in grass/clover and nitrogen-fertilized grass swards. *Grass and Forage Science* 49, 192–202.

Binnie, R.C., Harrington, F.J. and Murdoch, J.C. (1974) The effect of cutting height and nitrogen level on the yield, *in vitro* digestibility and chemical composition of Italian ryegrass swards. *Journal of the British Grassland Society* 29, 57–62.

Bird, P.R., Watson, M.J. and Cayley, J.W.D. (1978) The sulphur requirements of ruminants and the sulphur and nitrogen status of perennial pastures in Southern Australia. *Proceedings of Symposium, Sulphur in Forages.* An Foras Taluntais, Dublin, pp. 228–250.

Black, A.L. and Wight, J.R. (1979) Range fertilization: nitrogen and phosphorus uptake and recovery over time. *Journal of Range Management* 32, 349–353.

Black, A.S., Sherlock, R.R., Cameron, K.C., Smith, N.P. and Goh, K.M. (1985a) Comparison of three field methods for measuring ammonia volatilization from urea granules broadcast on to pasture. *Journal of Soil Science* 36, 271–280.

Black, A.S., Sherlock, R.R. and Smith, N.P. (1987) Effect of timing of simulated rainfall on ammonia volatilization from urea, applied to soil of varying moisture content. *Journal of Soil Science* 38, 679–687.

Black, A.S., Sherlock, R.R., Smith, N.P., Cameron, K.C. and Goh, K.M. (1984) Effect of previous urine application on ammonia volatilisation from 3 nitrogen fertilisers. *New Zealand Journal of Agricultural Research* 27, 413–416.

Black, A.S., Sherlock, R.R., Smith, N.P., Cameron, K.C. and Goh, K.M. (1985b) Effects of form of nitrogen, season, and urea application rate on ammonia volatilisation from pastures. *New Zealand Journal of Agricultural Research* 28, 469–474.

Black, W.J.M. and Richards, R.I.W.A. (1965) Grassland fertilizer practice and hypomagnesaemia. *Journal of the British Grassland Society* 20, 110–117.

Blackman, G.E. (1936) The influence of temperature and available nitrogen supply on the growth of pasture in spring. *Journal of Agricultural Science, Cambridge* 36, 620–647.

Blackman, G.E. and Templeman, W.G. (1938) The interaction of light intensity and nitrogen supply in the growth and metabolism of grasses and clover (*Trifolium repens*). 2. The influence of light intensity and nitrogen supply on the leaf production of frequently defoliated plants. *Annals of Botany* 2, 765–791.

Blackmer, A.M., Bremner, J.M. and Schmidt, E.L. (1980) Production of nitrous oxide by ammonia-oxidising chemoautotrophic micro-organisms in soil. *Applied and Environmental Microbiology* 40, 1060–1066.

Bland, B.F. (1967) The effect of cutting frequency and root segregation on the yield from perennial ryegrass–white clover associations. *Journal of Agricultural Science., Cambridge* 69, 391–397.

Blaser, R.E. (1964) Symposium on forage utilization: effects of fertility levels and stage of maturity on forage nutritive values. *Journal of Animal Science* 23, 246–253.

Blaser, R.E. and Brady, N.C. (1950) Nutrient competition in plant associations. *Agronomy Journal* 42, 128–135.

Blaxter, K.L., Wainman, F.W., Dewey, P.J.S., Davidson, J., Denerley, H. and Gunn, J.B. (1971) The effects of nitrogenous fertilizer on the nutritive value of artificially dried grass. *Journal of Agricultural Science, Cambridge* 76, 307–319.

Bolan, N.S., Hedley, M.J. and White, R.E. (1991) Processes of soil acidification during nitrogen cycling with emphasis on legume based pastures. *Plant and Soil* 134, 53–63.

Boller, B.C. and Nösberger, J. (1987) Symbiotically fixed nitrogen from field-grown white and red clover mixed with ryegrasses at low levels of ^{15}N fertilization. *Plant and Soil* 104, 219–226.

Boot, R.G.A. and Mensink, M. (1990) Size and morphology of root systems of perennial grasses from contrasting habitats as affected by nitrogen supply. *Plant and Soil* 129, 291–299.

Boote, K.J. (1976) Root:shoot relationships. *Proceedings Soil and Crop Science Society of Florida* 36, 15–23.

Bouldin, D.R., Klausner, S.D. and Reid, W.S. (1984) Use of nitrogen from manure. In: Hauck, R.D. (ed.) *Nitrogen in Crop Production*. American Society of Agronomy, Madison, Wisconsin, pp. 221–245.

Bouwman, A.F. (1989) Exchange of greenhouse gases between terrestrial ecosystems and the atmosphere. In: Bouwman, A.F. (ed.) *Soils and the Greenhouse Effect*. John Wiley and Sons, Chichester, pp. 61–127.

Bouwmeester, R.J.B., Vlek, P.L.G. and Stumpe, J.M. (1985) Effect of environmental factors on ammonia volatilization from a urea-fertilized soil. *Soil Science Society of America Journal* 49, 376–381.

Bowman, P.J., White, D.H., Cayley, J.W.D. and Bird, P.R. (1982) Predicting rates of pasture growth, senescence and decomposition. *Proceedings of the Australian Society of Animal Production* 14, 36–37.

Bremner, J.M. (1965) Nitrogen availability indexes. In: Black, C.A. (ed.) *Methods of Soil Analysis. Part 2. Chemical and Microbiological Properties*. American Society of Agronomy. Madison, Wisconsin, pp. 1324–1345.

Bremner, J.M. and McCarty, G.W. (1993) Inhibition of nitrification in soil by allelochemicals derived from plants and plant residues. In: Bollag, J-M. and Stotzky, G. (eds.) *Soil Biochemistry* Vol. 8. Marcel Dekker, New York, pp. 181–218.

Bremner, J.M. and Mulvaney, R.L. (1978) Urease activity in soils. In: Burns, R.G. (ed.) *Soil Enzymes*. Academic Press, London, pp. 149–196.

Bristow, A.W. and Jarvis, S.C. (1991) Effects of grazing and nitrogen fertiliser on the soil microbial biomass under permanent pasture. *Journal of the Science of Food and Agriculture* 54, 9–21.

Bristow, A.W., Ryden, J.C. and Whitehead, D.C. (1987) The fate at several time intervals of ^{15}N-labelled ammonium nitrate applied to an established grass sward. *Journal of Soil Science* 38, 245–254.

Bristow, A.W., Whitehead, D.C. and Cockburn, J.E. (1992) Nitrogenous constituents in the urine of cattle, sheep and goats. *Journal of the Science of Food and Agriculture* 59, 387–394.

Broadbent, F.E., Hill, G.N. and Tyler, K.B. (1958) Transformations and movement of urea in soils. *Proceedings of the Soil Science Society of America* 22, 303–307.

Broadbent, F.E., Nakashima, T. and Chang, G.Y. (1982) Estimation of nitrogen fixation by isotope dilution in field and greenhouse experiments. *Agronomy Journal* 74, 625–628.

Brockman, J.S. (1974) Quantity and timing of fertiliser N for grass and grass/clover swards. *Proceedings of the Fertiliser Society, London*, 142, 5–13.

Brockman, J.S. and Wolton, K.M. (1963) The use of nitrogen on grass/white clover swards. *Journal of the British Grassland Society* 18, 7–13.

Brookes, P.C., Landman, A., Pruden, G. and Jenkinson, D.S. (1985) Chloroform fumigation and the release of soil nitrogen: a rapid direct extraction method to

measure microbial biomass nitrogen in soil. *Soil Biology and Biochemistry* 17, 837–842.
Brophy, L.S., Heichel, G.H. and Russelle, M.P. (1987) Nitrogen transfer from forage legumes to grass in a systematic planting design. *Crop Science* 27, 753–758.
Brougham, R.W., Ball, P.R. and Williams, W.M. (1978) The ecology and management of white clover-based pastures. In: Wilson, J.R. (ed.) *Plant Relations in Pastures.* CSIRO, Melbourne, pp. 309–324.
Browne, D. (1966) Nitrogen use on grassland. 1. Effect of applied nitrogen on animal production from a ley. *Irish Journal of Agricultural Research* 5, 89–102.
Brunke, R., Alvo, P., Schuepp, P. and Gordon, R. (1988) Effect of meteorological parameters on ammonia loss from manure in the field. *Journal of Environmental Quality* 17, 431–436.
Bryant, A.M. and Ulyatt, M.J. (1965) Effects of nitrogenous fertilizer on the chemical composition of short-rotation ryegrass and its subsequent digestion by sheep. *New Zealand Journal of Agricultural Research* 8, 109–117.
Buijsman, E., Maas, H.F.M. and Asman, W.A.H. (1987) Anthropogenic NH_3 emissions in Europe. *Atmospheric Environment* 21, 1009–1022.
Burch, J.A. and Fox, R.H. (1989) The effect of temperature and initial soil moisture content on the volatilization of ammonia from surface-applied urea. *Soil Science* 147, 311–318.
Burden, R.J. (1982) Nitrate contamination of New Zealand aquifers: a review. *New Zealand Journal of Science* 25, 205–220.
Burford, J.R. (1976) Effect of the application of cow slurry to grassland on the composition of the soil atmosphere. *Journal of the Science of Food and Agriculture* 27, 115–126.
Burford, J.R. and Bremner, J.M. (1975) Relationships between the denitrification capacities of soils and total, water-soluble and readily decomposable soil organic matter. *Soil Biology and Biochemistry* 7, 389–394.
Burford, J.R., Greenland, D.J. and Pain, B.F. (1976) Effects of heavy dressings of slurry and inorganic fertilizers applied to grassland on the composition of drainage waters and the soil atmosphere. In: MAFF *Agriculture and Water Quality, MAFF Technical Bulletin* 32, 432–443.
Burns, I.G. (1979) Nitrate movement in soil and its agricultural significance. *Outlook on Agriculture* 9, 144–148.
Burton, D.L. and Beauchamp, E.G. (1985) Denitrification rate relationships with soil parameters in the field. *Communications in Soil Science and Plant Analysis* 16, 539–549.
Burns, L.C., Stevens, R.J., Smith, R.V. and Cooper, J.E. (1995) The occurrence and possible sources of nitrite in a grazed, fertilized, grassland soil. Soil *Biology and Biochemistry* 27, 47–59.
Burton, G.W. and Jackson, J.E. (1962) Effect of rate and frequency of applying six nitrogen sources on Coastal Bermudagrass. *Agronomy Journal* 54, 40–43.
Burton, G.W., Southwell, B.L. and Johnson, J.C. (1956) The palatability of Coastal Bermudagrass (*Cynodon dactylon* (L) Pers.) as influenced by nitrogen level and age. *Agronomy Journal* 48, 360–362.
Bussink, D.W. (1992) Ammonia volatilization from grassland receiving nitrogen fertilizer and rotationally grazed by dairy cattle. *Fertilizer Research* 33, 257–265.

Bussink, D.W. (1994) Relationships between ammonia volatilization and nitrogen fertilizer application rate, intake and excretion of herbage nitrogen by cattle on grazed swards. *Fertilizer Research* 38, 111–121.

Butler, G.W. and Bathurst, N.O. (1957) The underground transference of nitrogen from clover to associated grass. *Proceedings 7th International Grassland Congress, New Zealand,* 1956, 168–178.

Butler, G.W., Greenwood, R.M. and Soper, K. (1959) Effects of shading and defoliation on the turnover of root and nodule tissue of plants of *Trifolium repens, Trifolium pratense* and *Lotus uliginosus. New Zealand Journal of Agricultural Research* 2, 415–426.

Buttery, P.J. (1976) Protein synthesis in the rumen: its implication in the feeding of non-protein nitrogen to ruminants. In: Swan, H. and Broster, W.H. (eds) *Principles of Cattle Production.* Butterworths, London, pp. 145–168.

Cameron, C.D.T. (1966) The effects of nitrogen fertilizer application rates to grass on forage yields, body weight gains, feed utilization and vitamin A status of steers. *Canadian Journal of Animal Science* 46, 19–24.

Cameron, C.D.T. (1967) Intake and digestibility of nitrogen-fertilized grass hays by wethers. *Canadian Journal of Animal Science* 47, 123–125.

Cameron K.C. and Scotter, D.R. (1988) Nitrate leaching losses from temperate agricultural systems. In: Bacon, P.E., Evans, J., Storrier, R.R. and Taylor, A.C. (eds) *Nitrogen Cycling in Temperate Agricultural Systems.* Australian Society of Soil Science, Riverina, Vol. II, pp 251–281.

Cameron, K.C. and Wild, A. (1984) Potential aquifer pollution from nitrate leaching following the plowing of temporary grassland. *Journal of Environmental Quality* 13, 274–278.

Cameron, R.S. and Posner, A.M. (1979) Mineralisable organic nitrogen in soil fractionated according to particle size. *Journal of Soil Science* 30, 565–577.

Campbell, C.A. (1978) Soil organic carbon, nitrogen and fertility. In: Schnitzer, M. and Khan, S.U. (eds) *Soil Organic Matter.* Elsevier, Amsterdam, pp. 173–271.

Campbell, C.A. and Souster, W. (1982) Loss of organic matter and potentially mineralizable nitrogen from Saskatchewan soils due to cropping. *Canadian Journal of Soil Science* 62, 651–656.

Campbell, G.W., Atkins, D.H.F., Bower, J.S., Irwin, J.G., Simpson, D. and Williams, M.L. (1990) The spatial distribution of the deposition of sulphur and nitrogen in the UK. *Journal of the Science of Food and Agriculture* 53, 427–428.

Caradus, J.R. (1990) The structure and function of white clover root systems. *Advances in Agronomy* 43, 1–46.

Caradus, J.R. and Chapman, D.F. (1991) Variability of stolon characteristics and response to shading of two cultivars of white clover (*Trifolium repens* L.). *New Zealand Journal of Agricultural Research* 34, 239–247.

Caradus, J.R., Pinxterhuis, J.B., Hay, R.J.M., Lyons, T. and Hoglund, J.H. (1993) Response of white clover cultivars to fertiliser nitrogen. *New Zealand Journal of Agricultural Research* 36, 285–295.

Carlier, L. and Verbruggen I. (1994) Herbage production from grass seeds inoculated with *Azospirillum. Proceedings 15th General Meeting, European Grassland Federation,* Wageningen, Netherlands, pp. 145–147.

Carr, A.J.H. and Catherall, P.L. (1964) The assessment of disease in herbage crops. *Report Welsh Plant Breeding Station, 1963*, pp. 94–100.

Carran, R.A. (1983) Changes in soil nitrogen during pasture-crop sequences – a review. *Proceedings of the Agronomy Society of New Zealand* 13, 29–32.

Carran, R.A., Ball, P.R., Theobald, P.W. and Collins, M.E.G. (1982) Soil nitrogen balances in urine-affected areas under two moisture regimes in Southland. *New Zealand Journal of Experimental Agriculture* 10, 377–381.

Carroll, J.C. (1943) Effects of drought, temperature and nitrogen on turf grasses. *Plant Physiology* 18, 19–36.

Castellanos, J.Z. and Pratt, P.F. (1981) Mineralization of manure nitrogen – correlation with laboratory indexes. *Soil Science Society of America Journal* 45, 354–357.

Castle, M.E. (1965) Some recent grassland experiments and their significance in British grassland farming. *Agricultural Progress* 40, 35–41.

Castle, M.E. and Drysdale, A.D. (1962) Liquid manure as a grassland fertilizer. 1. The response to liquid manure and to dry fertilizer. *Journal of Agricultural Science, Cambridge* 58, 165–171.

Castle, M.E. and Holmes, W. (1960) The intensive production of herbage for crop drying. 7. The effect of further continued massive applications of nitrogen with and without phosphate and potash on the yield of grassland herbage. *Journal of Agricultural Science, Cambridge* 55, 251–260.

Castle, M.E. and Reid, D. (1968) The effects of single compared with split applications of fertilizer nitrogen on the yield and seasonal production of a pure grass sward. *Journal of Agricultural Science, Cambridge* 70, 383–389.

Castle, M.E. and Reid, D. (1987) A comparison of the effects of liquid manure (urine and water) and nitrogen fertilizers applied to a grass-clover sward on soils of different pH value. *Journal of Agricultural Science, Cambridge* 108, 17–23.

Castle, M.E., Reid, D. and Heddle, R.G. (1965) The effect of varying the date of application of fertilizer nitrogen on the yield and seasonal productivity of grassland. *Journal of Agricultural Science, Cambridge* 64, 177–184.

Chalmers, A.G. and Leech, P.K. (1987) *Survey of Fertilizer Practice: Fertilizer Use on Farm Crops in England and Wales, 1986.* MAFF, London.

Chalmers, A.G., Kershaw, C.D. and Leech, P.K. (1990) *Survey of Fertilizer Practice: Fertilizer Use on Farm Crops in England and Wales, 1989.* MAFF, London.

Chang, F.H. and Broadbent, F.E. (1982) Influence of trace metals on some soil nitrogen transformations. *Journal of Environmental Quality* 11, 1–4.

Chaney, K. and Paulson, G.A. (1988) Field experiments comparing ammonium nitrate and urea top-dressing for winter cereals and grassland in the UK. *Journal of Agricultural Science, Cambridge* 110, 285–299.

Chaussod, R., Catroux, G. and Juste, C. (1986) Effects of anaerobic digestion of organic wastes on carbon and nitrogen mineralization rates: laboratory and field experiments. In: Dam Kofoed, A., Williams, J.H. and L'Hermite, P. (eds) *Efficient Land Use of Sludge and Manure*. Elsevier, London, pp. 24–36.

Chestnutt, D.M.B. (1965) The effects of nitrogen, potash and phosphate on the yield and composition of a grass/clover sward. *Record of Agricultural Research (Northern Ireland)* 14, (1), 71–81.

Chestnutt, D.M.B. (1972) The effects of white clover and applied nitrogen on the nitrogen content of various grass/clover mixtures. *Journal of the British Grassland Society* 27, 211–216.

Christensen, S. (1983) Nitrous oxide emission from a soil under permanent grass: seasonal and diurnal fluctuations as influenced by manuring and fertilization. *Soil Biology and Biochemistry* 15, 531–536.

Christensen, S. and Tiedje, J.M. (1990) Brief and vigorous N_2O production by soil at spring thaw. *Journal of Soil Science* 41, 1–4.

Christie, P. (1987) Some long-term effects of slurry on grassland. *Journal of Agricultural Science, Cambridge* 108, 529–541.

Christie, P. and Beattie, J.A.M. (1989) Grassland soil microbial biomass and accumulation of potentially toxic metals from long-term slurry application. *Journal of Applied Ecology* 26, 597–612.

Chu, A.C.P. and Robertson, A.G. (1974) The effects of shading and defoliation on nodulation and nitrogen fixation by white clover. *Plant and Soil* 41, 509–519.

Church, D.C. (1976) *Digestive Physiology and Nutrition of Ruminants Volume 1. Digestive Physiology* 2nd edn. O and B Books Inc., Corvallis, Oregon.

Clarholm, M. (1985) Interactions of bacteria, protozoa and plants leading to mineralization of soil nitrogen. *Soil Biology and Biochemistry* 17, 181–187.

Clark, D.A. and Ulyatt, M.J. (1985) Utilization of forage legumes in ruminant livestock production in New Zealand. In: Barnes, R.F., Ball, P.R., Brougham, R.W., Marten, G.C. and Minson, D.J. (eds) *Forage Legumes for Energy Efficient Animal Production*. USDA, Washington, pp. 197–203.

Clark, F.E. (1977) Internal cycling of ^{15}nitrogen in shortgrass prairie. *Ecology* 58, 1322–1333.

Clark, F.E. and Paul, E.A. (1970) The microflora of grassland. *Advances in Agronomy* 22, 375–435.

Clarkson, D.T., Hopper, M.J. and Jones, L.H.P. (1986) The effect of root temperature on the uptake of nitrogen and the relative size of the root system in *Lolium perenne*. I. Solutions containing both NH_4^+ and NO_3^-. *Plant, Cell and Environment* 9, 535–545.

Clement, C.R. (1961) Benefit of leys – structural improvement or nitrogen reserves. *Journal of the British Grassland Society* 16, 194–200.

Clement, C.R. and Back, H.L. (1969) Prediction of nitrogen requirements of arable crops following leys. In: Hooper, L.J. and Eagle, D.J. (eds) *Nitrogen and Soil Organic Matter*, MAFF Technical Bulletin 15, 61–70.

Clement, C.R. and Williams, T.E. (1962) An incubation technique for assessing the nitrogen status of soils newly ploughed from leys. *Journal of Soil Science* 13, 82–91.

Clement, C.R. and Williams, T.E. (1967) Leys and soil organic matter. 2. The accumulation of nitrogen in soils under different leys. *Journal of Agricultural Science, Cambridge* 69, 133–138.

Coker, E.G., Hall, J.E., Carlton-Smith, C.H. and Davis, R.D. (1987) Field investigations into the manurial value of liquid undigested sewage sludge when applied to grassland. *Journal of Agricultural Science, Cambridge* 109, 479–494.

Colbourn, P. and Dowdell, R.J. (1984) Denitrification in field soils. *Plant and Soil* 76, 213–226.

Colbourn, P., Iqbal, M.M. and Harper, I.W. (1984) Estimation of the total gaseous nitrogen losses from clay soils under laboratory and field conditions. *Journal of Soil Science* 35, 11–22.

Collins, M. (1991) Nitrogen effects on yield and forage quality of perennial ryegrass and tall fescue. *Agronomy Journal* 83, 588–595.

Collins, R.P., Glendining, M.J. and Rhodes, I. (1991) The relationship between stolon characteristics, winter survival and annual yields in white clover (*Trifolium repens* L.). *Grass and Forage Science* 46, 51–61.
Collins, R.P. and Rhodes, I. (1989) Yield of white clover populations in mixture with contrasting perennial ryegrasses. *Grass and Forage Science* 44, 111–115.
Collins, R.P. and Rhode, I. (1995) Stolon characteristicts related to winter survival in white clover. *Journal of Agricultural Science, Cambridge* 124, 11–16.
Comfort, S.D., Kelling, K.A., Keeney, D.R. and Converse, J.C. (1990) Nitrous oxide production from injected liquid dairy manure. *Soil Science Society of America Journal* 54, 421–427.
Conroy, E. (1961) Effects of heavy applications of nitrogen on the composition of herbage. *Irish Journal of Agricultural Research* 1, 67–71.
Cooke, G.W. (1967) *The Control of Soil Fertility*. Crosby Lockwood & Son, London.
Cooke, G.W. (1975) *Fertilizing for Maximum Yield*, 2nd edn. Crosby Lockwood Staples, London.
Coombe, N.B. and Hood, A.E.M. (1980) Fertilizer nitrogen: effects on dairy cow health and performance. *Fertilizer Research* 1, 157–176.
Cooper, C.S., Klager, M.G. and Schulz-Schaeffer, J. (1962) Performance of six grass species under different irrigation and nitrogen treatments. *Agronomy Journal* 54, 283–288.
Corrall, A.J. and Fenlon, J.S. (1978) A comparative method for describing the seasonal distribution of production from grasses. *Journal of Agricultural Science, Cambridge* 91, 61–67.
Cotton, D.C.F. and Curry, J.P. (1980) The effects of cattle and pig slurry fertilizers on earthworms in grassland managed for silage production. *Pedobiologia* 20, 181–188.
Coupland, R.T. and Van Dyne, G.M. (1979) Systems synthesis. In: Coupland, R.T. (ed.) *Grassland Ecosystems of the World: Analysis of Grasslands and Their Uses*. Cambridge University Press, Cambridge, pp. 97–106.
Court, M.N., Stephen, R.C. and Waid, J.S. (1964) Toxicity as a cause of the inefficiency of urea as a fertilizer. I. Review. *Journal of Soil Science* 15, 42–48.
Cowling, D.W. (1961a) The effect of white clover and nitrogenous fertilizer on the production of a sward. 1. Total annual production. *Journal of the British Grassland Society* 16, 281–290.
Cowling, D.W. (1961b) The effect of nitrogenous fertilizer on an established white clover sward. *Journal of the British Grassland Society* 16, 65–68.
Cowling, D.W. (1966) The effect of the early application of nitrogenous fertilizer and of the time of cutting in spring on the yield of ryegrass/white clover swards. *Journal of Agricultural Science, Cambridge* 66, 413–431.
Cowling, D.W. (1968) Ammonia as a source of nitrogen for grass swards. *Journal of the British Grassland Society* 23, 53–60.
Cowling, D.W. (1982) Biological nitrogen fixation and grassland production in the United Kingdom. *Philosophical Transactions of the Royal Society, London, B* 296, 397–404.
Cowling, D.W., Green, J.O. and Green, S.M. (1964) The effect of white clover and nitrogenous fertilizer on the production of a sward. 3. Statistical interpretation of their relative contributions. *Journal of the British Grassland Society* 19, 419–424.

Cowling, D.W. and Lockyer, D.R. (1965) A comparison of the reaction of different grass species to fertilizer nitrogen and to growth in association with white clover. 1. Yield of dry matter. *Journal of the British Grassland Society* 20, 197–204.
Cowling, D.W. and Lockyer, D.R. (1967) A comparison of the reaction of different grass species to fertilizer nitrogen and to growth in association with white clover. 2. Yield of nitrogen. *Journal of the British Grasland Society* 22, 53–61.
Cowling, D.W. and Lockyer, D.R. (1970) The response of perennial ryegrass to nitrogen in various periods of the growing season. *Journal of Agricultural Science, Cambridge* 75, 539–546.
Cowling, D.W. and Lockyer, D.R. (1981) Increased growth of ryegrass exposed to ammonia. *Nature* 292, 337–338.
Croll, B.T. and Hayes, C.R. (1988) Nitrate and water supplies in the United Kingdom. *Environmental Pollution* 50, 163–187.
Crouchley, G. (1979) Nitrogen fixation in pasture. 8. Wairarapa plains dryland, Masterton. *New Zealand Journal of Experimental Agriculture* 7, 31–33.
Crush, J.R. (1987) Nitrogen fixation. In: Baker, M.J. and Williams, W.M. (eds) *White Clover.* CAB International, Wallingford, pp. 185–201.
Crush, J.R., Cosgrove, G.P. and Broughan, R.W. (1982) The effect of nitrogen fertiliser on clover nitrogen fixation in an intensively grazed Manawatu pasture. *New Zealand Journal of Experimental Agriculture* 10, 395–399.
Crush, J.R. and Lowther, W.L. (1985) Nitrogen fixation by the legume-Rhizobium symbiosis: external factors influencing the symbiosis. In: Barnes, R.F., Ball, P.R., Brougham, R.W., Marten, G.C. and Minson, D.J. (eds) *Forage Legumes for Energy-Efficient Animal Production.* USDA, Washington, pp. 155–159.
Culleton, N. and Lemaire, G. (1988) The efficiency of autumn nitrogen for grassland. In: Jenkinson, D.S. and Smith, K.A. (eds) *Nitrogen Efficiency in Agricultural Soils.* Elsevier, London, pp. 207–219.
Cunningham, R.K. (1968) Cation–anion relationships in crop nutrition. VI. The effects of part, age and species of plant and some soil characteristics. *Journal of Agricultural Science, Cambridge* 70, 237–244.
Cunningham, R.K. and Nielsen, K.F. (1965) Cation–anion relationships in crop nutrition. 5. The effects of soil temperature, light intensity and soil-water tension. *Journal of Agricultural Science, Cambridge* 64, 379–386.
Curry, J.P. (1976) Some effects of animal manures on earthworms in grassland. *Pedobiologia* 16, S.425–438.
Cuttle, S.P. and Bourne, P.C. (1992) Nitrogen immobilisation and leaching in pasture soils. *Proceedings of the Fertiliser Society*, No. 325.
Cuttle, S.P. and Bourne, P.C. (1993) Uptake and leaching of nitrogen from artificial urine applied to grassland on different dates during the growing season. *Plant and Soil* 150, 77–86.
Cuttle, S.P., Hallard, M., Daniel, G. and Scurlock, R.V. (1992) Nitrate leaching from sheep-grazed grass/clover and fertilized grass pastures. *Journal of Agricultural Science, Cambridge* 119, 335–343.
Dabney, S.M. and Bouldin, D.R. (1985) Fluxes of ammonia over an alfalfa field. *Agronomy Journal* 77, 572–578.
Dabney, S.M. and Bouldin, D.R. (1990) Apparant deposition velocity and compensation point of ammonia inferred from gradient measurements above and through alfalfa. *Atmospheric Environment* 24A, 2655–2666.

Dahlman, R.C. and Kucera, C.L. (1965) Root productivity and turnover in native prairie. *Ecology* 46, 84–89.

Dale, W.R. (1961) Some effects of sheep urine on pasture. *Proceedings of the New Zealand Grassland Association* 1961, 118–123.

Daly, G.T. (1990) The grasslands of New Zealand. In: Langer, R.H.M. (ed.) *Pastures, Their Ecology and Management*. Oxford University Press, Auckland, pp. 1–38.

Daly, M. and Mackenzie, G.H. (1988) Comparison of permanent pasture and a reseeded sward under grazing at different nitrogen levels. *Research and Development in Agriculture* 5, 89–92.

D'Aoust, M.J. and Tayler, R.S. (1968) The interaction between nitrogen and water in the growth of grass swards. *Journal of Agricultural Science, Cambridge* 70, 11–17.

Darwinkel, A. (1975) Aspects of assimilation and accumulation of nitrate in some cultivated plants. *Agricultural Research Reports, Wageningen* 843, pp. 64.

Date, R.A. and Brockwell, J. (1978) *Rhizobium* strain competition and host interaction for nodulation. In: Wilson, J.R. (ed.) *Plant Relations in Pastures*. CSIRO, Melbourne, pp. 202–216.

Davidson, E.A., Stark, J.M. and Firestone, M.K. (1990) Microbial production and consumption of nitrate in an annual grassland. *Ecology* 71, 1968–1975.

Davidson, R.L. (1969) Effects of soil nutrients and moisture on root/shoot ratios in *Lolium perenne* L. and *Trifolium repens* L. *Annals of Botany* 33, 571–577.

Davies, A. (1971) Changes in growth rate and morphology of perennial ryegrass swards at high and low nitrogen levels. *Journal of Agricultural Science, Cambridge* 77, 123–134.

Davies, A. (1988) The regrowth of grass swards. In: Jones, M.B. and Lazenby, A. (eds) *The Grass Crop*. Chapman and Hall, London, pp. 85–127.

Davies, A. and Evans, M.E. (1990) Effects of spring defoliation and fertilizer nitrogen on the growth of white clover in ryegrass/clover swards. *Grass and Forage Science* 45, 345–356.

Davies, D.A. and Fothergill, M. (1990) Productivity and persistence of white clover grown with three perennial ryegrass varieties and continuously stocked with sheep. *Proceedings 13th General Meeting, European Grassland Federation,* Banska Bystrica, pp. 157–162.

Davies, D.A., Munro, J.M.M. and Morgan, T.E.H. (1984) Potential pasture production in the uplands of Wales. 6. The relative performance of sown species. *Grass and Forage Science* 39, 229–238.

Davies, W. (1960) *The Grass Crop*, 2nd edn. E. and F.N. Spon Ltd, London.

Dawson, K.P. and Ryden, J.C. (1985) Uptake of fertilizer and soil nitrogen by ryegrass swards during spring and mid-season. *Fertilizer Research* 6, 177–188.

Day, T.A. and Detling, J.K. (1990) Grassland patch dynamics and herbivore grazing preference following urine deposition. *Ecology* 71, 180–188.

De Bode, M.J.C. (1991) Odour and ammonia emissions from manure storage. In: Nielsen, V.C, Voorburg, J.H. and L'Hermite, P. (eds) *Odour and Ammonia Emissions from Livestock Farming*. Elsevier, London, pp. 59–66.

Deenen, P.J.A.G. and Lantinga, E.A. (1993) Herbage and animal production responses to fertilizer nitrogen in perennial ryegrass swards. 1. Continuous grazing and cutting. *Netherlands Journal of Agricultural Science* 41, 179–203.

Deenen, P.J.A.G. and Middelkoop, N. (1992) Effects of cattle dung and urine on nitrogen uptake and yield of perennial ryegrass. *Netherlands Journal of Agricultural Science* 40, 469–482.

De Groot, T. (1963) The influence of heavy nitrogen fertilization on the health of livestock. *Journal of the British Grassland Society* 18, 112–118.

De Haan, S. (1986) Nitrogen in drainage water from containers with soils treated with different types of sewage sludge or municipal waste compost, including substrates consisting only of these products. In: Dam Kofoed, A., Williams, J.H. and L'Hermite, P. (eds) *Efficient Land Use of Sludge and Manure*. Elsevier, London, pp. 128–141.

Deinum, B. (1966) Climate, nitrogen and grass. Research into the influence of light intensity, water supply and nitrogen on the production and chemical composition of grass. *Mededelingen Landbouwhogeschool, Wageningen,* 66.11, pp. 91.

Deinum, B. (1985) Root mass of grass swards in different grazing systems. *Netherlands Journal of Agricultural Science* 33, 377–384.

Deinum, B. and Sibma, L. (1980) Nitrate content of herbage in relation to nitrogen fertilization and management. In: Prins, W.H. and Arnold, G.H. (eds) *The Role of Nitrogen in Intensive Grassland Production*. PUDOC, Wageningen, pp. 95–102.

De Klein, C.A.M. (1994) Denitrification and N_2O emission from urine-affected grassland soil in The Netherlands. *Proceedings 15th General Meeting, European Grassland Federation,* Wageningen, Netherlands, 392–396.

De La Lande Cremer, L.C.N. (1986) Dutch experience with slurry injection. In: Dam Kofoed, A., Williams, J.H. and L'Hermite, P. (eds) *Efficient Land Use of Sludge and Manure*. Elsevier, London, pp. 99–105.

De Vries, D.M. and Kruijne, A.A. (1960) The influence of nitrogen fertilization on the botanical composition of permanent grassland. *Stikstof* (English edn) 4, 26–36.

Delwiche, C.C. and Wijler, J. (1956) Non-symbiotic nitrogen fixation in soil. *Plant and Soil* 7, 113–129.

Denmead, O.T. (1983) Micrometeorological methods for measuring gaseous losses of nitrogen in the field. In: Freney, J.R. and Simpson, J.R. (eds.) *Gaseous Loss of Nitrogen from Plant-Soil Systems*. Martinus Nijhoff/Dr W Junk Publications, The Hague, pp. 133–157.

Denmead, O.T. (1990) An ammonia budget for Australia. *Australian Journal of Soil Research* 28, 887–900.

Denmead, O.T., Freney, J.R. and Simpson, J.R. (1976) A closed ammonia cycle within a plant canopy. *Soil Biology and Biochemistry* 8, 161–164.

Dennis, W.D. and Woledge, J. (1982) Photosynthesis by white clover leaves in mixed clover/ryegrass swards. *Annals of Botany* 49, 627–635.

Dennis, W.D. and Woledge, J. (1983) The effect of shade during leaf expansion on photosynthesis by white clover leaves. *Annals of Botany* 51, 111–118.

Dennis, W.D. and Woledge, J. (1985) The effect of nitrogenous fertilizer on the photosynthesis and growth of white clover/perennial ryegrass swards. *Annals of Botany* 55, 171–178.

Dent, J.W. and Aldrich, D.T.A. (1968) Systematic testing of quality in grass varieties. 2. The effect of cutting dates, season and environment. *Journal of the British Grassland Society* 23, 13–19.

Devine, J.R. and Holmes, M.R.J. (1963) Field experiments comparing ammonium nitrate, ammonium sulphate and urea applied repetitively to grassland. *Journal of Agricultural Science, Cambridge* 60, 297–304.

Devine, J.R. and Holmes, M.R.J. (1964) Field experiments comparing ammonium nitrate and ammonium sulphate as topdressings for winter wheat and grassland. *Journal of Agricultural Science, Cambridge* 62, 377–379.

Devine, J.R. and Holmes, M.R.J. (1965) Field experiments comparing winter and spring applications of ammonium sulphate, ammonium nitrate, calcium nitrate and urea to grassland. *Journal of Agricultural Science, Cambridge* 64, 101–107.

Dewes, T., Schmitt, L., Valentin, U. and Ahrens, E. (1990) Nitrogen losses during the storage of liquid livestock manures. *Biological Wastes* 31, 241–250.

De Willigen, P. and Neeteson, J.J. (1985) Comparison of six simulation models for the nitrogen cycle in the soil. *Fertilizer Research* 8, 157–171.

De Wit, C.T., Dijkshoorn, W. and Noggle, J.C. (1963) Ionic balance and growth of plants. *Verslagen van Landbouwkundige Onderzoekingen* 69.15, pp. 69.

Dibb, C. and Haggar, R.J. (1979) Evidence of effect of sward changes on yield. In: Charles, A.H. and Haggar, R.J. (eds) *Changes in Sward Composition and Productivity.* Occasional Symposium No. 10. British Grassland Society, Hurley, pp. 11–20.

Dickinson, C.H. and Craig, G. (1990) Effects of water on the decomposition and release of nutrients from cow pats. *New Phytologist* 115, 139–147.

Dickinson, C.H., Underhay, V.S.H. and Ross, V. (1981) Effect of season, soil fauna and water content on the decomposition of cattle dung pats. *New Phytologist* 88, 129–141.

Dijkshoorn, W. (1958) Nitrogen, chlorine and potassium in perennial ryegrass and their relation to the mineral balance. *Netherlands Journal of Agricultural Science* 6, 131–138.

Dilz, K. (1988) Efficiency of uptake and utilization of fertilizer nitrogen by plants. In: Jenkinson, D.S. and Smith, K.A. (eds) *Nitrogen Efficiency in Agricultural Soils.* Elsevier, London, pp. 1–26.

Dilz, K. and Mulder, E.G. (1962) The effect of soil pH, stable manure and fertilizer nitrogen on the growth of red clover and of red clover associations with perennial ryegrass. *Netherlands Journal of Agricultural Science* 10, 1–22.

Doak, B.W. (1952) Some chemical changes in the nitrogenous constituents of urine when voided on pasture. *Journal of Agricultural Science, Cambridge* 42, 162–171.

Donald, C.M. (1963) Competition among crop and pasture plants. *Advances in Agronomy* 15, 1–118.

Dormaar, J.F. and Willms, W.D. (1993) Decomposition of blue grama and rough fescue roots in prairie soils. *Journal of Range Management* 46, 207–213.

Douglas, B.F. and Magdoff, F.R. (1991) An evaluation of nitrogen mineralization indices for organic residues. *Journal of Environment Quality* 20, 368–372.

Dowdell, R.J. (1986) Environmental aspects of grassland manuring. In: Cooper, J.P. and Raymond, W.F. (eds) *Grassland Manuring,* Occasional Symposium No. 20. British Grassland Society, Hurley, pp. 46–54.

Dowdell, R.J., Morrison, J. and Hood, A.E.M. (1980) The fate of fertiliser nitrogen applied to grassland: uptake by plants, immobilisation into soil organic matter and losses by leaching and denitrification. In: Prins, W.H. and Arnold, G.H. (eds) *The*

Role of Nitrogen in Intensive Grassland Production. PUDOC, Wageningen, pp. 129–136.
Dowdell, R.J. and Webster, C.P. (1980) A lysimeter study using nitrogen-15 on the uptake of fertilizer nitrogen by perennial ryegrass swards and losses by leaching. *Journal of Soil Science* 31, 65–75.
Dowdell, R.J. and Webster, C.P. (1984) Effect of drought and irrigation on the fate of nitrogen applied to cut permanent grass swards in lysimeters: experimental design and crop uptake. *Journal of the Science of Food and Agriculture* 35, 1092–1104.
Doyle, C.J. and Elliott, J.G. (1983) Putting an economic value on increases in grass production. *Grass and Forage Science* 38, 169–177.
Draaijers, G.P.J., Ivens, W.P.M.F., Bos, M.M. and Bleuten, W. (1989) The contribution of ammonia emissions from agriculture to the deposition of acidifying and eutrophying compounds onto forests. *Environmental Pollution* 60, 55–66.
Draycott, A.P., Hodgson, D.R. and Holliday, R. (1967) Recent research on the value of fertilizers in solution. *Agricultural Progress* 42, 68–81.
Drury, C.F., McKenney, D.J. and Findlay, W.I. (1991) Relationships between denitrification, microbial biomass and indigenous soil properties. *Soil Biology and Biochemistry* 23, 751–755.
Drysdale, A.D. (1965) Liquid manure as a grassland fertilizer. 3. The effect of liquid manure on the yield and botanical composition of pasture and its interaction with nitrogen, phosphate and potash fertilizers. *Journal of Agricultural Science, Cambridge* 65, 333–340.
Drysdale, A.D. (1966) A comparison of two sources of nitrogen for grassland, with special reference to the grass/clover ratio. *Proceedings of 10th International Grassland Congress, Helsinki,* pp. 255–258.
DuBois, J.D. and Kapustka, L.A. (1983) Biological nitrogen influx in an Ohio relict prairie. *American Journal of Botany* 70, 8–16.
Dunlop, J. and Hart, A.L. (1987) Mineral nutrition. In: Baker, M.J. and Williams, W.M. (eds) *White Clover.* CAB International, Wallingford, pp. 153–183.
Eagles, C.F. and Othman, O.B. (1989) Yield and persistency of contrasting white clover populations grown in pure swards and in mixed swards with S.23 perennial ryegrass. *Annals of Applied Biology* 114, 545–557.
Eason, W.R. and Newman, E.I. (1990) Rapid cycling of nitrogen and phosphorus from dying roots of *Lolium perenne. Oecologia* 82, 432–436.
Edmeades, D.C., Rys, G., Smart, C.E. and Wheeler, D.M. (1986) Effect of lime on soil nitrogen uptake by a ryegrass-white clover pasture. *New Zealand Journal of Agricultural Research* 29, 49–53.
Edwards, A.C., Creasey, J. and Cresser, M.S. (1985) Factors influencing nitrogen inputs and outputs in two Scottish upland catchments. *Soil Use and Management* 1, 83–87.
Edwards, A.C. and Killham, K. (1986) The effect of freeze/thaw on gaseous nitrogen loss from upland soils. *Soil Use and Management* 2, 86–90.
Edwards, C.A. and Lofty, J.R. (1977) *Biology of Earthworms,* 2nd edn. Chapman and Hall, London.
Eggington, G.M. and Smith, K.A. (1986) Losses of nitrogen by denitrification from a grassland soil fertilized with cattle slurry and calcium nitrate. *Journal of Soil Science* 37, 69–80.

Eichner, M.J. (1990) Nitrous oxide emissions from fertilized soils: summary of available data. *Journal of Environmental Quality* 19, 272–280.

Elliott, P.W., Knight, D. and Anderson, J.M. (1990) Denitrification in earthworm casts and soil from pastures under different fertilizer and drainage regimes. *Soil Biology and Biochemistry* 22, 601–605.

Elliott, P.W., Knight, D. and Anderson, J.M. (1991) Variables controlling denitrification from earthworm casts and soil in permanent pastures. *Biology and Fertility of Soils* 11, 24–29.

Ellis, J.R., Mielke, L.N. and Schuman, G.E. (1975) The nitrogen status beneath beef cattle feedlots in Eastern Nebraska. *Proceedings of the Soil Science Society of America* 39, 107–111.

Engel, M.S. and Alexander, M. (1960) Autotrophic oxidation of ammonium and hydroxylamine. *Proceedings of the Soil Science Society of America* 24, 48–50.

Engin, M. and Sprent, J.I. (1973) Effects of water stress on growth and nitrogen-fixing activity of *Trifolium repens*. *New Phytologist* 72, 117–126.

Ennik, G.C., Gillet, M. and Sibma, L. (1980) Effect of high nitrogen supply on sward deterioration and root mass. In: Prins, W.H. and Arnold, G. H. (eds) *The Role of Nitrogen in Intensive Grassland Production*. PUDOC, Wageningen, pp. 67–76.

Epstein, E. (1972) *Mineral Nutrition of Plants: Principles and Perspectives.* Wiley, New York.

Erisman, J.W., Vermetten, A.W.M., Pinksterboer, E.F., Asman, W.A.H., Waijers-Ypelaan, A. and Slanina, J. (1987) In: Asman, W.A.H. and Diederen, S.M.A. (eds) *Ammonia and Acidification.* Proceedings of Symposium of the European Association for the Science of Air Pollution. Bilthoven, Netherlands, pp. 59–77.

Ernst, J.W. and Massey, H.F. (1960) The effects of several factors on volatilization of ammonia formed from urea in the soil. *Soil Science Society of America Proceedings* 24, 87–90.

Evans, D.R., Williams, T.A. and Mason, S.A. (1990) Contribution of white clover varieties to total sward production under typical farm management. *Grass and Forage Science* 45, 129–134.

Evans, P.S. (1973) The effect of repeated defoliation to three different levels on root growth of five pasture species. *New Zealand Journal of Agricultural Research* 16, 31–34.

Evans, P.S. (1977) Comparative root morphology of some pasture grasses and clovers. *New Zealand Journal of Agricultural Research* 20, 331–335.

Evans, P.S. (1978) Plant root distribution and water use patterns of some pasture and crop species. *New Zealand Journal of Agricultural Research* 21, 261–265.

Fangmeier, A., Hadwinger-Fangmeier, A., Van der Eerden, L. and Jäger, H-J. (1994) Effects of atmospheric ammonia on vegetation – a review. *Environmental Pollution* 86, 43–82.

Farquhar, G.D., Firth, P.M., Wetselaar, R. and Weir, B. (1980) On the gaseous exchange of ammonia between leaves and the environment: determination of the ammonia compensation point. *Plant Physiology* 66, 710–714.

Farquhar, G.D., Wetselaar, R. and Firth, P.M. (1979) Ammonia volatilization from senescing leaves of maize. *Science* 203, 1257–1258.

Feller, U. and Keist, M. (1986) Senescence and nitrogen metabolism in annual plants. In: Lambers, H., Neeteson, J.J. and Stulen, I. (eds) *Fundamental, Ecological and*

Agricultural Aspects of Nitrogen Metabolism in Higher Plants. Martinus Nijhoff, Dordrecht, pp. 219–234.

Fenn, L.B. and Hossner, L.R. (1985) Ammonia volatilization from ammonium or ammonium-forming nitrogen fertilizers. *Advances in Soil Science* 1, 123–169.

Fenn, L.B., Malstrom, H.L. and Wu, E. (1987) Ammonia losses from surface-applied urea as related to urea application rates, plant residue and calcium chloride addition. *Fertilizer Research* 12, 219–227.

Ferguson, R.B. and Kissel, D.E. (1986) Effects of soil drying on ammonia volatilization from surface-applied urea. *Soil Science Society of America Journal* 50, 485–490.

Field, A.C., Sykes, A.R. and Gunn, R.G. (1974) Effects of age and state of incisor dentition on faecal output of dry matter and on faecal and urinary output of nitrogen and minerals, of sheep grazing hill pastures. *Journal of Agricultural Science, Cambridge* 83, 151–160.

Field, T.R.O. and Ball, P.R. (1982) Nitrogen balance in an intensively utilised dairy-farm system. *Proceedings of the New Zealand Grassland Association* 43, 64–69.

Field, T.R.O., Ball, P.R. and Theobald, P.W. (1985) Leaching of nitrate from sheep-grazed pastures. *Proceedings of the New Zealand Grassland Association* 46, 209–214.

Firestone, M.K. (1982) Biological denitrification. In: Stevenson, F.J. (ed.) *Nitrogen in Agricultural Soils.* American Society of Agronomy, Madison, Wisconsin, pp. 289–326.

Fleisher, Z. and Hagin, J. (1981) Lowering ammonia volatilization losses from urea application by activation of nitrification process. *Fertilizer Research* 2, 101–107.

Floate, M.J.S. (1970) Decomposition of organic materials from hill soils and pastures. II. Comparative studies on the mineralization of carbon, nitrogen and phosphorus from plant materials and sheep faeces. *Soil Biology and Biochemistry* 2, 173–185.

Floate, M.J.S. (1987) Nitrogen cycling in managed grasslands. In: Snaydon, R.W. (ed.) *Managed Grasslands: Analytical Studies.* Elsevier, Amsterdam, pp. 163–172.

Floate, M.J.S. and Torrance, C.J.W. (1970) Decomposition of the organic materials from hill soils and pastures. 1. Incubation method for studying the mineralization of carbon, nitrogen and phosphorus. *Journal of the Science of Food and Agriculture* 21, 116–120.

Flowers, T.H. and Arnold, P.W. (1983) Immobilization and mineralization of nitrogen in soils incubated with pig slurry or ammonium sulphate. *Soil Biology and Biochemistry* 15, 329–335.

Follett, R.F., Power, J.F., Grunes, D.L. and Klein, C.A. (1977) Effect of N, K and P fertilization, N source, and clipping on potential tetany hazard of bromegrass. *Plant and Soil* 48, 485–508.

Follett, R.F. and Wilkinson, S.R. (1985) Soil fertility and fertilization of forages. In: Heath, M.E., Barnes, R.F. and Metcalfe, D.S. (eds) *Forages: The Science of Grassland Agriculture,* 4th edn. Iowa State University Press, Ames, pp. 304–317.

Forbes, T.D.A. and Hodgson, J. (1985) The reaction of grazing sheep and cattle to the presence of dung from the same or other species. *Grass and Forage Science* 40, 177–182.

Foster, S.S.D., Cripps, A.C. and Smith-Carington, A. (1982) Nitrate leaching to groundwater. *Philosophical Transactions of the Royal Society, London, B* 296, 477–489.

Fox, R.H., Myers, R.J.K. and Vallis, I. (1990) The nitrogen mineralization rate of legume residues in soil as influenced by their polyphenol, lignin and nitrogen contents. *Plant and Soil* 129, 251–259.

Frame, J. (1971) Fundamentals of grassland management. 10. The grazing animal. *Scottish Agriculture* 50, 28–44.

Frame, J. (1987) The role of white clover in United Kingdom pastures. *Outlook on Agriculture* 16, 28–34.

Frame, J. (1990) Herbage productivity of a range of grass species in association with white clover. *Grass and Forage Science* 45, 57–64.

Frame, J. (1991) Herbage production and quality of a range of secondary grass species at five rates of fertilizer nitrogen application. *Grass and Forage Science* 46, 139–151.

Frame, J. (1992) *Improved Grassland Management.* Farming Press, Ipswich.

Frame, J. and Boyd, A.G. (1987a) The effect of fertilizer nitrogen rate, white clover variety and closeness of cutting on herbage productivity from perennial ryegrass/white clover swards. *Grass and Forage Science* 42, 85–96.

Frame, J. and Boyd, A.G. (1987b) The effect of strategic use of fertilizer nitrogen in spring and/or autumn on the productivity of a perennial ryegrass/white clover sward. *Grass and Forage Science* 42, 429–438.

Frame, J., Harkess, R.D. and Talbot, M. (1989) The effect of cutting frequency and fertiliser nitrogen rate on herbage productivity from perennial ryegrass. *Research and Development in Agriculture* 6, 99–105.

Frame, J. and Newbould, P. (1986) Agronomy of white clover. *Advances in Agronomy* 40, 1–88.

France, G.D. and Thompson, D.C. (1993) An overview of efficient manufacturing processes. *Proceedings of the Fertiliser Society* No. 337.

Francis, G.S., Haynes, R.J., Sparling, G.P., Ross, D.J. and Williams, P.H. (1992) Nitrogen mineralization, nitrate leaching and crop growth following cultivation of a temporary leguminous pasture in autumn and winter. *Fertilizer Research* 33, 59–70.

Freney, J.R. and Black, A.S. (1988) Importance of ammonia volatilization as a loss process. In: Wilson, J.R. (ed.) *Advances in Nitrogen Cycling in Agricultural Ecosystems.* CAB International, Wallingford, pp. 156–173.

Freney, J.R., Simpson, J.R. and Denmead, O.T. (1981) Ammonia volatilization. In: Clark, F.E. and Rosswall, T. (eds) *Terrestrial Nitrogen Cycles, Ecological Bulletin* (Stockholm) 33, 291–302.

Froment, M.A., Chalmers, A.G. and Smith, K.A. (1992) Nitrate leaching from autumn and winter application of animal manures to grassland. In: Archer, J.R. and others (eds) *Nitrate and Farming Systems.* Association of Applied Biologists, Warwick, pp. 153–156.

Frost, J.P. (1994) Effect of spreading method, application rate and dilution on ammonia volatilization from slurry. *Grass and Forage Science* 49, 391–400.

Frost, J.P. and Stevens, R.J. (1991) Slurry utilisation. *Agricultural Research Institute of Northern Ireland, 64th Annual Report 1990–91*, 20–30.

Frost, J.P., Stevens, R.J. and Laughlin, R.J. (1990) Effect of separation and acidification of cattle slurry on ammonia volatilization and on the efficiency of slurry nitrogen for herbage production. *Journal of Agricultural Science, Cambridge* 115, 49–56.

Gardner, D.N. (1986) Nitrogen limitation on dairy farms. *Report, Milk Marketing Board,* No. 54, pp. 46.

Gardner, E.H., Jackson, T.L., Webster, G.R. and Turley, R.H. (1960) Some effects of fertilization on the yield, botanical and chemical composition of irrigated grass, and grass-clover pasture swards. *Canadian Journal of Plant Science* 40, 546–562.

Garstang, J.R. (1981) The long term effects of the application of different amounts of nitrogenous fertiliser on dry matter yield, sward composition and chemical composition of a permanent grass sward. *Experimental Husbandry* 37, 117–132.

Garwood, E.A. (1967a) Some effects of soil water conditions and soil temperature on the roots of grasses. 1. The effect of irrigation on the weight of root material under various swards. *Journal of the British Grassland Society* 22, 176–181.

Garwood, E.A. (1967b) Studies on the roots of grasses. *Annual Report 1966*, Grassland Research Institute, Hurley, 72–79.

Garwood, E.A., Clement, C.R. and Williams, T.E. (1972) Leys and soil organic matter. 3. The accumulation of macro-organic matter in the soil under different swards. *Journal of Agricultural Science, Cambridge* 78, 333–341.

Garwood, E.A. and Ryden, J.C. (1986) Nitrate loss through leaching and surface runoff from grassland: effects of water supply, soil type and management. In: van der Meer, H.G., Ryden, J.C. and Ennik, G.C. (eds) *Nitrogen Fluxes in Intensive Grassland Systems*. Martinus Nijhoff, Dordrecht, pp. 99–113.

Garwood, E.A., Salette, J. and Lemaire, G. (1980) The influence of water supply to grass on the response to fertilizer nitrogen and nitrogen recovery. In: Prins, W.H. and Arnold, G.H. (eds) *The Role of Nitrogen in Intensive Grassland Production*. PUDOC, Wageningen, pp. 59–65.

Garwood, E.A. and Sinclair, J. (1979) Use of water by six grass species. 2. Root distribution and use of soil water. *Journal of Agricultural Science, Cambridge* 93, 25–35.

Garwood, E.A. and Tyson, K.C. (1973) Losses of nitrogen and other plant nutrients to drainage from soil under grass. *Journal of Agricultural Science, Cambridge* 80, 303–312.

Garwood, E.A., Tyson, K.C. and Clement, C.R. (1977) A comparison of yield and soil conditions during 20 years of grazed grass and arable cropping. *Technical Report* 21, Grassland Research Institute, Hurley.

Garwood, E.A. and Williams, T.E. (1967) Growth, water use and nutrient uptake from the subsoil by grass swards. *Journal of Agricultural Science, Cambridge* 69, 125–130.

Gasser, J.K.R. (1964) Urea as a fertilizer. *Soils and Fertilizers* 27, 175–180.

George, J.R., Rhykerd, C.L., Noller, C.H., Dillon, J.E. and Burns, J.C. (1973) Effect of N fertilization on dry matter yield, total N, N recovery, and nitrate-N concentration of three cool-season forage grass species. *Agronomy Journal* 65, 211–216.

Gething, P.A. (1963) The effects of nitrogen, phosphate and potash on yields of herbage cut for conservation. *Proceedings of 1st Regional Conference, International Potash Institute, Wexford (Ireland)* 1963, pp. 83–96.

Geurink, J.H., Malestein, A., Kemp, A., Korzeniowski, A. and van't Klooster, A.T. (1982) Nitrate poisoning in cattle. 7. Prevention. *Netherlands Journal of Agricultural Science* 30, 105–113.

Gianello, C. and Bremner, J.M. (1988) A rapid steam distillation method of assessing potentially available organic nitrogen in soil. *Communications in Soil Science and Plant Analysis* 19, 1551–1568.

Gill, M., Beever, D.E. and Osbourn, D.F. (1989) The feeding value of grass and grass products. In: Holmes, W. (ed.) *Grass: its Production and Utilization*, 2nd edn. Blackwell, Oxford, pp. 89–129.
Gillet, M. (1982) Carbon and nitrogen relationships in plants: some practical consequences for grass. In: Corrall, A.J. (ed.) *Efficient Grassland Farming.* Occasional Symposium No. 14. British Grassland Society, Hurley, pp. 43–47.
Gisiger, L. (1950) Organic manuring of grassland. *Journal of British Grassland Society* 5, 63–79.
Gmur, N.F., Evans, L.S. and Cunningham, E.A. (1983) Effects of ammonium sulfate aerosols on vegetation. II. Mode of entry and responses of vegetation. *Atmospheric Environment* 17, 715–721.
Goedewaagen, M.A.J. and Schuurman, J.J. (1950) Root production by agricultural crops on arable land and on grassland as a source of organic matter in the soil. *Transactions of the International Society of Soil Science* 2, 28–31.
Goodman, P.J. (1988) Nitrogen fixation, transfer and turnover in upland and lowland grass-clover swards, using ^{15}N isotope dilution. *Plant and Soil* 112, 247–254.
Gordon, F.J. (1974) The use of nitrogen fertilisers on grassland for milk production. *Proceedings of the Fertiliser Society, London*, 142, 14–27.
Gostick, K.G. (1981) Accumulation of organic nitrogen in soils after application to grassland: some studies in England and Wales. In: Brogan, J.C. (ed.) *Nitrogen Losses and Surface Run-off from Landspreading of Manures.* Martinus Nijhoff, The Hague, pp. 369–372.
Goswami, A.K. and Willcox, J.S. (1969) Effect of applying increasing levels of nitrogen to ryegrass. *Journal of the Science of Food and Agriculture* 20, 592–599.
Gould, W. D., Hagedorn, C. and McCready, R.G.L. (1986) Urea transformations and fertilizer efficiency in soil. *Advances in Agronomy* 40, 209–238.
Goulding, K.W.T. (1990) Nitrogen deposition to land from the atmosphere. *Soil Use and Management* 6, 61–62.
Goulding, K.W.T. and Webster, C.P. (1992) Methods for measuring nitrate leaching. In: Archer, J.R. *et al.* (eds) *Nitrate and Farming Systems.* Association of Applied Biologists, Warwick, pp. 63–70.
Gracey, H.I. (1979) Nutrient content of cattle slurry and losses of nitrogen during storage. *Experimental Husbandry* 35, 47–51.
Granli, T. and Bockman, O.C. (1994) Nitrous oxide from agriculture. *Norwegian Journal of Agricultural Sciences,* Supplement No. 12, 1–128.
Grant, S.A., Torvell, L., Sim, E.M. and Small, J. (1991) The effect of stolon burial and defoliation early in the growing season on white clover performance. *Grass and Forage Science* 46, 173–182.
Green, C.J., Blackmer, A.M. and Yang, N.C. (1994) Release of fixed ammonium during nitrification in soils. *Soil Science Society of America Journal* 58, 1411–1415.
Green, J.O. and Baker, R.D. (1981) Classification, distribution and productivity of UK grasslands. In: Jollans, J.L. (ed.) *Grassland in the British Economy.* Centre for Agricultural Strategy, University of Reading, Paper No. 10, pp. 237–247.
Green, J.O. and Cowling, D.W. (1961) The nitrogen nutrition of grassland. *Proceedings 8th International Grassland Congress, Reading, 1960,* pp. 126–129.
Greenhalgh, J.F.D. (1981) A forward look at technical possibilities for grassland. In: Jollans, J.L. (ed.) *Grassland in the British Economy.* Centre for Agricultural Strategy, University of Reading, Paper No. 10, pp. 51–63.

Greenwood, D.J. (1975) The measurement of soil aeration. *MAFF Technical Bulletin,* 29, 261.

Greenwood, D.J. (1990) Production or productivity: the nitrate problem? *Annals of Applied Biology* 117, 209–231.

Griffith, W.K and Teel, M.R. (1965) Effect of nitrogen and potassium fertilization, stubble height and clipping frequency on yield and persistence of orchardgrass. *Agronomy Journal* 57, 147–149.

Griffith, W.K., Teel, M.R. and Parker, H.E. (1964) Influence of nitrogen and potassium on the yield and chemical composition of orchardgrass. *Agronomy Journal* 56, 473–475.

Grunes, D.L., Stout, P.R. and Brownell, J.R. (1970) Grass tetany of ruminants. *Advances in Agronomy* 22, 331–374.

Hageman, R.H. (1984) Ammonium versus nitrate nutrition of higher plants. In: Hauck, R.D. (ed.) *Nitrogen in Crop Production*. American Society of Agronomy, Madison, pp. 67–85.

Haggar, R.J. (1976) The seasonal productivity, quality and response to nitrogen of four indigenous grasses compared with *Lolium perenne*. *Journal of the British Grassland Society* 31, 197–207.

Haigh, R.A. and White, R.E. (1986) Nitrate leaching from a small, underdrained, grassland, clay catchment. *Soil Use and Management* 2, 65–70.

Hall, J.E. (1986) Soil injection research in the UK. In: Dam Kofoed, A., Williams, J.H. and L'Hermite, P. (eds) *Efficient Land Use of Sludge and Manure*. Elsevier, London, pp. 78–89.

Hallberg, G.R. (1989) Nitrate in ground water in the United States. In: Follet, R.F. (ed.) *Nitrogen Management and Ground Water Protection*. Elsevier, Amsterdam, pp. 35–74.

Halliday, J. and Pate, J.S. (1976) The acetylene reduction assay as a means of studying nitrogen fixation in white clover under sward and laboratory conditions. *Journal of British Grassland Society* 31, 29–35.

Hancock, J. (1953) Grazing behaviour of cattle. *Animal Breeding Abstracts* 21, 1–13.

Hansen, E.A. and Harris, A.R. (1975) Validity of soil-water samples collected with porous ceramic cups. *Soil Science Society of America, Proceedings* 39, 528–536.

Hargrove, W.L. (1989) Soil environmental and management factors influencing ammonia volatilization under field conditions. In: Bock, B.R. and Kissel, D.E. (eds) *Ammonia Volatilization from Urea Fertilizers,* Bulletin Y-206, National Fertilizer Development Center, Tennessee Valley Authority. Tennessee Valley Authority, Muscle Shoals, pp. 17–36.

Hargrove, W.L., Kissel, D.E. and Fenn, L.B. (1977) Field measurements of ammonia volatilization from surface applications of ammonium salts to a calcareous soil. *Agronomy Journal* 69, 473–476.

Harkess, R.D. and Frame, J. (1986) Efficient use of fertilizer nitrogen on grass swards: effects of timing, cutting management and secondary grasses. In: van der Meer, H.G., Ryden, J.C. and Ennik, G.C. (eds) *Nitrogen Fluxes in Intensive Grassland Systems.* Martinus Nijhoff, Dordrecht, pp. 29–37.

Harper, J.L., Jones, M. and Sackville-Hamilton, N.R. (1991) The evolution of roots and problems of analysing their behaviour. In: Atkinson, D. (ed.) *Plant Root Growth: An Ecological Perspective.* Blackwell Scientific Publications, London, pp. 3–22.

Harper, L.A., Catchpoole, V.R., Davis, R. and Weir, K.L. (1983) Ammonia volatilization: soil, plant and microclimate effects on diurnal and seasonal fluctuations. *Agronomy Journal* 75, 212–218.
Harris, W. (1978) Defoliation as a determinant of the growth, persistence and composition of pasture. In: Wilson, J.R. (ed.) *Plant Relations in Pastures.* CSIRO, Melbourne, pp. 67–85.
Harrison, A.F., Taylor, K., Hatton, J.C. and Howard, D.M. (1994) Role of nitrogen in herbage production by *Agrostis–Festuca* hill grassland. *Journal of Applied Ecology* 31, 351–360.
Hart, M.L. t' (1957) The influence of nitrogen fertilising on the botanical composition of grassland. *Stikstof* (English edn), 1, 33–37.
Hartmans, J. (1974) Factors affecting the herbage iodine content. *Netherlands Journal of Agricultural Science* 22, 195–206.
Hartung, J. (1991) Influence of housing and livestock on ammonia release from buildings. In: Nielsen, V.C., Voorburg, J.H. and L'Hermite, P. (eds) *Odour and Ammonia Emissions from Livestock Farming.* Elsevier, London, pp. 22–30.
Hassink, J. (1992a) Effect of grassland management on N mineralization potential, microbial biomass and N yield in the following year. *Netherlands Journal of Agricultural Science* 40, 173–185.
Hassink, J. (1992b) Effects of soil texture and structure on carbon and nitrogen mineralization in grassland soils. *Biology and Fertility of Soils* 14, 126–134.
Hassink, J. and Neeteson, J.J. (1991) Effect of grassland management on the amounts of soil organic N and C. *Netherlands Journal of Agricultural Science* 39, 225–236.
Hassink, J., Neutel, A.M. and De Ruiter, P.C. (1994) C and N mineralization in sandy and loamy grassland soils: the role of microbes and microfauna. *Soil Biology and Biochemistry* 26, 1565–1571.
Hassink, J., Scholefield, D. and Blantern, P. (1990) Nitrogen mineralisation in grassland soils. *Proceedings of 13th General Meeting, European Grasland Federation, Banska Bystrica, Czechoslovakia,* Vol. II, pp. 25–32.
Hatch, D.J., Jarvis, S.C. and Dollard, G.J. (1990a) Measurements of ammonia emission from grazed grassland. *Environmental Pollution* 65, 333–346.
Hatch, D.J., Jarvis, S.C. and Philipps, L. (1990b) Field measurement of nitrogen mineralization using soil core incubation and acetylene inhibition of nitrification. *Plant and Soil* 124, 97–107.
Hauck, R.D. (1984a) Significance of nitrogen fertilizer microsite reactions in soil. In: Hauck, R.D. (ed.) *Nitrogen in Crop Production.* American Society of Agronomy, Madison, pp. 507–519.
Hauck, R.D. (1984b) Atmospheric nitrogen chemistry, nitrification, denitrification and their inter-relationships. In: Hutzinger, O. (ed.) *Handbook of Environmental Chemistry, Vol. I, Part C.* Springer-Verlag, pp. 105–125.
Hauck, R.D. (1988) A human ecosphere perspective of agricultural nitrogen cycling. In: Wilson, J.R. (ed.) *Advances in Nitrogen Cycling in Agricultural Ecosystems.* CAB International, Wallingford, pp. 3–19.
Hauck, R.D. and Bremner, J.M. (1976) Use of tracers for soil and fertilizer nitrogen research. *Advances in Agronomy* 28, 219–266.
Havre, G.N. and Dishington, I.W. (1962) The mineral composition of pasture as influenced by various types of heavy nitrogen dressings. *Acta Agriculturae Scandinavica* 12, 298–308.

Hawkins, S.W. and Rose, P.H. (1979) The relationship between the rate of fertilizer nitrogen applied to grassland and milk production; an analysis of recorded farm data. *Grass and Forage Science* 34, 203–208.

Haydock, K.P. and Norris, D.O. (1966) Opposed curves for nitrogen per cent on dry weight given by Rhizobium dependent and nitrate dependent legumes. *Australian Journal of Science* 29, 426–427.

Haynes, R.J. (1986a) Uptake and assimilation of mineral nitrogen by plants. In: Haynes, R.J. (ed.) *Mineral Nitrogen in the Plant–Soil System.* Academic Press, London, pp. 303–378.

Haynes, R.J. (1986b) The decomposition process: mineralization, immobilization, humus formation and degradation. In: Haynes, R.J. (ed.) *Mineral Nitrogen in the Plant–Soil System.* Academic Press, London, pp. 52–126.

Haynes, R.J. (1986c) Nitrification. In: Haynes, R.J. (ed.) *Mineral Nitrogen in the Plant–Soil System.* Academic Press, London, pp. 127–165.

Haynes, R.J. and Sherlock, R.R. (1986) Gaseous losses of nitrogen. In Haynes, R.J. (ed.), *Mineral Nitrogen in the Plant–Soil System.* Academic Press, London, pp. 242–302.

Haynes, R.J. and Swift, R.S. (1990) Stability of soil aggregates in relation to organic constituents and soil water content. *Journal of Soil Science* 41, 73–83.

Haynes, R.J. and Williams, P.H. (1992) Changes in soil solution composition and pH in urine-affected areas of pasture. *Journal of Soil Science* 43, 323–334.

Haynes, R.J. and Williams, P.H. (1993) Nutrient cycling and soil fertility in the grazed pasture ecosystem. *Advances in Agronomy* 49, 119–199.

Haystead, A., King, J. and Lamb, W.I.C. (1979) Photosynthesis, respiration and nitrogen fixation in white clover. *Grass and Forage Science* 34, 125–130.

Haystead, A., Malajczuk, N. and Grove, T.S. (1988) Underground transfer of nitrogen between pasture plants infected with vesicular-arbuscular mycorrhizal fungi. *New Phytologist* 108, 417–423.

Haystead, A. and Marriott, C. (1978) Fixation and transfer of nitrogen in a white clover-grass sward under hill conditions. *Annals of Applied Biology* 88, 453–457.

Haystead, A. and Marriott, C. (1979) Transfer of legume nitrogen to associated grass. *Soil Biology and Biochemistry* 11, 99–104.

Head, M.J. and Rook, J.A.F. (1955) Hypomagnesaemia in dairy cattle and its possible relationship to ruminal ammonia production. *Nature, London* 176, 262–263.

Heath, M.E., Barnes, R.F. and Metcalfe, D.S. (eds) (1985) *Forages: The Science of Grassland Agriculture*, 4th edn. Iowa State University Press, Ames.

Heathwaite, A.L., Burt, T.P. and Trudgill, S.T. (1993) Overview – the nitrate issue. In: Burt, T.P., Heathwaite, A.L. and Trudgill, S.T. (eds) *Nitrate: Processes, Patterns and Management.* John Wiley and Sons, Chichester, pp. 3–21.

Heddle, R.G. (1967) Long-term effects of fertilizers on herbage production. I. Yields and botanical composition. *Journal of Agricultural Science, Cambridge* 69, 425–431.

Heddle, R.G. (1968) Nitrogenous fertilization of Italian ryegrass in spring. *Journal of the British Grassland Society* 23, 69–74.

Heddle, R.G. and Crooks, P. (1967) Long-term effects of fertilizers on herbage production. II. Chemical composition. *Journal of Agricultural Science, Cambridge* 69, 433–441.

Heichel, G.H. (1985) Forages and legumes. In: Heath, M.E., Barnes, R.F. and Metcalfe, D.S. (eds) *Forages: The Science of Grassland Agriculture*. Iowa State University Press, Ames, pp. 66–67.
Heichel, G.H. (1989) Dinitrogen fixation and nitrogen transfer in temperate legume-grass communities. *Proceedings 16th International Grassland Congress, Nice*, pp. 131–132.
Heichel, G.H., Barnes, D.K., Vance, C.P. and Henjum, K.I. (1984) N_2 fixation, and N and dry matter partitioning during a 4-year alfalfa stand. *Crop Science* 24, 811–815.
Heil, G.W., Werger, M.J.A., De Mol, W., van Dam, D. and Heijne, B. (1988) Capture of atmospheric ammonium by grassland canopies. *Science* 239, 764–765.
Hemingway, R.G. (1961) Magnesium, potassium, sodium and calcium contents of herbage as influenced by fertilizer treatments over a three-year period. *Journal of the British Grassland Society* 16, 106–116.
Hemingway, R.G. (1962) Copper, molybdenum, manganese and iron contents of herbage as influenced by fertilizer treatments over a three-year period. *Journal of the British Grassland Society* 17, 182–187.
Hemingway, R.G. (1963) Soil and herbage potasium levels in relation to yield. *Journal of the Science of Food and Agriculture* 14, 188–195.
Henry, C.M. and Deacon, J.W. (1981) Natural (non-pathogenic) death of the cortex of wheat and barley seminal roots, as evidenced by nuclear staining with acridine orange. *Plant and Soil* 60, 255–274.
Henzell, E.F., Fergus, I.F. and Martin, A.E. (1966) Accumulation of soil nitrogen and carbon under a *Desmodium uncinatum* pasture. *Australian Journal of Experimental Agriculture and Animal Husbandry* 6, 157–160.
Henzell, E.F. and Ross, P.J. (1973) The nitrogen cycle of pasture ecosystems. In: Butler, G.W. and Bailey, R.W. (eds) *Chemistry and Biochemistry of Herbage.* Vol. 2. Academic Press, London, pp. 227–246.
Herlihy, M. (1978) Dry matter response of ryegrass to ammonium and nitrate sources of nitrogen as a function of soil texture and moisture. *Plant and Soil* 50, 633–646.
Herlihy, M. (1979) Nitrogen mineralisation in soils of varying texture, moisture and organic matter. *Plant and Soil* 53, 255–267.
Herman, W.A., McGill, W.B. and Dormaar, J.F. (1977) Effects of initial chemical composition on decomposition of roots of three grass species. *Canadian Journal of Soil Science* 57, 205–215.
Herriott, J.B.D. and Wells, D.A. (1960) Clover nitrogen and sward productivity. *Journal of the British Grassland Society* 15, 63–69.
Herriott, J.B.D. and Wells, D.A. (1962) Gülle as a grassland fertilizer. *Journal of the British Grassland Society* 17, 167–170.
Herriott, J.B.D. and Wells, D.A. (1963) The grazing animal and sward productivity. *Journal of Agricultural Science, Cambridge* 61, 89–99.
Hides, D.H. (1978) Winter hardiness in *Lolium multiflorum* Lam. I. The effect of nitrogen fertilizer and autumn cutting managements in the field. *Journal of the British Grassland Society* 33, 99–105.
Hilder, E.J. (1966) Distribution of excreta by sheep at pasture. *Proceedings 10th International Grassland Congress,* Helsinki, pp. 977–981.
Hill, J. (1980) The remobilization of nutrients from leaves. *Journal of Plant Nutrition* 2, 407–444.

Hobbs, J.A. and Thompson, C.A. (1971) Effect of cultivation on the nitrogen and organic carbon contents of a Kansas Arguistoll (Chernozem). *Agronomy Journal* 63, 66–68.

Hobson, R.D. and Richards, I.R. (1978) The relationship between grass N yield and the total N supply to cut grass-white clover swards. *Journal of Agricultural Science, Cambridge* 90, 229–232.

Hodgson, D.R. and Draycott, A.P. (1968) Aqueous ammonia compared with other nitrogenous fertilizers as solids and solutions on grass. *Journal of Agricultural Science, Cambridge* 71, 195–203.

Hodgson, J. and Spedding, C.R.W. (1966) The health and performance of the grazing animal in relation to fertilizer nitrogen usage. 1. Calves. *Journal of Agricultural Science, Cambridge* 67, 155–167.

Hoglund, J.H. and Brock, J.L. (1987) Nitrogen fixation in managed graslands. In: Snaydon, R.W. (ed.) *Managed Grasslands: Analytical Studies*. Elsevier, Amsterdam, pp. 187–196.

Hoglund, J.H., Crush, J.R., Brock, J.L., Ball, R. and Carran, R.A. (1979) Nitrogen fixation in pasture. 12. General discussion. *New Zealand Journal of Experimental Agriculture* 7, 45–51.

Holland, P.T. and During, C. (1977) Movement of nitrate-N and transformations of urea-N under field conditions. *New Zealand Journal of Agricultural Research* 20, 479–488.

Holliday, R. and Wilman, D. (1962) The effect of white clover, fertilizer nitrogen and simulated animal residues on yield of grassland herbage. *Journal of the British Grassland Society* 17, 206–213.

Holliday, R. and Wilman, D. (1965) The effect of fertilizer nitrogen and frequency of defoliation on yield of grassland herbage. *Journal of the British Grassland Society* 20, 32–40.

Hollington, P.A. and Wilman, D. (1985) Effects of white clover and fertilizer nitrogen on clover and grass leaf dimensions, percentage cover and numbers of leaves and tillers. *Journal of Agricultural Science, Cambridge* 104, 595–607.

Holmes, J.C. and Lang, R.W. (1963) Effects of fertilizer nitrogen and herbage drymatter content on herbage intake and digestibility in bullocks. *Animal Production* 5, 17–26.

Holmes, W. (1949) The intensive production of herbage for crop-drying. Part 2. A study of the effect of massive dressings of nitrogenous fertilizer and of the time of their application on the yield, chemical and botanical composition of two grass leys. *Journal of Agricultural Science, Cambridge* 39, 128–141.

Holmes, W. (1968) The use of nitrogen in the management of pasture for cattle. *Herbage Abstracts* 38, 265–277.

Holmes, W. (1970) Animals for food. *Proceedings of the Nutrition Society* 29, 237–244.

Holmes, W. (1974) The role of nitrogen fertiliser in the production of beef from grass. *Proceedings of the Fertiliser Society*, London, 142, 57–69.

Holmes, W. (1989) *Grass: its Production and Utilization*, 2nd edn. Blackwell Scientific Publications, Oxford.

Holmes, W. and MacLusky, D.S. (1954) The intensive production of herbage for crop-drying. Part 5. The effect of continued massive applications of nitrogen with and without phosphate and potash on the yield of grassland herbage. *Journal of Agricultural Science, Cambridge* 45, 129–139.

Holter, P. (1983) Effect of earthworms on the disappearance rate of cattle droppings. In: Satchell, J.E. (ed.) *Earthworm Ecology*. Chapman & Hall, London, pp. 49–57.
Hoogerkamp, M. (1973) Accumulation of organic matter under grassland and its effects on grassland and on arable crops. *Agricultural Research Reports* 806, PUDOC, Wageningen, Netherlands, pp. 1–24.
Hoogerkamp, M. (1979) Avoiding the lean years. In: Charles, A.H. and Haggar, R.J. (eds) *Changes in Sward Composition and Productivity*. British Grassland Society Occasional Symposium No. 10. British Grassland Society, Hurley, pp. 199–205.
Hoogerkamp, M., Rogaar, H. and Eijsackers, H.J.P. (1983) Effect of earthworms on grassland on recently reclaimed polder soils in the Netherlands. In: Satchell, J.E. (ed.) *Earthworm Ecology*. Chapman and Hall, London, pp. 85–104.
Hopkins, A. (1986) Botanical composition of permanent grassland in England and Wales in relation to soil, environment and management factors. *Grass and Forage Science* 41, 237–246.
Hopkins, A. (1987) Distribution and management of grassland in the United Kingdom. In: Park, J.R. (ed.) *Environmental Management in Agriculture*. Bellhaven Press, London, pp. 178–185.
Hopkins, A., Adamson, A.H. and Bowling, P.J. (1994) Response of permanent and reseeded grassland to fertilizer nitrogen. 2. Effects on concentrations of Ca, Mg, K, Na, S, P, Mn, Zn, Cu, Co and Mo in herbage at a range of sites. *Grass and Forage Science* 49, 9–20.
Hopkins, A., Gilbey, J., Dibb, C., Bowling, P.J. and Murray, P.J. (1990) Response of permanent and reseeded grassland to fertilizer nitrogen. 1. Herbage production and herbage quality. *Grass and Forage Science* 45, 43–55.
Hopkins, A., Murray, P.J., Bowling, P.J., Rook, A.J. and Johnson, J. (1995) Productivity and nitrogen uptake of ageing and newly sown swards of perennial ryegrass (*Lolium perenne* L.) at different sites and with different nitrogen fertilizer treatments. *European Journal of Agronomy* 4, 65–75.
Hoult, E.H. and McGarity, J.W. (1986) The measurement and distribution of urease activity in a pasture system. *Plant and Soil* 93, 359–366.
Hoult, E.H. and McGarity, J.W. (1987) The influence of sward mass, defoliation and watering on ammonia volatilization losses from an Italian ryegrass sward top-dressed with urea. *Fertilizer Research* 13, 199–207.
Hungate, R.E. (1966) *The Rumen and its Microbes.* Academic Press, New York.
Hunt, I.V. (1966a) The effect of utilization of herbage on the response to fertilizer nitrogen. *Proceedings 9th International Grassland Congress, Sao Paulo,* 1965, 1113–1119.
Hunt, I.V. (1966b) The effect of age of sward on the yield and response of grass species to fertilizer nitrogen. *Proceedings 10th International Grassland Congress, Helsinki,* 1966, 249–254.
Hunt, L.A. (1965) Some implications of death and decay in pasture production. *Journal of the British Grassland Society* 20, 27–31.
Hunt, W.F. (1983) Nitrogen cycling through senescent leaves and litter in swards of Ruanui and Nui ryegrass with high and low nitrogen inputs. *New Zealand Journal of Agricultural Research* 26, 461–471.
Hutchings, N.J. (1984) The availability of nitrogen in liquid sewage sludges applied to grassland. *Journal of Agricultural Science, Cambridge* 102, 703–709.

Hutchinson, G.L. and Davidson, E.A. (1993) Processes for production and consumption of gaseous nitrogen oxides in soil. In: American Society of Agronomy, *Agricultural Ecosystem Effects on Trace Gases and Global Climate Change.* ASA Special Publication No. 55, pp. 79–93.

Hutchinson, G.L., Mosier, A.R. and Andre, C.E. (1982) Ammonia and amine emissions from a large cattle feedlot. *Journal of Environmental Quality* 11, 288–293.

Hutchinson, G.L. and Viets, F.G. (1969) Nitrogen enrichment of surface water by absorption of ammonia volatilized from cattle feedlots. *Science* 166, 514–515.

Hylton, L.O., Williams, D.E., Ulrich, A. and Cornelius, D.R. (1964) Critical nitrate levels for growth of Italian ryegrass. *Crop Science* 4, 16–19.

ICI Ltd. (1966) *Jealott's Hill Research Station; Guide to Field Experiments.* ICI, Bracknell, p. 90.

IPCC (Intergovernmental Panel on Climate Change) (1992) Climate change 1992. *Supplementary Report to the IPCC Scientific Assessment* (Houghton, J.T., Callander, B.A. and Varney, S.K., eds). Cambridge University Press, Cambridge.

Irvin, J.G. and Williams M.L. (1988) Acid rain: chemistry and transport. *Environmental Pollution* 50, 29–59.

Jackson, M.V. and Williams, T.E. (1979) Response of grass swards to fertilizer N under cutting or grazing. *Journal of Agricultural Science, Cambridge* 92, 549–562.

Jagtenberg, W.D. (1970) Predicting the best time to apply nitrogen to grassland in spring. *Journal of the British Grassland Society* 25, 266–271.

Jameson, H.R. (1959) Liquid nitrogenous fertilizers. *Journal of Agricultural Science, Cambridge* 53, 333–338.

Jansson, S.L. and Persson, J. (1982) Mineralization and immobilization of soil nitrogen. In: Stevenson, F.J. (ed.) *Nitrogen in Agricultural Soils.* American Society of Agronomy, Madison, Wisconsin, pp. 229–252.

Janzen, H.H. (1990) Deposition of nitrogen into the rhizosphere by wheat roots. *Soil Biology and Biochemistry* 22, 1155–1160.

Jaramillo, V.J. and Detling, J.K. (1992) Small-scale heterogeneity in a semi-arid North American grassland. I. Tillering, N uptake and retranslocation in simulated urine patches. *Journal of Applied Ecology* 29, 1–8.

Jarrige, R. and Alderman, G. (1987) *Feed Evaluation and Protein Requirement Systems for Ruminants.* C.E.C., Luxembourg.

Jarvis, S.C. (1987) The effects of low, regulated supplies of nitrate and ammonium nitrogen on the growth and composition of perennial ryegrass. *Plant and Soil* 100, 99–112.

Jarvis, S.C. and Barraclough, D. (1991) Variation in mineral nitrogen under grazed grassland swards. *Plant and Soil* 138, 177–188.

Jarvis, S.C., Barraclough, D., Williams, J. and Rook, A.J. (1991) Patterns of denitrification loss from grazed grassland: effects of N fertilizer inputs at different sites. *Plant and Soil* 131, 77–88.

Jarvis, S.C. and Hatch, D.J. (1985) The effects of aluminium on the growth of white clover dependent upon fixation of atmospheric nitrogen. *Journal of Experimental Botany* 36, 1075–1086.

Jarvis, S.C. and Hatch, D.J. (1994) Potential for denitrification at depth below long-term grass swards. *Soil Biology and Biochemistry* 26, 1629–1636.

Jarvis, S.C., Hatch, D.J. and Lockyer, D.R. (1989a) Ammonia fluxes from grazed grassland: annual losses from cattle production systems and their relation to nitrogen inputs. *Journal of Agricultural Science, Cambridge* 113, 99–108.

Jarvis, S.C., Hatch, D.J. and Roberts, D.H. (1989b) The effects of grassland management on nitrogen losses from grazed swards through ammonia volatilization: the relationship to excretal N returns from cattle. *Journal of Agricultural Science, Cambridge* 112, 205–216.

Jarvis, S.C. and Macduff, J.H. (1989) Nitrate nutrition of grasses from steady-state supplies in flowing solution culture following nitrate deprivation and/or defoliation. *Journal of Experimental Botany* 40, 965–975.

Jarvis, S.C., Macduff, J.H., Williams, J.R. and Hatch, D.J. (1990) Balances of forms of mineral N in grazed grassland soils: impact of N losses. *Proceedings 16th International Grassland Congress, Nice,* 1989, 151–152.

Jarvis, S.C. and Pain, B.F. (1990) Ammonia volatilisation from agricultural land. *Proceedings of the Fertiliser Society*, No. 298.

Jeater, R.S.L. (1967) Comparisons of liquefied (anhydrous) ammonia and ammonium nitrate as nitrogenous fertilizers for grassland. *Journal of the British Grassland Society* 22, 225–229.

Jenkinson, D.S. (1982) The supply of nitrogen from the soil. *MAFF Reference Book*, 385. HMSO, London, pp. 79–93.

Jenkinson, D.S. (1986) Nitrogen in UK arable agriculture. *Journal of the Royal Agricultural Society of England* 147, 178–189.

Jenkinson, D.S. (1990) Leaks in the nitrogen cycle. In: Merckx, R., Vereecken, H. and Vlassak, K. (eds) *Fertilization and the Environment.* Leuven University Press, pp. 35–39.

Jenkinson, D.S., Fox, R.H. and Rayner, J.H. (1985) Interactions between fertilizer nitrogen and soil nitrogen – the so called 'priming' effect. *Journal of Soil Science* 36, 425–444.

Jenkinson, D.S. and Rayner, J.H. (1977) The turnover of soil organic matter in some of the Rothamsted classical experiments. *Soil Science* 123, 298–305.

Jensen, H.L. (1947) Nitrogen fixation in leguminous plants. 7. The nitrogen fixing activity of root nodule tissue in *Medicago* and *Trifolium. Proceedings of Linnean Society of New South Wales* 72, 265–291.

Jensen, H.L. (1965) Nonsymbiotic nitrogen fixation. In: Bartholomew, W.V., and Clark, F.E. (eds) *Soil Nitrogen.* Agronomy No. 10. American Society of Agronomy, Madison, Wisconsin, pp. 436–480.

Johnes, P.J. and Burt, T.P. (1993) Nitrate in surface waters. In: Burt, T.P., Heathwaite, A.L. and Trudgill, S.T. (eds) *Nitrate: Processes, Patterns and Management.* John Wiley and Sons, Chichester, pp. 269–317.

Johnston, A.E. (1991) Soil fertility and soil organic matter. In: Wilson, W.S. (ed.) *Advances in Soil Organic Matter Research: The Impact on Agriculture and the Environment.* Royal Society of Chemistry, Cambridge, pp. 299–314.

Johnston, A.E., McEwen, J., Lane, P.W., Hewitt, M.V., Poulton, P.R. and Yeoman, D.P. (1994) Effects of one to six year old ryegrass-clover leys on soil nitrogen and on the subsequent yields and fertilizer nitrogen requirements of the arable sequence winter wheat, potatoes, winter wheat, winter beans (*Vicia faba*) grown on a sandy loam soil. *Journal of Agricultural Science, Cambridge* 122, 73–89.

Jollans, J.L. (ed.) (1981) *Grassland in the British Economy*. Centre for Agricultural Strategy, University of Reading, Paper No. 10.

Jollans, J.L. (1985) *Fertilizers in UK Farming*. Centre for Agricultural Strategy, University of Reading, Report No. 9.

Jones, D. and Haggar, R.J. (1993) Impact of nitrogen and organic manures on yield and botanical diversity of a grassland field margin. In: Hopkins, A. and Younie, D. (eds) *Forward with Grass into Europe*. Occasional Symposium No. 27. British Grassland Society, Hurley, pp. 135–138.

Jones, D.I.H., Griffith, G. ap. and Walters, R.J.K. (1965) The effect of nitrogen fertilizers on the water-soluble carbohydrate content of grasses. *Journal of Agricultural Science, Cambridge* 64, 323–328.

Jones, D.J.C. (1960) The effect of sulphate of ammonia applications on the sulphur content of various grass and clover mixtures. *Journal of Agricultural Science, Cambridge* 54, 188–194.

Jordan, C. (1989) The effect of fertilizer type and application rate on denitrification losses from cut grassland in Northern Ireland. *Fertilizer Research* 19, 45–55.

Jordan, C. and Smith, R.V. (1985) Factors affecting leaching of nutrients from an intensively managed grassland in County Antrim, Northern Ireland. *Journal of Environmental Management* 20, 1–15.

Karraker, P.E., Bortner, C.E. and Fergus, E.N. (1950) Nitrogen balance in lysimeters as affected by growing Kentucky bluegrass and certain legumes separately and together. *Bulletin 557 Kentucky Agriculture Experimental Station*, pp. 16.

Kaupenjohann, M., Dohler, H. and Bauer, M. (1989) Effects of N-emmissions on nutrient status and vitality of *Pinus sylvestris* near a hen-house. *Plant and Soil* 113, 279–282.

Keatinge, J.D.H., Steward, R.H. and Garrett, M.K. (1979) The influence of temperature and soil water potential on the leaf extension rate of perennial ryegrass in Northern Ireland. *Journal of Agricultural Science, Cambridge* 92, 175–183.

Keeney, D.R. (1982) Nitrogen – availability indices. In: Page, A.L. (ed.) *Methods of Soil Analysis, Part 2. Chemical and Microbiological Properties*. American Society of Agronomy, Madison, Wisconsin, pp. 711–733.

Keeney, D.R. and MacGregor, A.N. (1978) Short-term cycling of ^{15}N-urea in a ryegrass-white clover pasture. *New Zealand Journal of Agricultural Research* 21, 443–448.

Kemp, A. (1960) Hypomagnesaemia in milking cows : the response of serum magnesium to alterations in herbage composition resulting from potash and nitrogen dressings on pasture. *Netherlands Journal of Agricultural Science* 8, 281–304.

Kemp, A., Deijs, W.B. and Kluvers, E. (1966) Influences of higher fatty acids on the availability of magnesium in milking cows. *Netherlands Journal of Agricultural Science* 14, 290–295.

Kemp, A., Hemkes, O.J. and van Steenbergen, T. (1979) The crude protein production of grassland and the utilization by milking cows. *Netherlands Journal of Agricultural Science* 27, 36–47.

Kennedy, I.R. (1992) *Acid Soil and Acid Rain*, 2nd edn. Research Studies Press Ltd., Taunton.

Kershaw, E.S. (1963) The crude protein and nitrate nitrogen relationship in S22 in response to nitrogen and potash fertilizer treatments. *Journal of the British Grassland Society* 18, 323–327.

Kershaw, E.S. and Banton, C.L. (1965) The mineral content of S22 ryegrass on calcareous loam soil in response to fertilizer treatments. *Journal of the Science of Food and Agriculture* 16, 698–701.

Ketelaars, J.J.M.H. and Rap, H. (1994) Ammonia volatilization from urine applied to the floor of a dairy cow barn. *Proceedings 15th General Meeting, European Grassland Federation,* Wageningen, Netherlands, pp. 413–417.

Kilian, K.C., Attoe, O.J. and Engelbert, L.E. (1966) Urea-formaldehyde as a slowly available form of nitrogen for Kentucky bluegrass. *Agronomy Journal* 58, 204–206.

Kirchmann, H. (1985) Losses, plant uptake and utilisation of manure nitrogen during a production cycle. *Acta Agriculturae Scandinavica Supplementum* 24, 1–77.

Kirchmann, H. (1991) Carbon and nitrogen mineralization of fresh, aerobic and anaerobic animal manures during incubation with soil. *Swedish Journal of Agricultural Research* 21, 165–173.

Kirchmann, H. and Witter, E. (1992) Composition of fresh, aerobic and anaerobic farm animal dungs. *Bioresource Technology* 40, 137–142.

Kirkham, F.W. and Wilkins R.J. (1994) The productivity and response to inorganic fertilizers of species-rich wetland hay meadows on the Somerset Moors: the effect of nitrogen, phosphorus and potassium on herbage production. *Grass and Forage Science* 49, 163–175.

Kissel, D.E., Brewer, H.L. and Arkin, G.F. (1977) Design and test of a field sampler for ammonia volatilization. *Soil Science Society of America Journal* 41, 1133–1138.

Kissel, D.E. and Cabrera, M.L. (1989) Factors affecting urea hydrolysis. In: Bock, B.R. and Kissel, D.E. (eds) *Ammonia Volatilization from Urea Fertilizers,* Bulletin Y-206. National Fertilizer Development Center, Tennessee Valley Authority, Muscle Shoals, Alabama, pp. 53–66.

Klarenbeek, J.V. and Bruins, M.A. (1987) Ammonia emissions from livestock buildings and slurry spreading in The Netherlands. In: Nielsen, V.C., Voorburg, J.H. and L'Hermite, P. (eds) *Volatile Emissions from Livestock Farming and Sewage Operations.* Elsevier, London, pp. 73–83.

Klarenbeek, J.V. and Bruins, M.A. (1991) Ammonia emissions after land spreading of animal slurries. In: Nielsen, V.C., Voorburg, J.H. and L'Hermite, P. (eds) *Odour and Ammonia Emissions from Livestock Farming.* Elsevier, London, pp. 107–115.

Klasink, A., Steffens, G. and Kowalewsky, H-H. (1991) Odour and ammonia emissions from grassland and arable land. In: Nielsen, V.C., Voorburg, J.H. and L'Hermite, P. (eds) *Odour and Ammonia Emissions from Livestock Farming.* Elsevier, London, pp. 170–176.

Kleter, H. J. (1968) Influence of weather and nitrogen fertilization on white clover percentage of permanent grassland. *Netherlands Journal of Agricultural Science* 16, 43–52.

Knight, D., Elliott, P.W. and Anderson, J.M. (1989) Effects of earthworms upon transformations and movement of nitrogen from organic matter applied to agricultural soils. In: Hansen, J.A. and Henriksen, K. (eds) *Nitrogen in Organic Wastes Applied to Soils.* Academic Press, London, pp. 59–80.

Knowles, R. (1981) Denitrification. In: Paul, E.A. and Ladd, J. (eds) *Soil Biochemistry,* Vol 5. Marcel Dekker, New York, pp. 323–369.

Knowles, R. (1982) Denitrification. *Microbiological Reviews* 46, 43–70.

Kolenbrander, G.J. (1977) The nitrogen balance sheet. In: Voorburg, J.H. (ed.) *Utilisation of Manure by Land Spreading.* CEC, Luxembourg, pp. 673–698.

Kolenbrander, G.J. (1981) Leaching of nitrogen in agriculture. In: Brogan, J.C. (ed.) *Nitrogen Losses and Surface Run-off from Landspreading of Manures*. Martinus Nijhoff/Dr Junk, The Hague, pp. 199–216.

Kowalenko, C.G. (1978) Organic nitrogen, phosphorus and sulfur in soils. In: Schnitzer, M. and Khan, S.U. (eds) *Soil Organic Matter*. Elsevier, Amsterdam, pp. 95–136.

Kreil, W., Wacker, G., Kaltofen, H. and Hey, E. (1966) Heavy nitrogen fertilizing to pasture. *Proceedings of 9th International Grassland Congress, Sao Paulo, 1965*, pp. 1093–1098.

Kresge, C.B. and Decker, A.M. (1966) Nutrient balance in Midland Bermudagrass as affected by differential nitrogen and potassium fertilization. 1. Forage yields and persistence. *Proceedings of 9th International Grassland Congress, Sao Paulo*, 1965, pp. 671–674.

Kresge, C.B. and Satchell, D.P. (1960) Gaseous loss of ammonia from nitrogen fertilizers applied to soils. *Agronomy Journal* 52, 104–107.

Kreula, M., Rauramaa, A. and Ettala, T. (1978) The effect of feeding on the hippuric acid content of cow's urine. *Journal of the Scientific Agricultural Society of Finland* 50, 372–377.

Kruse, M., ApSimon, H.M. and Bell, J.N.B. (1989) Validity and uncertainty in the calculation of an emission inventory for ammonia arising from agriculture in Great Britain. *Environmental Pollution* 56, 237–257.

Kuntze, H. (1964) [Decomposition of grass and clover roots in soil.] *Zeitschrift für Planzenernährung und Bodenkunde* 120, 383–400.

Laidlaw, A.S. (1980) The effects of nitrogen fertilizer applied in spring on swards of perennial ryegrass sown with four cultivars of white clover. *Grass and Forage Science* 35, 295–299.

Laidlaw, A.S. (1984) Quantifying the effect of nitrogen fertilizer applications in spring on white clover content in perennial ryegrass-white clover swards. *Grass and Forage Science* 39, 317–321.

Laidlaw, A.S. (1988) The contribution of different white clover cultivars to the nitrogen yield of mixed swards. *Grass and Forage Science* 43, 347–350.

Laidlaw, A.S. and Steen, R.W.J. (1989) Turnover of grass laminae and white clover leaves in mixed swards continuously grazed with steers at a high- and low-N fertilizer level. *Grass and Forage Science* 44, 249–258.

Laidlaw, A.S. and Stewart, T.A. (1987) Clover development in the sixth to ninth year of a grass/clover sward as affected by out-of-season management and spring fertilizer nitrogen application. *Research and Development in Agriculture* 4, 155–160.

Lambert, D.A. (1964) The effect of level of nitrogen and cutting treatment on leaf area in swards of S48 timothy (*Phleum pratense* L.) and S215 meadow fescue (*Festuca pratensis* L.). *Journal of the British Grassland Society* 19, 396–402.

Lambert, J. and Toussaint, B. (1978) An investigation of the factors influencing the phosphorus content of herbage. *Phosphorus in Agriculture* 73, 1–12.

Lambert, M.G., Renton, S.W. and Grant, D.A. (1982) Nitrogen balance studies in some North Island hill pastures. In: Gandar, P.W. (ed.) *Nitrogen Balances in New Zealand Ecosystems*. DSIR, Palmerston North, pp. 35–39.

Langer, R.H.M. (1959) Growth and nutrition of timothy (*Phleum pratense* L.). 4. The effect of nitrogen, phosphorus and potassium supply on growth during the first year. *Annals of Applied Biology* 47, 211–221.

Langer, R.H.M. (1963) Tillering in herbage grasses. *Herbage Abstracts* 33, 141–148.

Lantinga, E.A., Keuning, J.A., Groenwold, J. and Deenen, P.A.G. (1987) Distribution of excreted nitrogen by grazing cattle and its effects on sward quality, herbage production and utilization. In: van der Meer, H.G., Unwin, R.J., van Dijk, T.A. and Ennik, G.C. (eds) *Animal Manure on Grassland and Fodder Crops: Fertilizer or Waste*. Martinus Nijhoff, Dordrecht, pp. 103–117.

Large, R.V. and Spedding, C.R.W. (1966) The health and performance of the grazing animal in relation to fertilizer nitrogen usage. 2. Weaned lambs. *Journal of Agricultural Science, Cambridge* 67, 41–52.

Lauer, D.A., Bouldin, D.R. and Klausner, S.D. (1976) Ammonia volatilization from dairy manure spread on the soil surface. *Journal of Environmental Quality* 5, 134–141.

Lawlor, D.W. (1991) Concepts of nutrition in relation to cellular processes and environment. In: Porter, J.R. and Lawlor, D.W. (eds) *Plant Growth: Interactions with Nutrition and Environment.* Cambridge University Press, Cambridge, pp. 1–32.

Laws, J.A. and Pain, B.F. (1994) Defining and evaluating grassland management systems associated with four contrasting nitrogen inputs. *Proceedings 15th General Meeting, European Grassland Federation,* Wageningen, Netherlands, pp. 387–391.

Lazenby, A. (1981) British grasslands: past, present and future. *Grass and Forage Science* 36, 243–266.

Lazenby, A. (1988) The grass crop in perspective: selection, plant performance and animal production. In: Jones, M.B. and Lazenby, A. (eds) *The Grass Crop.* Chapman and Hall, London, pp. 311–360.

Lazenby, A. and Rogers, H.H. (1965) Selection criteria in grass breeding. 5. Performance of *Lolium perenne* genotypes grown at different nitrogen levels and spacings. *Journal of Agricultural Science, Cambridge* 65, 79–89.

Leaver, J.D. (1976) Utilisation of grassland by dairy cows. In: Swan, H. and Broster, W.H. (eds) *Principles of Cattle Production*. Butterworths, London, pp. 307–327.

Leconte, D. (1987) [Response of *Trifolium repens* in a mixed grass sward in lower Normandy. II. Physiological studies in a pure sward.] *Fourrages* 109, 27–39. (*Soils and Fertilizers*, 1987, 50, 11424.)

Ledgard, S.F. (1991) Transfer of fixed nitrogen from white clover to associated grasses in swards grazed by dairy cows, estimated using ^{15}N methods. *Plant and Soil* 131, 215–223.

Ledgard, S.F., Brier, G.J. and Upsdell, M.P. (1990) Effect of clover cultivar on production and nitrogen fixation in clover-ryegrass swards under dairy cow grazing. *New Zealand Journal of Agricultural Research* 33, 243–249.

Ledgard, S.F. and Peoples, M.B. (1988) Measurement of nitrogen fixation in the field. In: Wilson, J.R. (ed.) *Advances in Nitrogen Cycling in Agricultural Ecosystems.* CAB International, Wallingford, pp. 351–367.

Ledgard, S.F., Steele, K.W. and Saunders, W.H.M. (1982) Effects of cow urine and its major constitutents on pasture properties. *New Zealand Journal of Agricultural Research* 25, 61–68.

Lee, D.S. and Dollard, G.J. (1994) Uncertainties in current estimates of emissions of ammonia in the United Kingdom. *Environmental Pollution* 86, 267–277.
Lee, G.R., Davies, L.H., Armitage, E.R. and Hood, A.E.M. (1977) The effects of rates of nitrogen application on seven perennial ryegrass varieties. *Journal of the British Grassland Society* 32, 83–87.
Lee, K.E. (1985) *Earthworms: Their Ecology and Relationships with Soils and Land Use.* Academic Press, Sydney.
Legg, J.O. and Meisinger, J.J. (1982) Soil nitrogen budgets. In: Stevenson, F.J. (ed.) *Nitrogen in Agricultural Soils.* American Society of Agronomy, Madison, Wisconsin, pp. 503–566.
Lemaire, G. and Denoix, A. (1987) Summer dry matter accumulation in *Festuca arundinacea* and *Dactylis glomerata* populations in Western France. II. Interaction between moisture level and nitrogen nutrition. *Agronomie* 7, 381–389 (*Soils and Fertilizers*, 1988, 51-8422.)
L'Estrange, J.L., Owen, J.B. and Wilman, D. (1967) Effects of a high level of nitrogenous fertilizer and date of cutting on the availability of the magnesium and calcium of herbage to sheep. *Journal of Agricultural Science, Cambridge* 68, 173–178.
Levi-Minzi, R., Riffaldi, R. and Saviozzi, A. (1986) Organic matter and nutrients in fresh and mature farmyard manure. *Agricultural Wastes* 16, 225–236.
Lightner, J.W., Mengel, D.B. and Rhykerd, C.L. (1990) Ammonia volatilization from nitrogen fertilizer surface applied to orchardgrass sod. *Soil Science Society of America Journal* 54, 1478–1482.
Limmer, A.W. and Steele, K.W. (1983) Effect of cow urine upon denitrification. *Soil Biology and Biochemistry* 15, 409–412.
Limmer, A.W., Steele, K.W. and Wilson, A.T. (1982) Direct field measurement of N_2 and N_2O evolution from soil. *Journal of Soil Science* 33, 499–507.
Linehan, P.A. and Lowe, J. (1961) Yielding capacity and grass/clover ratio of herbage swards as influenced by fertilizer treatments. *Proceedings 8th International Grassland Congress,* Reading 1960, 133–137.
Lloyd, A. (1992a) Nitrate leaching following the break-up of grassland for arable cropping. In: Archer, J.R. *et al.* (eds) *Nitrate and Farming Systems.* Association of Applied Biologists, Wellesbourne, Warwick, pp. 243–247.
Lloyd, A. (1992b) Urea as a nitrogen fertilizer for grass cut for silage. *Journal of Agricultural Science, Cambridge* 119, 373–381.
Lloyd, D. (1993) Aerobic denitrification in soils and sediments: from fallacies to facts. *Trends in Ecology and Evolution* 8, 352–356.
Lockyer, D.R. (1984) A system for the measurement in the field of losses of ammonia through volatilization. *Journal of the Science of Food and Agriculture* 35, 837–848.
Lockyer, D.R. and Cowling, D.W. (1977) Non-symbiotic nitrogen fixation in some soils of England and Wales. *Journal of the British Grassland Society* 32, 7–11.
Lockyer, D.R. and Whitehead, D.C. (1990) Volatilization of ammonia from cattle urine applied to grassland. *Soil Biology and Biochemistry* 22, 1137–1142.
Long, F.N.J. and Gracey, H.I. (1990a) Herbage production and nitrogen recovery from slurry injection and fertilizer nitrogen application. *Grass and Forage Science* 45, 77–82.

Long, F.N.J. and Gracey, H.I. (1990b) Effect of fertilizer nitrogen source and cattle slurry on herbage production and nitrogen utilization. *Grass and Forage Science* 45, 431–442.

Lotero, J., Woodhouse, W.W. and Petersen, R.G. (1966) Local effect on fertility of urine voided by grazing cattle. *Agronomy Journal* 58, 262–265.

Low, A.J. (1972) The effect of cultivation on the structure and other physical characteristics of grassland and arable soils (1945–70). *Journal of Soil Science* 23, 363–380.

Low, A.J. (1973) Nitrate and ammonium nitrogen concentration in water draining through soil monoliths in lysimeters cropped with grass or clover or uncropped. *Journal of the Science of Food and Agriculture* 24, 1489–1495.

Low, A.J. and Armitage, E.R. (1959) Irrigation of grassland. *Journal of Agricultural Science, Cambridge* 52, 256–262.

Low, A.J. and Armitage, E.R. (1970) The composition of the leachate through cropped and uncropped soils in lysimeters compared with that of the rain. *Plant and Soil* 33, 393–411.

Low, A.J. and Piper, F.J. (1961) Urea as a fertilizer. Laboratory and pot-culture studies. *Journal of Agricultural Science, Cambridge* 57, 249–255.

Lowe, J. (1966) Output of pastures under a clover nitrogen regime in Northern Ireland. *Proceedings of 10th International Grassland Congress, Helsinki, 1966,* pp. 187–191.

Lowe, J. (1967) Botanical development and output of a sward seeded with perennial ryegrass and white clover under stated fertilizer treatments. *Record of Agricultural Research (Northern Ireland)* 16(1), 75–91.

Loynachan, T.E., Bartholomew, W.V. and Wollum, A.G. (1976) Nitrogen transformations in aerated swine manure slurries. *Journal of Environmental Quality* 5, 293–297.

Luebs, R.E., Davis, K.R. and Laag, A.E. (1974) Diurnal fluctuation and movement of atmospheric ammonia and related gases from dairies. *Journal of Environmental Quality* 3, 265–269.

Lüscher, A. and Nösberger, J. (1992) Overwintering and spring growth of white clover. *Proceedings 14th General Meeting, European Grassland Federation,* Lahti, Finland, 167–171.

Lycklama, J.C. (1963) The absorption of ammonium and nitrate by perennial ryegrass. *Acta Botanica Neerlandica* 12, 361–423.

Lyster, S., Morgan, M.A. and O'Toole, P. (1980) Ammonia volatilization from soils fertilized with urea and ammonium nitrate. *Journal of Life Sciences, Royal Dublin Society* 1, 167–176.

Lyttleton, J.W. (1973) Proteins and nucleic acids. In: Butler, G.W. and Bailey, R.W. (eds) *Chemistry and Biochemistry of Herbage*. Vol. 1. Academic Press, London, pp. 63–103.

Ma, W.-C., Brussaard, L. and de Riddler, J.A. (1990) Long-term effects of nitrogenous fertilizers on grassland earthworms (Oligochaeta: Lumbricidae): their relation to soil acidification. *Agriculture, Ecosystems and Environment* 30, 71–80.

McAllister, J.S.V. (1966a) A study of the responses to alternative sources of nitrogen in Northern Ireland. 1. Effects of time of application on yield and mineral composition of Italian ryegrass. *Record of Agricultural Research (Northern Ireland)* 15(2), 67–87.

McAllister, J.S.V. (1966b) A study of the responses to alternative sources of nitrogen in Northern Ireland. 2. Effects of moderate and of heavy dressings on the yield and mineral composition of Italian ryegrass. *Record of Agricultural Research (Northern Ireland)* 15(2), 89–110.

McAllister, J.S.V. and McConaghy, S. (1960) The application of heavy dressings of nitrogen to pasture. *Research Experiment Record, Ministry of Agriculture, Northern Ireland* 10, 87–104.

McAllister, J.S.V., McConaghy, S., Coey, W.E. and Kerr, J.A.M. (1965) The effects of different nitrogen treatments on the spring growth of ryegrass. 1. Effects on yield and nitrogen content of the herbage. *Record of Agricultural Research (Northern Ireland)* 14(2), 15–29.

McAuliffe, C., Chamblee, D.S., Uribe-Arango, H. and Woodhouse, W.W. (1958) Influence of inorganic nitrogen on nitrogen fixation by legumes as revealed by N^{15}. *Agronomy Journal* 50, 334–337.

McCarrick, R.B. and Wilson, R.K. (1966) Effects of nitrogen fertilization of mixed swards on herbage yield, dry matter digestibility and voluntary food intake of the conserved herbages. *Journal of the British Grassland Society* 21, 195–199.

McCarty, G.W. and Bremner, J.M. (1991) Production of urease by microbial activity in soils under aerobic and anaerobic conditions. *Biology and Fertility of Soils* 11, 228–230.

McConaghy, S., Stewart, J.W.B. and Lowe, J. (1962) The effect on soils and herbage of a nitrogenous fertilizer containing ammonium nitrate, applied regularly at varying levels. *Research Experiment Record, Ministry of Agriculture, Northern Ireland* 12, 71–92.

MacDiarmid, B.N. and Watkin, B.R. (1971) The cattle dung patch. 1. Effect of dung patches on yield and botanical composition of surrounding and underlying pasture. *Journal of the British Grassland Society* 26, 239–245.

MacDiarmid, B.N. and Watkin, B.R. (1972a) The cattle dung patch. 2. Effect of a dung patch on the chemical status of the soil and ammonia nitrogen losses from the patch. *Journal of the British Grassland Society* 27, 43–48.

MacDiarmid, B.N. and Watkin, B.R. (1972b) The cattle dung patch. 3. Distribution and rate of decay of dung patches and their influence on grazing behaviour. *Journal of the British Grassland Society* 27, 48–54.

Macduff, J.H., Steenvoorden, J.H.A.M., Scholefield, D. and Cuttle, S.P. (1990) Nitrate leaching losses from grazed grassland. *Proceedings of the 13th General Meeting, European Grassland Federation*, Banska Bystrica, Czechoslovakia 2, 18–24.

Macduff, J.H. and White, R.E. (1984) Components of the nitrogen cycle measured for cropped and grassland soil-plant systems. *Plant and Soil* 76, 35–47.

McGarry, S.J., O'Toole, P. and Morgan, M.A. (1987) Effects of soil temperature and moisture content on ammonia volatilization from urea-treated pasture and tillage soils. *Irish Journal of Agricultural Research* 26, 173–182.

McGill, W.B., Hunt, H.W., Woodmansee, R.G. and Reuss, J.O. (1981) Phoenix, a model of the dynamics of carbon and nitrogen in grassland soils. In: Clark, F.E. and Rosswall, T. (eds) *Terrestrial Nitrogen Cycles. Ecological Bulletins (Stockholm)* 33, 49–115.

McInnes, K.J., Kissel, D.E. and Kanemasu, E.T. (1985) Estimating ammonia flux: a comparison between the integrated horizontal flux method and theoretical solutions of the diffusion profile. *Agronomy Journal* 77, 884–889.

Mackenzie, G.H. and Daly, M. (1982) Nitrogen use on perennial ryegrass-white clover swards. *Grass and Forage Science* 37, 181–183.

McLean, E.O., Adams, D. and Franklin, R.E. (1956) Cation exchange capacities of plant roots as related to their nitrogen contents. *Proceedings of Soil Science Society of America* 20, 345–347.

MacLeod, L.B. (1965) Effect of nitrogen and potassium on the yield, botanical composition and competition for nutrients in three alfalfa-grass associations. *Agronomy Journal* 57, 129–134.

MacLusky, D.S. (1960) Some estimates of the areas of pasture fouled by the excreta of dairy cows. *Journal of the British Grassland Society* 15, 181–188.

MacLusky, D.S. and Morris, D.W. (1964) Grazing methods, stocking rate and grassland production. *Agricultural Progress* 39, 97–108.

McNaught, K.J. and Christoffels, P.J.E. (1961) Effect of sulphur deficiency on sulphur and nitrogen levels in pastures and lucerne. *New Zealand Journal of Agricultural Research* 4, 177–196.

MAFF/DOE (1993) Review of the rules for sewage sludge application to agricultural land: soil fertility aspects of potentially toxic elements. *Report by the Independent Scientific Committee on Soil Fertility Aspects of Sewage Sludge Use in Agriculture.* MAFF, London, 91 pp.

Mahler, H.R. and Cordes, E.H. (1971) *Biological Chemistry,* 2nd edn. Harper and Row, New York.

Mahoney, A.W. and Poulton, B.R. (1962) Effects of nitrogen fertilization and date of harvest on the acceptibility of timothy forage. *Journal of Dairy Science* 45, 1575.

Mallarino, A.P., Wedin, W.F., Perdomo, C.H., Goyenola, R.S. and West, C.P. (1990) Nitrogen transfer from white clover, red clover, and birdsfoot trefoil to associated grass. *Agronomy Journal* 82, 790–795.

Mangan, J.L. (1982) The nitrogenous constituents of fresh forages. In: Thomson, D.J., Beever, D.E. and Gunn, R.G. (eds) *Forage Protein in Ruminant Animal Production.* British Society of Animal Production Occasional Publication No. 6, pp. 25–40.

Mannetje, L.'t (1994) Towards sustainable grassland management in The Netherlands. *Proceedings 15th General Meeting, European Grassland Federation,* Wageningen, Netherlands, pp. 3–18.

Marriott, C.A. (1988) Seasonal variation in white clover content and nitrogen fixing (acetylene reducing) activity in a cut upland sward. *Grass and Forage Science* 43, 253–262.

Marriott, C. and Smith, M. (1992) Senescence and decomposition of white clover stolons in grazed upland grass/clover swards. *Plant and Soil* 139, 219–227.

Marriott, C.A. and Haystead, A. (1990) The effect of defoliation on the nitrogen economy of white clover: regrowth and the remobilization of plant organic nitrogen. *Annals of Botany* 66, 465–474.

Marriott, C.A., Smith, M.A. and Baird, M.A. (1987) The effect of sheep urine on clover performance in a grazed upland sward. *Journal of Agricultural Science, Cambridge* 109, 177–185.

Marsh, R. and Campling, R.C. (1970) Fouling of pastures by dung. *Herbage Abstracts* 40, 123–130.

Marstorp, H. and Kirchmann, H. (1991) Carbon and nitrogen mineralization and crop uptake of nitrogen from six green manure legumes decomposing in soil. *Acta Agriculturae Scandinavica* 41, 243–252.

Martin-Smith, M. (1963) Uricolytic enzymes in soil. *Nature, London* 197, 361–362.

Mason, V.C., Kessank, P., Ononiwu, J.C. and Narang, M.P. (1981) Factors influencing faecal nitrogen excretion in sheep. 2. Carbohydrate fermentation in the caecum and large intenstine. *Zeitschrift für Tierphysiologie, Tierernährung und Futtermittelkunde* 45, 174–184.

Masterton, C.L. and Murphy, P.M. (1976) Application of the acetylene reduction technique to the study of nitrogen fixation by white clover in the field. In: Nutman, P.S. (ed.) *Symbiotic Nitrogen Fixation in Plants.* Cambridge University Press, Cambridge, pp. 299–316.

Masterton, C.L. and Sherwood, M.T. (1970) White clover/Rhizobium symbiosis: a review. In: Lowe, J. (ed.) *White Clover Research.* Occasional Symposium No. 6, British Grassland Society, pp. 11–39.

Mayne, C.S. and Wright, I.A. (1988) Herbage intake and utilization by the grazing dairy cow. In: Garnsworthy, P.C. (ed.) *Nutrition and Lactation in the Diary Cow.* Butterworths, Guildford, pp. 280–288.

Metson, A.J. and Hurst, F.B. (1953) Effects of sheep dung and urine on a soil under pasture at Lincoln, Canterbury, with particular reference to potassium and nitrogen equilibria. *New Zealand Journal of Science and Technology, Section A* 35, 327–359.

Metson, A.J. and Saunders, W.M.H. (1978) Seasonal variations in chemical composition of pasture. II. Nitrogen, sulphur and soluble carbohydrate. *New Zealand Journal of Agricultural Research* 21, 355–364.

Metson, A.J., Saunders, W.M.H., Collie, T.W. and Graham, V.W. (1966) Chemical composition of pastures in relation to grass tetany in beef breeding cows. *New Zealand Journal of Agricultural Research* 9, 410–436.

Meyer, R.D. and Jarvis, S.C. (1989) The effects of fertiliser/urine interactions on NH_3-N losses from grassland soils. *Proceedings 16th International Grassland Congress*, Nice, pp. 155–156.

Middelkoop, N. and Deenen, P.J.A.G. (1990) The local influence of cattle dung and urine and its interactions with fertilizer nitrogen on herbage dry matter production. *Proceedings 13th General Meeting, European Grassland Federation,* pp. 67–70.

Miles, D.G. and Williams, I.G. (1964) Winter hardiness in pasture varieties. *Report of Welsh Plant Breeding Station 1963*, pp. 70–71.

Millard, P., Sharp, G.S. and Scott, N.M. (1985) The effect of sulphur deficiency on the uptake and incorporation of nitrogen in ryegrass. *Journal of Agricultural Science, Cambridge* 105, 501–504.

Millard, P., Thomas, R.J. and Buckland, S.T. (1990) Nitrogen supply affects the remobilization of nitrogen for the regrowth of defoliated *Lolium perenne* L. *Journal of Experimental Botany* 41, 941–947.

Miller, W.J., Adams, W.E., Nussbaumer, R., McCreery, R.A. and Perkins, H.F. (1964) Zinc content of Coastal Bermudagrass as influenced by frequency and season of harvest, location, and level of N and lime. *Agronomy Journal* 56, 198–201.

Minchin, F.R. and Pate, J.S. (1973) The carbon balance of a legume and the functional economy of its root nodules. *Journal of Experimental Botany* 24, 259–271.

Ministry of Agriculture, Fisheries and Food (1993) *Agriculture in the UK, 1992*. HMSO, London.
Ministry of Agriculture, Fisheries and Food (1994) *Fertiliser Recommendations for Agricultural and Horticultural Crops*. Reference Book 209, 6th edn. HMSO, London.
Minson, D.J. (1990) *Forage in Ruminant Nutrition*. Academic Press, London.
Molloy, S.P. and Tunney, H. (1983) A laboratory study of ammonia volatilization from cattle and pig slurry. *Irish Journal of Agricultural Research* 22, 37–45.
Moloney, D. and Murphy, W.E. (1963) The effect of different levels of nitrogen on a grass clover sward under grazing conditions. I. Animal output. *Irish Journal of Agricultural Research* 2, 1–12.
Monaghan, R.M. and Barraclough, D. (1992) Some chemical and physical factors affecting the rate and dynamics of nitrification in urine-affected soil. *Plant and Soil* 143, 11–18.
Monaghan, R.M. and Barraclough, D. (1993) Nitrous oxide and dinitrogen emissions from urine-affected soil under controlled conditions. *Plant and Soil* 151, 127–138.
Moore, D.R.E. and Waid, J.S. (1971) The influence of washings of living roots on nitrification. *Soil Biology and Biochemistry* 3, 69–83.
Morrill, L.G. and Dawson, J.E. (1967) Patterns observed for the oxidation of ammonium to nitrate by soil organisms. *Proceedings Soil Science Society of America* 31, 757–760.
Morris, D.R., Zuberer, D.A. and Weaver, R.W. (1985) Nitrogen fixation by intact grass-soil cores using $^{15}N_2$ and acetylene reduction. *Soil Biology and Biochemistry* 17, 87–91.
Morrison, J. (1980) The influence of climate and soil on the yield of grass and its response to fertilizer nitrogen. In: Prins, W.H. and Arnold, G.H. (eds) *The Role of Nitrogen in Intensive Grassland Production*. PUDOC, Wageningen, pp. 51–57.
Morrison, J. (1987) Effects of nitrogen fertilizer. In: Snaydon, R.W. (ed.) *Managed Grassland: Analytical Studies*. Elsevier, Amsterdam, pp. 61–70.
Morrison, J., Denehy, H. and Chapman, P.F. (1983) Possibilities for the strategic use of fertilizer N on white clover/grass swards. *Proceedings 9th General Meeting, European Grassland Federation.* British Grassland Society Occasional Symposium No. 14, 227–231.
Morrison, J., Jackson, M.V. and Sparrow, P.E. (1980) *The Response of Perennial Ryegrass to Fertilizer Nitrogen in Relation to Climate and Soil.* Technical Report 27, Grassland Research Institute, Hurley.
Mortensen, W.P., Baker, A.S. and Dermanis, P. (1964) Effects of cutting frequency of orchardgrass and nitrogen rate on yield, plant nutrient composition and removal. *Agronomy Journal* 56, 316–320.
Morton, J.D. and Baird, D.B. (1990) Spatial distribution of dung patches under sheep grazing. *New Zealand Journal of Agricultural Research* 33, 285–294.
Mosier, A.R. (1980) Acetylene inhibition of ammonium oxidation in soil. *Soil Biology and Biochemistry* 12, 443–444.
Mosier, A.R. (1994) Nitrous oxide emissions from agricultural soils. *Fertilizer Research* 37, 191–200.
Mouat, M.C.H. and Walker, T.W. (1959) Competition for nutrients between grasses and white clover. 1. Effect of grass species and nitrogen supply. *Plant and Soil* 11, 30–40.

Mountford, J.O., Lakhani, K.H. and Kirkham, F.W. (1993) Experimental assessment of the effects of nitrogen addition under hay-cutting and aftermath grazing on the vegetation of meadows on a Somerset peat moor. *Journal of Applied Ecology* 30, 321–332.

Muck, R.E. (1981) Urease activity in bovine feces. *Journal of Dairy Science* 65, 2157–2163.

Mudd, A.J. (1970) The influence of heavily fertilized grass on mineral metabolism of dairy cows. *Journal of Agricultural Science, Cambridge* 74, 11–21.

Mullaly, J.V., McPherson, J.B., Mann, A.P. and Rooney, D.R. (1967) The effect of length of legume and non-legume leys on gravimetric soil nitrogen at some locations in the Victorian wheat areas. *Australian Journal of Experimental Agriculture and Animal Husbandry* 7, 568–571.

Mulvaney, R.L. and Bremner, J.M. (1981) Control of urea transformations in soils. In: Paul, E.A. and Ladd, J.N. (eds) *Soil Biochemistry*, Volume 5. Marcel Dekker, New York, pp. 153–196.

Mummey, D.L., Smith, J.L. and Bolton, H. (1994) Nitrous oxide flux from a shrub-steppe ecosystem: sources and regulation. *Soil Biology and Biochemistry* 26, 279–286.

Munro, I.A. (1958) Irrigation of grassland. The influence of irrigation and nitrogen treatments on the yield and utilization of a riverside meadow. *Journal of the British Grassland Society* 13, 213–221.

Munro, J.M.M. and Davies, D.A. (1974) Potential pasture production in the uplands of Wales. 5. The nitrogen contribution of white clover. *Journal of the British Grassland Society* 29, 213–223.

Munro, P.E. (1966) Inhibition of nitrifiers by grass root extracts. *Journal of Applied Ecology* 3, 231–238.

Murdock, L.W. and Frye, W.W. (1985) Comparison of urea and urea-ammonium polyphosphate with ammonium nitrate in production of tall fescue. *Agronomy Journal* 77, 630–633.

Murphy, M.D. and O'Donnell, T. (1989) Sulphur deficiency in herbage in Ireland. *Irish Journal of Agricultural Research* 28, 79–90.

Murphy, P.M. (1987) The contribution of white clover in grassland. *Irish Grassland and Animal Production Association Journal* 1987, 3–10.

Murphy, P.M., Turner, S. and Murphy, M. (1986) Effect of spring applied urea and calcium ammonium nitrate on white clover (*Trifolium repens*) performance in a grazed ryegrass-clover pasture. *Irish Journal of Agricultural Research* 25, 251–259.

Myers, R.J.K. (1987) Modelling the behaviour of nitrogen in soil-plant systems. In: Bacon, P.E., Evans, J., Storrier, R.R. and Taylor, A.C. (eds) *Nitrogen Cycling in Temperate Agricultural Systems*. Australian Society of Soil Science, Riverina, pp. 397–425.

Mytton, L.R. (1975) Plant genotype × rhizobium strain interactions in white clover. *Annals of Applied Biology* 80, 103–107.

National Research Council (Panel on Nitrates), USA (1978) *Nitrates: an Environmental Assessment*. National Academy of Sciences, Washington D.C.

National Research Council (Subcommittee on Nitrogen Usage in Ruminants), USA (1985) *Ruminant N Usage*. National Academy of Sciences, Washington D.C.

Nehring, K., Zelik, U. and Schiemann, R. (1965) [On the content of organic compounds in the urine of cattle, sheep and pigs.] *Archiv für Tierernährung* 15, 45–52.

Nelson, D.W. (1982) Gaseous losses of nitrogen other than through denitrification. In: Stevenson, F.J. (ed.) *Nitrogen in Agricultural Soils.* American Society of Agronomy, Madison, Wisconsin, pp. 327–363.

Nelson, K.E., Turgeon, A.J. and Street, J.R. (1980) Thatch influence on mobility and transformation of nitrogen carriers applied to turf. *Agronomy Journal* 72, 487–492.

Nelson, S.H. and Sosulski, F.W. (1984) Amino acid and protein content of *Poa pratensis* as related to nitrogen application and colour. *Canadian Journal of Plant Science* 64, 691–697.

Nesheim, L. and Boller, B.C. (1991) Nitrogen fixation by white clover when competing with grasses at moderately low temperatures. *Plant and Soil* 133, 47–56.

Newbould, P. (1981) The potential of indigenous plant resources. In: Frame, J. (ed.) *The Effective Use of Forage and Animal Resources in the Hills and Uplands.* British Grassland Society Occasional Symposium, No. 12. British Grassland Society, Hurley, pp. 1–15.

Newbould, P. and others (1982) The effect of *Rhizobium* inoculation on white clover in improved hill soils in the United Kingdom. *Journal of Agricultural Science, Cambridge* 99, 591–610.

Newman, R.J., Allen, B.F. and Cook, M.G. (1962) The effect of nitrogen on winter pasture production in southern Victoria. *Australian Journal of Experimental Agriculture and Animal Husbandry* 2, 20–24.

Nielsen, K.F. and Cunningham, R.K. (1964) The effects of soil temperature and form and level of nitrogen on growth and chemical composition of Italian ryegrass. *Proceedings of the Soil Science Society of America* 28, 213–218.

Noller, C.H. and Rhykerd, C.L. (1974) Relationship of nitrogen fertilization and chemical composition of forage to animal health and performance. In: Mays, D.A. (ed.) *Forage Fertilization.* American Society of Agronomy, Madison, Wisconsin, pp. 363–394.

Nommick, H. and Vahtras, K. (1982) Retention and fixation of ammonium and ammonia in soils. In: Stevenson, F.J. (ed.) *Nitrogen in Agricultural Soils.* American Society of Agronomy, Madison, Wisconsin, pp. 123–171.

North of Scotland College of Agriculture (1965) Grassland Experimental Centre, Muchalls, Kincardine: *Guide to Experiments*, pp. 6–7.

Nowakowski, T.Z. (1961) The effect of different nitrogenous fertilizers, applied as solids or solutions, on the yield and nitrate-N content of established grass and newly-sown ryegrass. *Journal of Agricultural Science, Cambridge* 56, 287–292.

Nowakowski, T.Z. (1962) Effects of nitrogen fertilizers on total nitrogen, soluble nitrogen and soluble carbohydrate contents of grass. *Journal of Agricultural Science, Cambridge* 59, 387–392.

Nowakowski, T.Z. (1964) Mineral fertilisation and organic composition of herbage. *Proceedings 2nd Regional Conference International Potash Institute, Morat, Switzerland,* 1964, 63–73.

Nowakowski, T.Z. and Cunningham, R.K. (1966) Nitrogen fractions and soluble carbohydrates in Italian ryegrass. 2. Effects of light intensity, form and level of nitrogen. *Journal of the Science of Food and Agriculture* 17, 145–150.

Nowakowski, T.Z., Cunningham, R.K. and Nielsen, K.F. (1965) Nitrogen fractions and soluble carbohydrates in Italian ryegrass. 1. Effects of soil temperature, form and level of nitrogen. *Journal of the Science of Food and Agriculture* 16, 124–134.

Nowakowski, T.Z. and Gasser, J.K.R. (1967) The effect of a nitrification inhibitor on the concentration of nitrate in plants. *Journal of Agricultural Science, Cambridge* 68, 131–133.

Nyborg, M. and Hoyt, P.B. (1978) Effects of soil acidity and liming on mineralization of soil nitrogen. *Canadian Journal of Soil Science* 58, 331–338.

Oakes, D. B. (1982) Nitrate pollution of groundwater resources – mechanisms and modelling. In: Zwirnmann, K.H. (ed.) *Non-Point Nitrate Pollution of Municipal Water Supply Sources: Issues of Analysis and Control.* International Institute for Applied Systems Analysis, Collaborative Proceedings Series CP-82-S4, Laxenburg, Austria, pp. 207–230.

O'Brien, T.A. (1960) The influence of nitrogen on seedling and early growth of perennial ryegrass and cocksfoot. *New Zealand Journal of Agricultural Research* 3, 399–411.

O'Connor, K.F. (1974) Nitrogen in agrobiosystems and its environmental significance. *New Zealand Agricultural Science* 8, 137–148.

O'Connor, K.F. (1990) Pasture and soil fertility. In Langer, R.H.M. (ed.) *Pastures: their Ecology and Management.* Oxford University Press, Auckland, pp. 157–196.

Odhuba, E.K., Reid, R.L. and Jung, G.A. (1965) Nutritive evaluation of tall fescue pasture. *Journal of Animal Science* 24, 1216.

Oenema, O., Postmus, J., Prins, W.H. and Neeteson, J.J. (1989) Seasonal variations in soil mineral nitrogen and in the response of grassland to nitrogen fertilization. *Proceedings 16th International Grassland Congress, Nice*, pp. 159–160.

Oenema, O., Wopereis, F.A. and Ruitenberg, G.H. (1992) Developing new recommendations for nitrogen fertilisation of intensively managed grassland in the Netherlands. In: Archer, J.R. *et al.* (eds) *Nitrate and Farming Systems.* Association of Applied Biologists, Warwick, pp. 249–253.

Øien, A. and Selmer-Olsen, A.R. (1980) A laboratory method for evaluation of available nitrogen in soil. *Acta Agriculturae Scandinavica* 30, 149–156.

Okereke, G.U. and Meints, V.W. (1985) Immediate immobilization of labelled ammonium sulfate and urea nitrogen in soils. *Soil Science* 140, 105–109.

Opperman, M.H., Wood, M. and Harris, P.J. (1989) Changes in inorganic N following the application of cattle slurry to soil at two temperatures. *Soil Biology and Biochemistry* 21, 319–321.

O'Riordan, E.G., Dodd, V.A., Tunney, H. and Fleming, G.A. (1986) The chemical composition of Irish sewage sludges. *Irish Journal of Agricultural Research* 25, 223–229.

O'Riordan, E.G., Dodd, V.A., Tunney, H. and Fleming, G.A. (1987) The fertiliser nutrient value of an anaerobically digested sewage sludge under grassland field conditions. *Irish Journal of Agricultural Research* 26, 199–211.

O'Riordan, E.G., Dodd, V.A., Fleming, G.A. and Tunney, H. (1994) Repeated application of a metal-rich sewage sludge to grassland. 1. Effects on metal levels in soil. *Irish Journal of Agricultural and Food Research* 33, 41–51.

Orr, R.J., Parsons, A.J., Treacher, T.T. and Penning, P.D. (1988) Seasonal patterns of grass production under cutting or continuous stocking managements. *Grass and Forage Science* 43, 199–207.

Orr, R.J., Penning, P.D., Parsons, A.J and Champion, R.A. (1995) Herbage intake and N excretion by sheep grazing monocultures or a mixture of grass and white clover. *Grass and Forage Science* 50, 31–40.

Ørskov, E.R. (1992) *Protein Nutrition in Ruminants*, 2nd edn. Academic Press, London.

Oswalt, D.L., Bertrand, A.R. and Teel, M.R. (1959) Influence of nitrogen fertilization and clipping on grass roots. *Proceedings of Soil Science Society of America* 23, 228–230.

O'Toole, P., McGarry, S.J. and Morgan, M.A. (1985a) Ammonia volatilization from urea-treated pasture and tillage soils: effects of soil properties. *Journal of Soil Science* 36, 613–620.

O'Toole, P.O. and Morgan, M.A. (1988) Efficiency of fertilizer urea: the Irish experience. In: Jenkinson, D.S. and Smith, K.A. (eds) *Nitrogen Efficiency in Agricultural Soils*. Elsevier, London, pp. 191–206.

O'Toole, P., Morgan, M.A. and McAleese, D.M. (1982) Effects of soil properties, temperature and urea concentration on patterns and rates of urea hydrolysis in some Irish soils. *Irish Journal of Agricultural Research* 21, 185–197.

O'Toole, P., Morgan, M.A. and McGarry, S.J. (1985b) A comparative study of urease activities in pasture and tillage soils. *Communications in Soil and Plant Analysis* 16, 759–773.

Ourry, A., Boucaud, J. and Salette, J. (1988) Nitrogen mobilization from stubble and roots during regrowth of defoliated perennial ryegrass. *Journal of Experimental Botany* 39, 803–809.

Ourry, A., Boucaud, J. and Salette, J. (1990) Partitioning and remobilization of nitrogen during regrowth in nitrogen-deficient ryegrass. *Crop Science* 30, 1251–1254.

Overgaard, K. (1984) Trends in nitrate pollution of groundwater in Denmark. *Nordic Hydrology* 15, 177–184.

Overman, A.R., Wilkinson, S.R. and Evers, G.W. (1992) Yield response of Bermudagrass and Bahiagrass to applied nitrogen and overseeded clover. *Agronomy Journal* 84, 998–1001.

Overrein, L.N. and Moe, P.G. (1967) Factors affecting urea hydrolysis and ammonia volatilization in soil. *Soil Science Society of America Proceedings* 31, 57–61.

Owens, J.D., Evans, M.R., Thacker, F.E., Hisset, R. and Baines, S. (1973) Aerobic treatment of piggery waste. *Water Research* 7, 1745–1766.

Owens, L.B., Van Keuren, R.W. and Edwards, W.M. (1983) Nitrogen loss from a high-fertility, rotational pasture program. *Journal of Environmental Quality* 12, 346–350.

Pain, B.F. (1994) Reducing emissions from land associated with ruminant production. *Proceedings 15th General Meeting, European Grassland Federation,* Wageningen, Netherlands, pp. 290–301.

Pain, B.F., Misselbrook, T.H. and Rees, Y.J. (1994) Effects of nitrification inhibitor and acid addition to cattle slurry on nitrogen losses and herbage yields. *Grass and Forage Science* 49, 209–215.

Pain, B.F., Phillips, V.R., Clarkson, C.R. and Klarenbeek, J.V. (1989) Loss of nitrogen through ammonia volatilization during and following the application of pig or cattle slurry to grassland. *Journal of the Science of Food and Agriculture* 47, 1–12.

Pain, B.F., Smith, K.A. and Dyer, C.J. (1986) Factors affecting the response of cut grass to the nitrogen content of dairy cow slurry. *Agricultural Wastes* 17, 189–202.

Pain, B.F., Thompson, R.B., De La Lande Cremer, L.C.N.J. and Ten Holte, L. (1987) The use of additives in livestock slurries to improve their flow properties, conserve nitrogen and reduce odours. In: van der Meer, H.G., Unwin, R.J., van Dijk, T.A.

and Ennik, G.C. (eds) *Animal Manure on Grassland and Fodder Crops: Fertilizer or Waste*. Martinus Nijhoff, Dordrecht.

Pain, B.F., Thompson, R.B., Rees, Y.J. and Skinner, J.H. (1990) Reducing gaseous losses of nitrogen from cattle slurry applied to grassland by the use of additives. *Journal of the Science of Food and Agriculture* 50, 141–153.

Panditharatne, S., Allen, V.G., Fontenot, J.P. and McClure, W.H. (1986) Yield, chemical composition and digestibility by sheep of orchardgrass fertilized with different rates of nitrogen and sulphur or associated red clover. *Journal of Animal Science* 62, 813–821.

Pang, P.C., Cho, C.M. and Hedlin, R.A. (1975) Effect of pH and nitrifier population on nitrification of band-applied and homogeneously mixed urea nitrogen in soils. *Canadian Journal of Soil Science* 55, 15–21.

Parker, C.A. (1957) Non-symbiotic nitrogen-fixing bacteria in soil. 3. Total nitrogen changes in a field soil. *Journal of Soil Science* 8, 48–59.

Parkin, T.B. (1987) Soil microsites as a source of denitrification variability. *Soil Science Society of Americal Journal* 51, 1194–1199.

Parkin, T.B. and Berry, E.C. (1994) Nitrogen transformations associated with earthworm casts. *Soil Biology and Biochemistry* 26, 1233–1238.

Parkin, T.B., Kaspar, H.F., Sexstone, A.J. and Tiedje, J.M. (1984) A gas-flow soil core method to measure field denitrification rates. *Soil Biology and Biochemistry* 16, 323–330.

Parks, W.L. and Fisher, W.B. (1958) Influence of soil temperature and nitrogen on ryegrass growth and chemical composition. *Proceedings of the Soil Science Society of America* 22, 257–259.

Parr, J.F. (1967) Biochemical considerations for increasing the efficiency of nitrogen fertilizers. *Soils and Fertilizers* 30, 207–213.

Parsons, A.J. (1988) The effects of season and management on the growth of grass swards. In: Jones, M.B. and Lazenby, A. (eds) *The Grass Crop*. Chapman and Hall, London, pp. 129–177.

Parsons, A.J., Leafe, E.L., Collet, B., Penning, P.D. and Lewis, J. (1983) The physiology of grass production under grazing. II. Photosynthesis, crop growth and animal intake of continuously-grazed swards. *Journal of Applied Ecology* 20, 127–139.

Parsons, A.J., Orr, R.J., Penning, P.D. and Lockyer, D.R. (1991a) Uptake, cycling and fate of nitrogen in grass-clover swards continuously grazed by sheep. *Journal of Agricultural Science, Cambridge* 116, 47–61.

Parsons, J.W. and Tinsley, J. (1975) Nitrogen substances. In: Gieseking, J.E. (ed.) *Soil Components, Vol. 1. Organic Components*. Springer Verlag, New York, pp. 263–304.

Parsons, L.L., Murray, R.E. and Smith, M.S. (1991b) Soil denitrification dynamics: spatial and temporal variations of enzyme activity, populations and nitrogen gas loss. *Soil Science Society of America Journal* 55, 90–95.

Parton, W.J., Morgan, J.A., Altenhofen, J.M. and Harper, L.A. (1988a) Ammonia volatilization from spring wheat plants. *Agronomy Journal* 80, 419–425.

Parton, W.J., Mosier, A.R. and Schimel, D.S. (1988b) Rates and pathways of nitrous oxide production in a shortgrass steppe. *Biogeochemistry* 6, 45–58.

Patto, P.M., Clement, C.R. and Forbes, T.J. (1978) *Permanent Grassland Studies. 2. Grassland Poaching in England and Wales*. Grassland Research Institute, Hurley, 19 pp.

Paul, E.A. (1988) Towards the year 2000: directions for future nitrogen research. In: Wilson, J.R. (ed.) *Advances in Nitrogen Cycling in Agricultural Ecosystems*. CAB International, Wallingford, pp. 417–425.

Paul, E.A. and Clark, F.E. (1989) *Soil Microbiology and Biochemistry.* Academic Press, San Diego.

Paul, J.W. and Beauchamp, E.G. (1989) Effect of carbon constituents in manure on denitrification in soil. *Canadian Journal of Soil Science* 69, 49–61.

Pearse, P.J. and Wilman, D. (1984) Effects of applied nitrogen on grass leaf initiation, development and death in field swards. *Journal of Agricultural Science, Cambridge* 103, 405–413.

Peel, S. and Lloveras, J. (1994) New targets for sustainable forage production and utilization. *Proceedings 15th General Meeting European Grassland Federation,* Wageningen, Netherlands, pp. 35–47.

Penman, H.L. (1962) Woburn irrigation, 1951–59. 2. Results for grass. *Journal of Agricultural Science, Cambridge* 58, 349–364.

Penning, P.D., Parsons, A.J., Orr, R.J. and Treacher, T.T. (1991) Intake and behaviour responses by sheep to changes in sward characteristics under continuous stocking. *Grass and Forage Science* 46, 15–28.

Perez, C.B. and Story, C.D. (1960) The effect of nitrate in nitrogen-fertilized hays on fermentation *in vitro. Journal of Animal Science* 19, 1311.

Petersen, R.G., Woodhouse, W.W. and Lucas, H.L. (1956) The distribution of excreta by freely grazing cattle and its effect on pasture fertility. 2. Effect of returned excreta on the residual concentration of some fertilizer elements. *Agronomy Journal* 48, 444–449.

Phillips, V.R., Pain, B.F. and Klarenbeek, J.V. (1991) Factors influencing the odour and ammonia emissions during and after the land spreading of animal slurries. In: Nielsen, V.C., Voorburg, J.H. and L'Hermite, P. (eds) *Odour and Ammonia Emissions from Livestock Farming*. Elsevier, London, pp. 98–106.

Phipps, R.H. (1975) The effects on dairy cows of grazing pasture containing high levels of nitrate-nitrogen. *Journal of the British Grassland Society* 30, 45–49.

Porqueddu, C., Roggero, P.P., Sitzia, M. and Sulas, L. (1994) Soil conservation role of permanent pastures in the Mediterranean environment. *Proceedings 15th General Meeting, European Grassland Federation,* Wageningen, Netherlands, pp. 235–238.

Postmus, J. and Schepers, J.H. (1980) Temperature sum and date of spring application of nitrogen on grassland – results in the Netherlands. In: Prins, W.H. and Arnold, G.H. (eds) *The Role of Nitrogen in Intensive Grassland Production.* PUDOC, Wageningen, p. 159.

Poth, M. and Focht, D.D. (1985) ^{15}N kinetic analysis of N_2O production by *Nitrosomonas europaea:* an examination of nitrifier denitrification. *Applied and Environmental Microbiology* 49, 1134–1141.

Power, J.F. (1968) Mineralization of nitrogen in grass roots. *Proceedings of the Soil Science Society of America* 32, 673–674.

Power, J.F. (1981) Long-term recovery of fertilizer nitrogen applied to a native mixed prairie. *Soil Science Society of America Journal* 45, 782–786.

Power, J.F. (1983) Recovery of nitrogen and phosphorus after 17 years from various fertilizer materials applied to crested wheatgrass. *Agronomy Journal* 75, 249–254.

Power, J.F. (1985) Nitrogen- and water-use efficiency of several cool-season grasses receiving ammonium nitrate for 9 years. *Agronomy Journal* 77, 189–192.

Power, J.F. and Legg, J.O. (1984) Nitrogen-15 recovery for five years after application of ammonium nitrate to crested wheatgrass. *Soil Science Society of America Journal* 48, 322–326.

Prine, G.M. and Burton, G.W. (1956) The effect of N rate and clipping frequency upon the yield, protein content and certain morphological characteristics of Coastal Bermudagrass (*Cynodon dactylon* (L) Pers). *Agronomy Journal* 48, 296–301.

Prine, G.M., Gardner, F.P. and Willard, C.J. (1963) Irrigation and nitrogen treatment of forage crops. *Research Circular 119, Ohio Agricultural Experiment Station,* 35 pp.

Prins, W.H. (1980) Changes in quantity of mineral nitrogen in three grassland soils as affected by intensity of nitrogen fertilization. *Fertilizer Research* 1, 51–63.

Prins, W.H. (1983) Effect of a wide range of nitrogen applications on herbage nitrate content in long-term fertilizer trials on all-grass swards. *Fertilizer Research* 4, 101–113.

Prins, W.H., Dilz, K. and Neeteson, J.J. (1988a) Current recommendations for nitrogen fertilisation within the EEC in relation to nitrate leaching. *Proceedings of the Fertiliser Society, London*, No. 276.

Prins, W.H. and Neeteson, J.J. (1982) Grassland productivity as affected by intensity of nitrogen fertilization in preceding years. *Netherlands Journal of Agricultural Science* 30, 245–258.

Prins, W.H., Postmus, J., Reker, A.M. and Ruiter, B. (1988b) Nitrogen use on grassland in spring in the Netherlands and elsewhere in Europe: temperature sum, stage of growth, rate and source of nitrogen. *Netherlands Fertilizer Technical Bulletin* No. 17, 55 pp.

Prins, W.H. and Snijders, P.J.M. (1987) Negative effects of animal manures on grassland due to surface spreading and injection. In: van der Meer, H.G., Unwin, R.J., van Dijk, T. A. and Ennik, G.C. (eds) *Animal Manure on Grassland and Fodder Crops: Fertilizer or Waste*. Martinus Nijhoff, Dordrecht, pp. 129–135.

Prins, W.H., van Burg, P.F.J. and Wieling, H. (1980) The seasonal response of grassland to nitrogen at different intensities of nitrogen fertilization, with special reference to methods of response measurements. In: Prins, W.H. and Arnold, G.H. (eds) *The Role of Nitrogen in Intensive Grassland Production*. PUDOC, Wageningen, pp. 35–49.

Purchase, B.S. (1974) Evaluation of the claim that grass root exudates inhibit nitrification. *Plant and Soil* 41, 527–539.

Rachhpal-Singh, and Nye, P.H. (1984) The effect of soil pH and high urea concentrations on urease activity in soil. *Journal of Soil Science* 35, 519–527.

Rachhpal-Singh and Nye, P.H. (1986a) A model of ammonia volatilization from applied urea. 1. Development of the model. *Journal of Soil Science* 37, 9–20.

Rachhpal-Singh and Nye, P.H. (1986b) A model of ammonia volutalization from applied urea. III Sensitivity analysis, mechanisms and applications. *Journal of Soil Science* 37, 31–40.

Rahman, H., McDonald, P. and Simpson, K. (1960) Effect of nitrogen and potassium fertilizers on the mineral status of perennial ryegrass. 1. Mineral content. *Journal of the Science of Food and Agriculture* 11, 422–428.

Raison, R.J., Connell, M.J. and Khanna, P.K. (1987) Methodology for studying fluxes of soil mineral-N *in situ*. *Soil Biology and Biochemistry* 19, 521–530.

Ramage, C.H., Eby, C., Mather, R.E. and Purvis, E.R. (1958) Yield and chemical composition of grasses fertilized heavily with nitrogen. *Agronomy Journal* 50, 59–62.

Raymond, W.F. and Spedding, C.R.W. (1965) Nitrogenous fertilizers and the feed value of grass. *Proceedings 1st General Meeting European Grassland Federation*. pp. 151–160.

Reddy, K.R., Rao, P.S.C. and Jessup, R.E. (1982) The effect of carbon mineralization on denitrification kinetics in mineral and organic soils. *Soil Science Society of America Journal* 46, 62–68.

Reid, D. (1962) Studies on the cutting management of grass-clover swards. III. The effects of prolonged close and lax cutting on herbage yields and quality. *Journal of Agricultural Science, Cambridge* 59, 359–368.

Reid, D. (1966) The response of herbage yields and quality to a wide range of nitrogen application rates. *Proceedings of 10th International Grassland Congress*, Helsinki, 1966, pp. 209–213.

Reid, D. (1970) The effects of a wide range of nitrogen application rates on the yields from a perennial ryegrass sward with and without white clover. *Journal of Agricultural Science, Cambridge* 74, 227–240.

Reid, D. (1972) The effects of the long-term application of a wide range of nitrogen rates on the yields from perennial ryegrass swards with and without white clover. *Journal of Agricultural Science, Cambridge* 79, 291–301.

Reid, D. (1978) The effects of frequency of defoliation on the yield response of a perennial ryegrass sward to a wide range of nitrogen application rates. *Journal of Agricultural Science, Cambridge* 90, 447–457.

Reid, D. (1983) The combined use of fertilizer nitrogen and white clover as nitrogen sources for herbage growth. *Journal of Agricultural Science, Cambridge* 100, 613–623.

Reid, D. (1984) The seasonal distribution of nitrogen fertiliser dressings on pure perennial ryegrass swards. *Journal of Agricultural Science, Cambridge* 103, 659–669.

Reid, D. (1985) A comparison of the yield responses of four grasses to a wide range of nitrogen application rates. *Journal of Agricultural Science, Cambridge* 105, 381–387.

Reid, D. and Castle, M.E. (1965) The response of grass–clover and pure-grass leys to irrigation and fertilizer nitrogen treatment. 1. Irrigation effects. *Journal of Agricultural Science, Cambridge* 64, 185–194.

Reid, D. and Strachan, N.H. (1974) The effects of a wide range of nitrogen rates on some chemical constituents of herbage from perennial ryegrass swards with and without white clover. *Journal of Agricultural Science, Cambridge* 83, 393–401.

Reid, R.L. and Jung, G.A. (1965) The influence of fertilizer treatment on the intake, digestibility and palatability of tall fescue hay. *Journal of Animal Science* 24, 615–625.

Reid, R.L., Jung, G.A. and Kinsey, C.M. (1967) Nutritive value of nitrogen-fertilized orchardgrass pasture at different periods of the year. *Agronomy Journal* 59, 519–525.

Reid, R.L., Jung, G.A. and Murray, S.J. (1966) Nitrogen fertilization in relation to the palatibility and nutritive value of orchardgrass. *Journal of Animal Science* 25, 636–645.

Reid, R.L., Jung, G.A., Post, A.J., Horn, F.P., Kahle, E.B., Bubar, J.D. and Daniel, K. (1974) Effects of nitrogen and micro-element fertilization on quality of pasture and on the health, nutritional status and reproductive performance of sheep. *Journal of Animal Science* 38, 163–171.

Reisenauer, H.M., Clement, C.R. and Jones, L.H.P. (1982) Comparative efficacy of ammonium and nitrate for grasses. *Proceedings of International Plant Nutrition Colloquium* Vol. 2. Commonwealth Agricultural Bureaux, Farnham Royal, pp. 539–544.

Reith, J.W.S., Burridge, J.C., Berrow, M.L. and Caldwell, K.S. (1983) Effects of the application of fertilisers and trace elements on the cobalt content of herbage cut for conservation. *Journal of the Science of Food and Agriculture* 34, 1163–1170.

Reith, J.W.S., Inkson, R.H.E. and collaborators (1961) The effects of fertilizers on herbage production. 1. The effect of nitrogen, phosphate and potash on yield. *Journal of Agricultural Science, Cambridge* 56, 17–29.

Reith, J.W.S., Inkson, R.H.E. and collaborators (1964) The effects of fertilizers on herbage production. 2. The effect of nitrogen, phosphorus and potassium on botanical and chemical composition. *Journal of Agricultural Science, Cambridge* 63, 209–219.

Reith, J.W.S. and Mitchell, R.L. (1964) The effect of soil treatment on trace element uptake by plants. In: Bould, C., Prevot, P. and Magness, J.R. (eds) *Plant Analysis and Fertilizer Problems,* Vol. 4. American Society of Horticultural Science. East Lansing, Michigan, pp. 241–254.

Reynolds, C.M. and Wolf, D.C. (1987a) Effect of soil moisture and air relative humidity on ammonia volatilization from surface-applied urea. *Soil Science* 143, 144–152.

Reynolds, C.M. and Wolf, D.C. (1987b) Influence of urease activity and soil properties on ammonia volatilization from urea. *Soil Science* 143, 418–425.

Reynolds, C.M., Wolf, D.C. and Armbruster, J.A. (1985) Factors related to urea hydrolysis in soils. *Soil Science Society of America Journal* 49, 104–108.

Richards, I.R. (1976) Nitrogen under grazing – response to fertilizer N and the role of white clover. In: Hodgson, J. and Jackson, D.K. (eds) *Pasture Utilization by the Grazing Animal.* Occasional Symposium, British Grassland Society No. 8, pp. 69–77.

Richards, I.R. (1977) Influence of soil and sward characteristics on the response to nitrogen. *Proceedings of an International Meeting on Animal Production from Temperate Grasslands,* Dublin, pp. 45–49.

Richards, I.R. and Wolton, K.M. (1976a) A note on the properties of urine excreted by grazing cattle. *Journal of the Science of Food and Agriculture* 27, 426–428.

Richards, I.R. and Wolton, K.M. (1976b) The spatial distribution of excreta under intensive cattle grazing. *Journal of the British Grassland Society* 31, 89–92.

Richardson, A.C. and Syers, J.K. (1985) Edaphic limitations and soil nutrient requirements of legume-based forage systems in temperate regions of New Zealand. In: Barnes, R.F., Ball, P.R., Brougham, R.W., Marten, G.C. and Minson, D.J. (eds) *Forage Legumes for Energy-Efficient Animal Production.* USDA, Washington, pp. 89–94.

Richter, G.M., Hoffman, A., Nieder, R. and Richter, J. (1989) Nitrogen mineralization in loamy arable soils after increasing the ploughing depth and ploughing grasslands. *Soil Use and Management* 5, 169–173.

Rinne, S-L., Sillanpää, M., Huokuna, E. and Hiivola, S-L. (1974a) Effects of heavy nitrogen fertilization on potassium, calcium, magnesium and phosphorus contents in ley grasses. *Annales Agriculturae Fenniae* 13, 96–108.

Rinne, S-L., Sillanpää, M., Huokuna, E. and Hiivola, S-L. (1974b) Effects of heavy nitrogen fertilization on iron, manganese, sodium, zinc, copper, strontium, molybdenum and cobalt contents in ley grasses. *Annales Agriculturae Fenniae* 13, 109–118.

Robbins, C.W. and Carter, D.L. (1980) Nitrate-nitrogen leached below the root zone during and following alfalfa. *Journal of Environmental Quality* 9, 447–450.

Roberts, A.M., Hudson, J.A. and Roberts, G. (1989) A comparison of nutrient losses following grassland improvement using two different techniques in an upland area of mid-Wales. *Soil Use and Management* 5, 174–179.

Robinson, R.R. and Sprague, V.G. (1947) The clover populations and yields of a Kentucky bluegrass sod as affected by nitrogen fertilization, clipping treatments and irrigation. *Agronomy Journal* 39, 107–116.

Robson, M.J. and Deacon, M.J. (1978) Nitrogen deficiency in small closed communities of S24 ryegrass. II. Changes in the weight and chemical composition of individual leaves during their growth and death. *Annals of Botany* 42, 1199–1213.

Robson, M.J. and Parsons, A.J. (1978) Nitrogen deficiency in small closed communities of S24 ryegrass. I. Photosynthesis, respiration, dry matter production and partition. *Annals of Botany* 42, 1185–1197.

Robson, M.J., Parsons, A.J. and Williams, T.E. (1989) Herbage production: grasses and legumes. In: Holmes, W. (ed.) *Grass: its Production and Utilization*. Blackwell Scientific Publications, Oxford, pp. 7–88.

Rodgers, G.A., Penny, A., Widdowson, F.V. and Hewitt, M.V. (1987) Tests of nitrification and of urease inhibitors, when applied with either solid or aqueous urea, on grass grown on a light sandy soil. *Journal of Agricultural Science, Cambridge* 108, 109–117.

Rodgers, G.A., Widdowson, F.V., Penny, A. and Hewitt, M.V. (1984) Comparison of the effects of aqueous and of prilled urea, used alone or with urease or nitrification inhibitors, with those of Nitro-chalk on ryegrass leys. *Journal of Agricultural Science, Cambridge* 103, 671–685.

Roelofs, J.G.M. (1986) The effect of airborne sulphur and nitrogen deposition on aquatic and terrestrial heathland vegetation. *Experientia* 42, 372–377.

Roelofs, J.G.M. and Houdijk, A.L.F.M. (1991) Ecological effects of ammonia. In: Nielsen, V.C., Voorburg, J.H. and L'Hermite, P. (eds) *Odour and Ammonia Emissions from Livestock Farming*. Elsevier, London, pp. 10–16.

Roelofs, J.G.M., Kempers, A.J., Houdijke, A.L.F.M. and Jansen, J. (1985) The effect of air-borne ammonium sulphate on *Pinus nigra* var. *maritima* in the Netherlands. *Plant and Soil* 84, 45–56.

Rogers, H.H. and Aneja, V.P. (1980) Uptake of atmospheric ammonia by selected plant species. *Environmental and Experimental Botany* 20, 251–257.

Rosen, K., Gundersen, P., Tegnhammar, L., Johansson, M. and Frogner, T. (1992) Nitrogen enrichment of Nordic forest ecosystems. *Ambio* 21, 364–368.

Ross, D.J. and Cairns, A. (1980) Nitrogen availability in some soils from tussock grasslands and introduced pastures. 5. Influence of standing dead material and roots from five tussock species on nitrogen mineralisation. *New Zealand Journal of Science* 23, 11–18.

Rowarth, J.S., Gillingham, A.G., Tillman, R.W. and Syers, J.K. (1985) Release of phosphorus from sheep faeces on grazed, hill country pastures. *New Zealand Journal of Agricultural Research* 28, 497–504.

Royal Society (1983) *The Nitrogen Cycle of the United Kingdom: Report of a Royal Society Study Group.* Royal Society, London.

Rus Jerez, B.E., Ball, P.R. and Tillman, R.W. (1988) The role of earthworms in nitrogen release from herbage residues. In: Jenkinson, D.S. and Smith, K.A. (eds) *Nitrogen Efficiency in Agricultural Soils.* Elsevier, London, pp. 355–370.

Russel, J.S. (1960) Soil fertility changes in the long-term experimental plots at Kybybolite, South Australia. 1. Changes in pH, total nitrogen, organic carbon and bulk density. *Australian Journal of Agricultural Research* 11, 902–926.

Ryden, J.C. (1983) Denitrification loss from a grassland soil in the field receiving different rates of nitrogen as ammonium nitrate. *Journal of Soil Science* 34, 355–365.

Ryden, J.C. (1984) The flow of nitrogen in grassland. *Proceedings of the Fertiliser Society, London.* No. 229.

Ryden, J.C. (1986) Gaseous losses of nitrogen from grassland. In: Van der Meer, H.G., Ryden, J.C. and Ennik, G.C. (eds) *Nitrogen Fluxes in Intensive Grassland Systems.* Martinus Nijhoff, Dordrecht, pp. 59–73.

Ryden, J.C., Ball, P.R. and Garwood, E.A. (1984) Nitrate leaching from grassland. *Nature, London* 311, 50–53.

Ryden, J.C. and Dawson, K.P. (1982) Evaluation of the acetylene-inhibition technique for the measurement of denitrification in grassland soils. *Journal of the Science of Food and Agriculture* 33, 1197–1206.

Ryden, J.C. and Lockyer, D.R. (1985) Evaluation of a system of wind tunnels for field studies of ammonia loss from grassland through volatilization. *Journal of the Science of Food and Agriculture* 36, 781–788.

Ryden, J.C., Skinner, J.H. and Nixon, D.J. (1987a) Soil core incubation system for the field measurement of denitrification using acetylene-inhibition. *Soil Biology and Biochemistry* 19, 753–757.

Ryden, J.C., Whitehead, D.C., Lockyer, D.R., Thompson, R.B., Skinner, J.H. and Garwood, E.A. (1987b) Ammonia emission from grassland and livestock production systems in the UK. *Environmental Pollution* 48, 173–184.

Ryle, G.J.A. (1970) Effects of two levels of applied nitrogen on the growth of S37 cocksfoot in small simulated swards in a controlled environment. *Journal of the British Grassland Society* 25, 20–29.

Ryle, G.J.A., Powell, C.E. and Gordon, A.J. (1979) The respiratory costs of nitrogen fixation in soyabean, cowpea and white clover. *Journal of Experimental Botany* 30, 145–153.

Sabine, J.R. (1983) Earthworms as a source of food and drugs. In: Satchell, J.E. (ed.) *Earthworm Ecology.* Chapman and Hall, London, pp. 285–296.

Safley, L.M., Barker, J.C. and Westerman, P.W. (1984) Characteristics of fresh dairy manure. *Transactions of the American Society of Agricultural Engineers 1984,* 1150–1162.

Sahrawat, K.L. and Keeney, D.R. (1986) Nitrous oxide emission from soils. *Advances in Soil Science* 4, 103–148.

Salt, P.D. (1965) An apparatus for measuring losses of ammonia from decomposing plant materials. *Chemistry and Industry*, 1965, 461–462.

Sanderson, M.A. and Wedin, W.F. (1989) Nitrogen in the detergent fibre fractions of temperate legumes and grasses. *Grass and Forage Science* 44, 159–168.

Sarathchandra, S.U. (1978) Nitrification activities of some New Zealand soils and the effect of some clay types on nitrification. *New Zealand Journal of Agricultural Research* 21, 615–621.

Satchell, J.E. (1967) Lumbricidae. In: Burges, A. and Raw, F. (eds) *Soil Biology*. Academic Press, London, pp. 259–322.

Schechtner, G., Tunney, H., Arnold, G.H. and Keuning, J.A. (1980) Positive and negative effects of cattle manure on grassland with special reference to high rates of application. In: Prins, W.H. and Arnold, G.H. (eds) *The Role of Nitrogen in Intensive Grassland Production*. PUDOC, Wageningen, pp. 77–93.

Schimel, D.S. and Parton, W.J. (1986) Microclimatic controls of nitrogen mineralization and nitrification in shortgrass steppe soils. *Plant and Soil* 93, 347–357.

Schjoerring, J.K. (1991) Ammonia emission from the foliage of growing plants. In: Sharkey, T.D., Holland, E.A. and Mooney, H.A. (eds), *Trace Gas Emissions by Plants*. Academic Press, London, pp. 267–292.

Schmidt, D.R. and Tenpas, G.H. (1965) Seasonal response of grasses fertilized with nitrogen compared to a legume-grass mixture. *Agronomy Journal* 57, 428–431.

Schmidt, E.L. (1982) Nitrification in soil. In: Stevenson, F.J. (ed.) *Nitrogen in Agricultural Soils*. American Society of Agronomy, Madison, Wisconsin, pp. 253–288.

Scholefield, D., Corrè, W.J., Colbourne, P., Jarvis, S.C., Hawkins, J. and De Klein, C.A.M. (1990) Denitrification in grazed grassland soils assessed using the core incubation technique with acetylene inhibition. *Proceedings 13th General Meeting, European Grassland Federation,* Banska Bystrica, Czechoslovakia, Vol. 2, pp. 8–12.

Scholefield, D., Garwood, E.A. and Titchen, N.M. (1988) The potential of management practices for reducing losses of nitrogen from grazed pastures. In: Jenkinson, D.S. and Smith, K.A. (eds) *Nitrogen Efficiency in Agricultural Soils*. Elsevier, London, pp. 220–230.

Scholefield, D., Lockyer, D.R., Whitehead, D.C. and Tyson, K.C. (1991) A model to predict transformations and losses of nitrogen in UK pastures grazed by beef cattle. *Plant and Soil* 132, 165–177.

Scholefield, D. and Titchen, N.M. (1995) Development of a rapid field test for soil mineral nitrogen and its application to grazed grassland. *Soil Use and Management* 11, 33–43.

Scholefield, D. and Tyson, K.C. (1992) Comparing the levels of nitrate leaching from grass/clover and N-fertilized grass swards grazed with beef cattle. *Proceedings 14th General Meeting, European Grassland Federation,* Lahti, Finland, pp. 530–533.

Scholefield, D., Tyson, K.C., Garwood, E.A., Armstrong, A.C., Hawkins, J. and Stone, A.C. (1993) Nitrate leaching from grazed grassland lysimeters: effects of fertilizer input, field drainage, age of sward and patterns of weather. *Journal of Soil Science*, 44, 601–613.

Schrøder, H. (1990) Agricultural production and the eutrophication of the Baltic and adjacent seas. In: Merckx, R., Verecken, H. and Vlassak, K. (eds) *Fertilization and the Environment*. Leuven University Press, Leuven, pp. 11–19.

Schuurkes, J.A.A.R., Maenen, M.M.J. and Roelofs, J.G.M. (1988) Chemical characteristics of precipitation in NH_3-affected areas. *Atmospheric Environment* 22, 1689–1698.

Schuurman, J.J. and Knot, L. (1974) The effect of nitrogen on the root and shoot development of *Lolium multiflorum* var. *westerwoldicum*. *Netherlands Journal of Agricultural Science* 22, 82–88.

Schwank, O., Blum, H. and Nösberger, J. (1986) The influence of irradiance distribution on the growth of white clover (*Trifolium repens* L.) in differently managed canopies of permanent grassland. *Annals of Botany* 57, 273–281.

Scott, R.S. (1977) Effects of animals on pasture production. II. Pasture production and N and K requirements of cattle and sheep pastures measured under a common method of defoliation. *New Zealand Journal of Agricultural Research* 20, 31–36.

Sears, P.D. (1953) Pasture growth and soil fertility. 7. General discussion of the experimental results and of their application to farming practice in New Zealand. *New Zealand Journal of Science and Technology Section A* 35, 221–236.

Sears, P.D., Goodall, V.C., Jackman, R.H. and Robinson, G.S. (1965) Pasture growth and soil fertility. VIII. The influence of grasses, white clover, fertilizers and the return of herbage clippings on pasture production of an impoverished soil. *New Zealand Journal of Agricultural Research* 8, 270–283.

Sears, P.D. and Newbold, R.P. (1942) The effect of sheep droppings on yield, botanical composition and chemical composition of pasture. *I. New Zealand Journal of Science and Technology Section A* 24, 36–61.

Sexstone, A.J., Parkin, T.B. and Tiedje, J.M. (1985) Temporal response of soil denitrification rates to rainfall and irrigation. *Soil Science Society of America Journal* 49, 99–103.

Shaw, P.G., Brockman, J.S. and Wolton, K.M. (1966) The effect of cutting and grazing on the response of grass/white clover swards to fertilizer nitrogen. *Proceedings 10th International Grassland Congress,* Helsinki, 1966, 240–244.

Sheehy, J.E. (1989) How much dinitrogen fixation is required in grazed grassland? *Annals of Botany* 64, 159–161.

Sheldrick, R.D., Lavender, R.H. and Martyn, T.M. (1990) Dry matter yield and response to nitrogen of an *Agrostis stolonifera* dominant sward. *Grass and Forage Science* 45, 203–213.

Sherlock, R.R., Black, A.S. and Smith, N.P. (1988) Micro-environment soil pH around broadcast urea granules and its relationship to ammonia volatilisation. In: Bacon, P.E., Evans, J., Storrier, R.R. and Black, A.C. (eds) *Nitrogen Cycling in Temperate Agricultural Systems.* Australian Society of Soil Science, Riverina, pp. 316–326.

Sherlock, R.R. and Goh, K.M. (1983) Initial emission of nitrous oxide from sheep urine applied to pasture soil. *Soil Biology and Biochemistry* 15, 615–617.

Sherlock, R.R. and Goh, K.M. (1984) Dynamics of ammonia volatilization from simulated urine patches and aqueous urea applied to pasture. I. Field experiments. *Fertilizer Research* 5, 181–195.

Sherlock, R.R. and Goh, K.M. (1985) Dynamics of ammonia volatilization from simulated urine patches and aqueous urea applied to pasture. III. Field verification of a simplified model. *Fertilizer Research* 6, 23–36.

Sherwood, M. (1986) Nitrate leaching following application of slurry and urine to field plots. In: Dam Kofoed, A., Williams, J.H. and L'Hermite, P. (eds) *Efficient Land Use of Sludge and Manure*. Elsevier, London, pp. 150–157.

Sherwood, M. and Fanning, A. (1981) Nutrient content of surface run-off water from land treated with animal wastes. In: Brogan, J.C. (ed.) *Nitrogen Losses and Surface Run-off from Landspreading of Manures*. Martinus Nijhoff/Dr. W. Junk, The Hague, pp. 5–17.

Sherwood, M. and Fanning, A. (1989) Leaching of nitrate from simulated urine patches. In: Germon, J.C. (ed.) *Management Systems to Reduce Impact of Nitrates*. Elsevier, London, pp. 32–44.

Sibma, L. and Alberda, Th. (1980) The effect of cutting frequency and nitrogen fertilizer rates on dry matter production, nitrogen uptake and herbage nitrate content. *Netherlands Journal of Agricultural Science* 28, 243–251.

Sillanpaa, M. and Rinne, S-L. (1975) The effect of heavy nitrogen fertilization on the uptake of nutrients and on some properties of soil cropped with grasses. *Annales Agriculturae Fenniae* 14, 210–226.

Silvertown, J. (1980) The dynamics of a grassland ecosystem: botanical equilibrium in the Park Grass experiment. *Journal of Applied Ecology* 17, 491–504.

Simpson, D., Wilman, D. and Adams, W.A. (1987) The distribution of white clover (*Trifolium repens* L.) and grasses within six sown hill swards. *Journal of Applied Ecology* 24, 201–216.

Simpson, D., Wilman, D. and Adams, W.A. (1988) Responses of white clover and grass to applications of potassium and nitrogen on a potassium-deficient hill soil. *Journal of Agricultural Science, Cambridge* 110, 159–167.

Simpson, J.R. (1968) Losses of urea nitrogen from the surface of pasture soils. *Transactions 9th International Congress of Soil Science, Adelaide*, Vol. 2, pp. 459–466.

Simpson, J.R. (1987) Nitrogen nutrition of pastures. In: Wheeler, J.L., Pearson, C.J. and Robards, G.E. (eds) *Temperate Pastures: Their Production, Use and Management*. CSIRO, Australia, pp. 143–154.

Simpson, J.R. and Freney, J.R. (1967) The fate of labelled mineral nitrogen after addition to three pasture soils of different organic matter contents. *Australian Journal of Agricultural Research* 18, 613–623.

Simpson, J.R. and Stobbs, J.H. (1981) Nitrogen supply and animal production from pastures. In: Morley, F.H.W. (ed.) *Grazing Animals*. Elsevier, Amsterdam, pp. 261–287.

Sims, P.L. and Coupland, R.T. (1979) Natural temperate grasslands: Producers. In: Coupland, R.T. (ed.) *Grassland Ecosystems of the World*. Cambridge University Press, Cambridge, pp. 49–72.

Sims, P.L. and Singh, J.S. (1978) The structure and function of ten western North American grasslands. III. Net primary production, turnover and efficiencies of energy capture and water use. *Journal of Ecology* 66, 573–597.

Skiba, U., Hargreaves, K.J., Fowler, D. and Smith, K.A. (1992) Fluxes of nitric and nitrous oxides from agricultural soils in a cool temperate climate. *Atmospheric Environment* 26A, 2477–2488.

Skinner, R.J. (1987) Growth responses in grass to sulphur fertilizer. *Proceedings International Symposium on Elemental Sulphur in Agriculture*, Vol. 2. Syndicat Francais du Soufre, Marseille, pp. 525–535.

Skrijka, P. (1987) Investigations of the fertilizer value of sheep excrements left on pasture. In: van der Meer, H.G., Unwin, R.J., van Dijk, T.A. and Ennik, G.C. (eds) *Animal Manure on Grassland and Fodder Crops, Fertilizer or Waste?* Martinus Nijhoff, Dordrecht, pp. 325–327.

Smith, C.J. (1987) Denitrification in the field. In: Wilson, J.R. (ed.) *Advances in Nitrogen Cycling in Agricultural Ecosystems.* CAB International, Wallingford, pp. 387–398.

Smith, D. (1973) The non-structural carbohydrates. In: Butler, G.W. and Bailey, R.W. (eds) *Chemistry and Biochemistry of Herbage,* Vol. 1. Academic Press, London, pp. 105–155.

Smith, D. and Jewiss, O.R. (1966) Effects of temperature and nitrogen supply on the growth of timothy (*Phleum pratense* L.). *Annals of Applied Biology* 58, 145–157.

Smith, G.S., Cornforth, I.S. and Henderson, H.V. (1985) Critical leaf concentrations for deficiencies of nitrogen, potassium, phosphorus, sulphur and magnesium in perennial ryegrass. *New Phytologist* 101, 393–409.

Smith, J.H. and Peterson, J.R. (1982) Recycling of nitrogen through land application of agricultural, food processing and municipal wastes. In: Stevenson, F.J. (ed.) *Nitrogen in Agricultural Soils.* American Society of Agronomy, Madison, Wisconsin, pp. 791–831.

Smith, K.A. and Arah, J.R.M. (1990) Losses of nitrogen by denitrification and emissions of nitrogen oxides from soils. *Proceedings of The Fertiliser Society, London*, No. 299.

Smith, K.A. and Chambers, B.J. (1993) Utilizing the nitrogen content of organic manures on farms – problems and practical solutions. *Soil Use and Management* 9, 105–112.

Smith, K.A., Crichton, I.J., McTaggart, I.P. and Lang, R.W. (1989) Inhibition of nitrification by dicyandiamide in cool temperate conditions. In: Hansen, J.A. and Henriksen, K. (eds) *Nitrogen in Organic Wastes Applied to Soils.* Academic Press, London, pp. 289–303.

Smith, L.P. (1984) *The Agricultural Climate of England and Wales.* MAFF Reference Book 435. HMSO, London.

Smith, M.S. and Tiedje, J.M. (1979) Phases of denitrification following oxygen depletion in soil. *Soil Biology and Biochemistry* 11, 261–267.

Smith, S.J., Chichester, F.W. and Kissel, D.E. (1978) Residual forms of fertilizer nitrogen in field soils. *Soil Science* 125, 165–169.

Smith, S.J. and Power, J.F. (1985) Residual forms of fertilizer nitrogen in a grassland soil. *Soil Science* 140, 362–367.

Smith S.J., Schepers, J.S. and Porter, L.K. (1990) Assessing and managing agricultural nitrogen losses to the environment. *Advances in Soil Science* 14, 1–43.

Smith, S.J. and Young, S.B. (1975) Distribution of nitrogen forms in virgin and cultivated soils. *Soil Science* 120, 354–360.

Smith, S.R., Tibbett, M. and Evans, T.D. (1992) Nitrate accumulation potential of sewage sludges applied to soil. In: Archer, J.R. *et al.* (eds) *Nitrate and Farming Systems.* Association of Applied Biologists, Warwick, pp. 157–161.

Sommer, S.G. and Christensen, B.T. (1991) Effect of dry matter content on ammonia loss from surface applied cattle slurry. In: Nielsen, V.C., Voorburg, J.H. and L'Hermite, P. (eds) *Odour and Ammonia Emissions from Livestock Farming.* Elsevier, London, pp. 141–147.

Sommer, S.G., Christensen, B.T., Nielsen, N.E. and Schjoerring, J.K. (1993) Ammonia volatilization during storage of cattle and pig slurry: effect of surface cover. *Journal of Agricultural Science, Cambridge* 121, 63–71.

Sommer, S.G. and Jensen, C. (1994) Ammonia volatilization from urea and ammoniacal fertilizers surface applied to winter wheat and grassland. *Fertilizer Research* 37, 85–92.

Sommer, S.G. and Jensen, E.S. (1991) Foliar absorption of atmospheric ammonia by ryegrass in the field. *Journal of Environmental Quality* 20, 153–156.

Sommer, S.G. and Olesen, J.E. (1991) Effects of dry matter content and temperature on ammonia loss from surface-applied cattle slurry. *Journal of Environmental Quality* 20, 679–683.

Sommer, S.G., Olesen, J.E. and Christensen, B.T. (1991) Effects of temperature, wind speed and air humidity on ammonia volatilization from surface applied cattle slurry. *Journal of Agricultural Science, Cambridge* 117, 91–100.

Sommers, L.E. (1977) Chemical composition of sewage sludges and analysis of their potential use as fertilizers. *Journal of Environmental Quality* 6, 225–232.

Sparling, G.P. (1985) The soil biomass. In: Vaughan, D. and Malcolm, R.E. (eds) *Soil Organic Matter and Biological Activity*. Martinus Nijhoff/Dr W. Junk, Dordrecht, pp. 223–262.

Spatz, G., Neuendorff, J., Pape, A. and Schröder, C. (1992) Nitrogen dynamics under excrement patches in pastures. *Zeitschrift für Pflanzenernahrung und Bodenkunde* 155, 301–305.

Spedding, C.R.W. (1971) *Grassland Ecology*. Clarendon Press, Oxford.

Speirs, R.B. and Frost, C.A. (1987) The enhanced acidification of a field soil by very low concentrations of atmospheric ammonia. *Research and Development in Agriculture* 4, 83–86.

Sprague, M.A. and Taylor, B.B. (1970) Forage composition and losses from orchardgrass silage as affected by maturity and nitrogen fertilization. *Agronomy Journal* 62, 749–753.

Sprague, V.G. and Garber, R.J. (1950) Effect of time and height of cutting and nitrogen fertilization on the persistence of the legume and production of orchardgrass–Ladino and bromegrass–Ladino associations. *Agronomy Journal* 42, 586–593.

Stanford, G. (1977) Evaluating the nitrogen-supplying capacities of soils. *Proceedings of International Seminar on Soil Environment and Fertility Management in Intensive Agriculture*, Tokyo, 1977, pp. 412–418.

Stanford, G. (1982) Assessment of soil nitrogen availability. In: Stevenson, F.J. (ed.) *Nitrogen in Agricultural Soils*. American Society of Agronomy, Madison, Wisconsin, pp. 651–688.

Stanford, G. and Epstein, E. (1974) Nitrogen mineralization-water relations in soils. *Soil Science Society of America Proceedings* 38, 103–107.

Stanford, G., Vander Pol, R.A. and Dzienia, S. (1975) Denitrification rates in relation to total and extractable soil carbon. *Soil Science Society of America Proceedings* 39, 284–289.

Steele, K.W. (1987) Nitrogen losses from managed grasslands. In: Snaydon, R.W. (ed.) *Ecosystems of the World. 17B. Managed Grasslands*. Elsevier, Amsterdam, pp. 197–204.

Steele, K.W. and Brock, J.L. (1985) Nitrogen cycling in legume-based forage production systems in New Zealand. In: Barnes, R.F., Ball, P.R., Brougham, R.W.,

Marten, C.G. and Minson, D.J. (eds) *Forage Legumes for Energy-Efficient Animal Production.* USDA, Washington, pp. 171–176.
Steele, K.W., Judd, M.J. and Shannon, P.W. (1984) Leaching of nitrate and other nutrients from a grazed pasture. *New Zealand Journal of Agricultural Research* 27, 5–11.
Steele, K.W. and Shannon, P. (1982) Concepts relating to the nitrogen economy of a Northland intensive beef farm. In: Gandar, P.W. (ed.) *Nitrogen Balances in New Zealand Ecosystems.* DSIR, Palmerston North, pp. 85–90.
Steele, K.W., Watson, R.N., Bonish, P.M., Littler, R.A. and Yeates, G.W. (1985) Effect of invertebrates on nitrogen fixation in temperate pastures. *Proceedings 15th International Grassland Congress*, Kyoto, 450–451.
Steele, K.W., Wilson, A.T. and Saunders, W.M.H. (1980) Nitrification activity in New Zealand grassland soils. *New Zealand Journal of Agricultural Research* 23, 249–256.
Steen, E. (1991) Usefulness of the mesh bag method in quantitative root studies. In: Atkinson, D. (ed.) *Plant Root Growth: An Ecological Perspective.* Blackwell Scientific Publications, London, pp. 75–86.
Steenbergen, T.V. (1977) Influence of type of soil and year on the effect of nitrogen fertilisation on the yield of grassland. *Stikstof* (English edn) 20, 29–35.
Steenvoorden, J.H.A.M. (1988) How to reduce nitrogen losses in intensive-grassland management. In: Eijsackers, H. and Quispel, A. (eds) *Ecological Implications of Contemporary Agriculture. Ecological Bulletins* 39, 126–130.
Steenvoorden, J.H.A.M. (1989) Agricultural practices to reduce nitrogen losses *via* leaching and surface runoff. In: Germon, J.C. (ed.) *Management Systems to Reduce Impact of Nitrates.* Elsevier, London, pp. 72–84.
Steenvoorden, J.H.A.M., Fonck, H. and Oosterom, H.P. (1986) Losses of nitrogen from intensive grassland systems by leaching and surface runoff. In Van der Meer, H.G., Ryden J.C. and Ennik, G.C. (eds) *Nitrogen Fluxes in Intensive Grassland Systems.* Martinus Nijhoff, Dordrecht, pp. 85–97.
Stevens, R.J. (1988) Some factors influencing the efficiency of fertiliser nitrogen for grass production in spring. In: Jenkinson, D.S. and Smith, K.A. (eds) *Nitrogen Efficiency in Agricultural Soils.* Elsevier, London, pp. 177–189.
Stevens, R.J., Gracey, H.I., Kilpatrick, D.J., Camlin, M.S., O'Neil, D.G. and McLaughlan, W. (1989a) Effect of date of application and form of nitrogen on herbage production in spring. *Journal of Agricultural Science, Cambridge* 112, 329–337.
Stevens, R.J. and Laughlin, R.J. (1988) The effects of times of application and chemical forms on the efficiencies of ^{15}N-labelled fertilizers for ryegrass at two contrasting field sites. *Journal of the Science of Food and Agriculture* 43, 9–16.
Stevens, R.J. and Laughlin, R.J. (1989) A microplot study of the fate of ^{15}N-labelled ammonium nitrate and urea applied at two rates to ryegrass in spring. *Fertilizer Research* 20, 33–39.
Stevens R.J., Laughlin, R.J., Atkins, G.J. and Prosser, S.J. (1993) Automated determination of nitrogen-15-labelled dinitrogen and nitrous oxide by mass spectrometry. *Soil Science Society of America Journal* 57, 981–988.
Stevens, R.J., Laughlin, R.J. and Frost, J.P. (1989b) Effect of acidification with sulphuric acid on the volatilization of ammonia from cow and pig slurries. *Journal of Agricultural Science, Cambridge* 113, 389–395.

Stevens, R.J., Laughlin, R.J. and Frost, J.P. (1992) Effects of separation, dilution, washing and acidification on ammonia volatilization from surface-applied cattle slurry. *Journal of Agricultural Science, Cambridge* 119, 383–389.
Stevens, R.J., Laughlin, R.J. and Kilpatrick, D.J. (1989c) Soil properties related to the dynamics of ammonia volatilization from urea applied to the surface of acidic soils. *Fertilizer Research* 20, 1–9.
Stevens, R.J. and Logan, H.J. (1987) Determination of the volatilization of ammonia from surface-applied cattle slurry by the micrometeorological mass balance method. *Journal of Agricultural Science, Cambridge* 109, 205–207.
Stevens, R.J. and Watson, C. (1986) The response of grass for silage to sulphur application at 20 sites in Northern Ireland. *Journal of Agricultural Science, Cambridge* 107, 565–571.
Stevenson, F.J. (1965) Origin and distribution of nitrogen in soil. In: Bartholomew, W.V. and Clark, F.E. (eds) *Soil Nitrogen*. American Society of Agronomy, Madison, Wisconsin, pp. 1–42.
Stevenson, F.J. (1982a) *Humus Chemistry: Genesis, Composition, Reactions.* Wiley, New York.
Stevenson, F.J. (1982b) Organic forms of soil nitrogen. In: Stevenson, F.J. (ed.) *Nitrogen in Agricultural Soils.* American Society of Agronomy, Madison, Wisconsin, pp. 67–122.
Stevenson, F.J. (1986) *Cycles of Soil: Carbon, Nitrogen, Phosphorus, Sulfur, Micronutrients.* Wiley, New York.
Stewart, A.B. and Holmes, W. (1953) Nitrogenous manuring of grassland. 1. Some effects of heavy dressings of nitrogen on the mineral composition. *Journal of the Science of Food and Agriculture* 4, 401–408.
Stiles, W. (1966) Ten years of irrigation experiments. *Annual Report, Grassland Research Institute, Hurley 1965*. pp. 57–66.
Stiles, W. and Williams, T.E. (1965) The response of a ryegrass/white clover sward to various irrigation regimes. *Journal of Agricultural Science, Cambridge* 65, 351–364.
Stillings, B.R., Bratzler, J.W., Marriott, L.F. and Miller, R.C. (1964) Utilization of magnesium and other minerals by ruminants consuming low and high nitrogen-containing forages and vitamin D. *Journal of Animal Science* 23, 1148–1154.
Stillwell, M.A. and Woodmansee, R.G. (1981) Chemical transformations of urea-nitrogen and movement of nitrogen in a shortgrass prairie soil. *Soil Science Society of America Journal* 45, 893–898.
Stockdill, S.M.J. (1982) Effects of introduced earthworms on the productivity of New Zealand pastures. *Pedobiologia* 24, 29–35.
Sugimoto, Y. and Ball, P.R. (1989) Nitrogen losses from cattle dung. *Proceedings 16th International Grassland Congress,* Nice, pp. 153–154.
Sutton, M.A., Moncrieff, J.B. and Fowler, D. (1992) Deposition of atmospheric ammonia to moorlands *Environmental Pollution* 75, 15–24.
Sutton, M.A., Pitcairn, C.E.R. and Fowler, D. (1993) The exchange of ammonia between the atmosphere and plant communities. *Advances in Ecological Research* 24, 301–393.
Swift, G., Cleland, A.T. and Franklin, M.F. (1988) A comparison of nitrogen fertilizers for spring and summer grass production. *Grass and Forage Science* 43, 297–303.

Swift, G., Vipond, J.E., McClelland, T.H., Cleland, A.T., Milne, J.A. and Hunter, E.A. (1993) A comparison of diploid and tetraploid perennial ryegrass and tetraploid ryegrass/white clover swards under continuous sheep stocking at controlled sward heights. I. Sward characteristics. *Grass and Forage Science* 48, 279–289.

Syers, J.K., Sharpley, A.N. and Keeney, D.R. (1979) Cycling of nitrogen by surface-casting earthworms in a pasture ecosystem. *Soil Biology and Biochemistry* 11, 181–185.

Syers, J.K., Skinner, R.J. and Curtin, D. (1987) Soil and fertiliser sulphur in UK agriculture. *Proceedings of the Fertiliser Society* No. 264, pp. 43.

Syers, J.K. and Springett, J.A. (1983) Earthworm ecology in grassland soils. In: Satchell, J.E. (ed.) *Earthworm Ecology*. Chapman and Hall, London, pp. 67–83.

Syers, J.K. and Springett, J.A. (1984) Earthworms and soil fertility. *Plant and Soil* 76, 93–104.

Syrjala-Qvist, L., Pekkarinen, E., Setala, J. and Kangasmaki, T. (1984) Effect of nitrogen fertilization on the protein quality of timothy grass and silage. *Journal of Agricultural Science in Finland* 56, 193–198.

Tayler, R.S. (1965) The irrigation of grassland. *Outlook on Agriculture* 4, 234–242.

Taylor, H.J. and Bell, J.N.B. (1988) Studies on the tolerance to SO_2 of grass populations in polluted areas. V. Investigations into the development of tolerance to SO_2 and NO_2 in combination and NO_2 alone. *New Phytologist* 110, 327–338.

Templeton, W.C. and Taylor, T.H. (1966) Some effects of nitrogen, phosphorus and potassium fertilization on botanical composition of a tall fescue-white clover sward. *Agronomy Journal* 58, 569–572.

Terman, G.L. (1979) Volatilization losses of nitrogen as ammonia from surface-applied fertilizers, organic amendments and crop residues. *Advances in Agronomy* 31, 189–223.

Terman, G.L. and Hunt, C.M. (1964) Volatilization losses of nitrogen from surface-applied fertilizers, as measured by crop response. *Soil Science Society of America Proceedings* 28, 667–672.

Theron, J.J. (1965) The influence of fertilizers on the organic matter content of the soil under natural veld. *South African Journal of Agricultural Science* 8, 525–534.

Thomas, B. and Thompson, A. (1948) The ash content of some grasses and herbs on the Palace Leas hay plots at Cockle Park. *Empire Journal of Experimental Agriculture* 16, 221–230.

Thomas, B., Thompson, A., Oyenuga, V.A. and Armstrong, R.H. (1952) The ash constituents of some herbage plants at different stages of maturity. *Empire Journal of Experimental Agriculture* 20, 10–22.

Thomas, C. and Chamberlain, D.G. (1990) Evaluation and prediction of the nutritive value of pastures and forages. In: Wiseman, J. and Cole, D.J.A. (eds) *Feedstuff Evaluation*. Butterworths, London, pp. 319–336.

Thomas, H. (1984) Effects of drought on growth and competitive ability of perennial ryegrass and white clover. *Journal of Applied Ecology* 21, 591–602.

Thomas, P.C., Chamberlain, D.G., Kelly, N.C. and Wait, M.K. (1980) The nutritive value of silages: digestion of nitrogenous constituents in sheep receiving diets of grass silage and grass silage and barley. *British Journal of Nutrition* 43, 469–479.

Thomas, R.J. (1992) The role of the legume in the nitrogen cycle of productive and sustainable pastures. *Grass and Forage Science* 47, 133–142.

Thomas, R.J., Logan, K.A.B., Ironside, A.D. and Bolton, G.R. (1988) Transformations and fate of sheep urine-N applied to an upland UK pasture at different times during the growing season. *Plant and Soil* 107, 173–181.

Thomas, R.J., Logan, K.A.B., Ironside, A.D. and Bolton, G.R. (1990) The effects of grazing with and without excretal returns on the accumulation of nitrogen by ryegrass in a continuously grazed upland sward. *Grass and Forage Science* 45, 65–75.

Thomas, R.J., Logan, K.A.B., Ironside, A.D. and Milne, J.A. (1986) Fate of sheep urine applied to an upland grass sward. *Plant and Soil* 91, 425–427.

Thompson, K.F. and Poppi, D.P. (1990) Livestock production from pasture. In Langer, R.H.M. (ed.) *Pastures, Their Ecology and Management.* Oxford University Press, Melbourne, pp. 263–283.

Thompson, R.B. (1989) Denitrification in slurry-treated soil: occurrence at low temperatures, relationship with soil nitrate and reduction by nitrification inhibitors. *Soil Biology and Biochemistry* 21, 875–882.

Thompson, R.B. and Pain, B.F. (1989) Denitrification from cattle slurry applied to grassland. In Hansen, J.A. and Henriksen, K. (eds) *Nitrogen in Organic Wastes Applied to Soils.* Academic Press, London, pp. 247–260.

Thompson, R.B., Pain, B.F. and Lockyer, D.R. (1990a) Ammonia volatilization from cattle slurry following surface application to grassland. 1. Influence of mechanical separation, changes in chemical composition during volatilization and the presence of the grass sward. *Plant and Soil* 125, 109–117.

Thompson, R.B., Pain, B.F. and Rees, Y.J. (1990b) Ammonia volatilization from cattle slurry following surface application to grassland. 2. Influence of application rate, wind speed and applying slurry in narrow bands. *Plant and Soil* 125, 119–128.

Thompson, R.B., Ryden, J.C. and Lockyer, D.R. (1987) Fate of nitrogen in cattle slurry following surface application or injection to grassland. *Journal of Soil Science* 38, 689–700.

Thomson, D.J. (1984) The nutritive value of white clover. In: Thomson, D.J. (ed.) *Forage Legumes.* British Grassland Society Occasional Symposium No. 16, pp. 78–92.

Thornton, R.H. (1982) Guidelines for future research. In: Gandar, P.W. (ed.) *Nitrogen Balances in New Zealand Ecosystems.* DSIR, Palmerston North, pp. 259–262.

Tiedje, J.M., Simkins, S. and Groffman, P.M. (1989) Perspectives on measurement of denitrification in the field including recommended protocols for acetylene based methods. *Plant and Soil* 115, 261–284.

Tiessen, H., Stewart, J.W.B. and Bettany, J.R. (1982) Cultivation effects on the amounts and concentration of carbon, nitrogen and phosphorus in grassland soils. *Agronomy Journal* 74, 831–835.

Titko, S., Street, J.R. and Logan, T.J. (1987) Volatilization of ammonia from granular and dissolved urea applied to turfgrass. *Agronomy Journal* 79, 535–540.

Toomre, R.I. (1966) Effect of high rates of nitrogenous fertilizers on the yield and chemical composition of grasses. *Proceedings 10th International Grassland Congress*, Helsinki, 1966, pp. 227–230.

Topps, J.H. and Elliott, R.C. (1966) Partition of nitrogen in the urine of African sheep given low-protein diets. *Proceedings of the Nutrition Society* 25, xix–xx.

Triboi, E. (1987) Recovery of mineral N fertilizer in herbage and soil organic matter in grasslands of the Massif Central, France. *Fertilizer Research* 13, 99–116.

Troughton, A. (1981) Length of life of grass roots. *Grass and Forage Science* 36, 117–120.

Tukey, H.B.J. and Morgan, T.V. (1964) The occurrence of leaching from above ground plant parts and the nature of the material leached. *Proceedings of 16th International Horticulture Congress*, 1962, Vol. 4, 153–160.

Tunney, H. and Molloy, S.P. (1986) Comparison of grass production with soil injected and surface spread cattle slurry. In: Dam Kofoed, A., Williams, J.H. and L'Hermite, P. (eds) *Efficient Land Use of Sludge and Manure*. Elsevier, London, pp. 90–98.

Tveitnes, S. and Haland, A. (1989) Influence of the nitrification inhibitor dicyandiamide (DCD) on the nitrogen efficiency of cattle slurry. *Norwegian Journal of Agricultural Sciences* 3, 343–350.

Tyson, K.C., Roberts, D.H., Clement, C.R. and Garwood, E.A. (1990) Comparison of crop yields and soil conditions during 30 years under annual tillage or grazed pasture. *Journal of Agricultural Science, Cambridge* 115, 29–40.

Underhay, V.H.S. and Dickinson, C.H. (1978) Water, mineral and energy fluctuations in decomposing cattle dung pats. *Journal of the British Grassland Society* 33, 189–196.

Unkovich, M.J., Pate, J.S., Sanford, P. and Armstrong, E.L. (1994) Potential precision of the ^{15}N natural abundance method in field estimates of nitrogen fixation by crop and pasture legumes in south-west Australia. *Australian Journal of Agricultural Research* 45, 119–132.

Unwin, R.J. (1986) Leaching of nitrate after application of organic manures: lysimeter studies. In: Dam Kofoed, A., Williams, J.H. and L'Hermite, P. (eds) *Efficient Land Use of Sludge and Manure*. Elsevier, London, pp. 158–167.

Unwin, R.J., Pain, B.F. and Whinham, W.N. (1986) The effect of rate and time of application of nitrogen in cow slurry on grass cut for silage. *Agricultural Wastes* 15, 253–268.

Unwin, R.J. and Vellinga, Th. V. (1994) Fertilizer recommendations for intensively managed grassland. *Proceedings 15th General Meeting, European Grassland Federation*, Wageningen, Netherlands, pp. 590–602.

Vallis, I. (1978) Nitrogen relationships in grass/legume mixtures. In: Wilson, J.R. (ed.) *Plant Relations in Pastures*. CSIRO, Melbourne, pp. 190–201.

Vallis, I., Harper, L.A., Catchpoole, V.R. and Weier, K.L. (1982) Volatilization of ammonia from urine patches in a subtropical pasture. *Australian Journal of Agricultural Research* 33, 97–107.

Van Breemen, N. and others (1982) Soil acidification from atmospheric ammonium sulphate in forest canopy throughfall. *Nature, London* 299, 548–550.

Van Breemen, N. and van Dijk, H.F.G. (1988) Ecosystem effects of atmospheric deposition of nitrogen in the Netherlands. *Environmental Pollution* 54, 249–274.

Van Burg, P.F.J. (1961) Nitrogen fertilization and the seasonal production of grassland herbage. *Proceedings 8th International Grassland Congress*, Reading, 1960, pp. 142–146.

Van Burg, P.F.J. (1964) Nitrogen fertilization of grassland: the effect of nitrogen solutions. *Stikstof* (English edn), No. 8, 28–32.

Van Burg, P.F.J. (1966) Nitrate as an indicator of the nitrogen-nutrition status of grass. *Proceedings 10th International Grassland Congress*, Helsinki, 1966, pp. 267–272.

Van Burg, P.F.J. (1968) Nitrogen fertilizing of grassland in spring. *Netherlands Nitrogen Technical Bulletin*, No. 6, pp. 45.

Van Burg, P.F.J. (1969) The agronomic value of anhydrous ammonia in Western Europe. *Outlook on Agriculture* 6, 55–59.

Van Burg, P.F.J. and van Brakel, G.D. (1965) The fertilizer value of anhydrous ammonia on permanent grassland. *Stikstof* (English edn), No. 9, 28–36.

Van Burg, P.F.J., van Brakel, G.D., and Schepers, J.H. (1967) The agricultural value of anhydrous ammonia on grassland: experiments 1963–1965. *Netherlands Nitrogen Technical Bulletin*, No. 2, pp. 31.

Van Burg, P.F.J., Prins, W.H., den Boer, D.J. and Sluiman, W.J. (1981) Nitrogen and intensification of livestock farming in EEC countries. *Proceedings of the Fertiliser Society, London* 199, 1–78.

Van Burg, P.F.J., Dilz, K. and Prins, W.H. (1982) Agricultural value of various nitrogen fertilizers. *Netherlands Nitrogen Technical Bulletin* No. 13, pp. 51.

Van der Eerden, L.J.M. (1982) Toxicity of ammonia to plants. *Agriculture and Environment* 7, 223–235.

Van der Meer, H.G. (1983) Effective use of nitrogen on grassland farms. *British Grassland Society, Occasional Symposium* No. 14, 61–68. Proceedings 9th General Meeting European Grassland Federation, Reading.

Van der Meer, H.G. (1994) Grassland and society. *Proceedings 15th General Meeting, European Grassland Federation,* Wageningen, Netherlands, pp. 19–32.

Van der Meer, H.G., Thompson, R.B., Snijders, P.J.M. and Geurink, J.H. (1987) Utilization of nitrogen from injected and surface-spread cattle slurry applied to grassland. In: van der Meer, H.G., Unwin, R.J., van Dijk, T.A. and Ennik, G.C. (eds) *Animal Manure on Grassland and Fodder Crops: Fertilizer or Waste?* Martinus Nijhoff, Dordrecht, pp. 47–71.

Van der Meer, H.G. and van Uum-van Lohuyzen, M.G. (1986) The relationship between inputs and outputs of nitrogen in intensive grassland systems. In: van der Meer, H.G., Ryden, J.C. and Ennik, G.C. (eds) *Nitrogen Fluxes in Intensive Grassland Systems*. Martinus Nijhoff, Dordrecht, pp. 1–18.

Van der Meer, H.G. and Wedin, W.F. (1989) Present and future role of grasslands and fodder crops in temperate countries with special reference to over-production and environment. *Proceedings 16th International Grassland Congress*, Nice, pp. 1711–1718.

Van der Molen, H. (1964) Hypomagnesaemia and grass fertilization in the Netherlands. *Outlook on Agriculture* 4, 55–63.

Van der Molen, J., Bussink, D.W., Vertregt, N., van Faassen, H.G. and den Boer, D.J. (1989) Ammonia volatilization from arable and grassland soils. In: Hansen, J.A. and Henriksen, K. (eds) *Nitrogen in Organic Wastes Applied to Soils*. Academic Press, London, pp. 185–201.

Van Dijk, T.A. and Sturm, H. (1983) Fertiliser value of animal manures on the continent. *Proceedings of the Fertiliser Society, London*, No. 220.

Van Straalen, W.M. and Tamminga, S. (1990) Protein degradation of ruminant diets. In: Wiseman, J. and Cole, D.J.A. (eds) *Feedstuff Evaluation*. Butterworths, London, pp. 55–72.

Van Strien, A.J., Melman, Th. C.P. and de Heiden, J.L.H. (1988) Extensification of dairy farming and floristic richness of peat grassland. *Netherlands Journal of Agricultural Science* 36, 339–355.

Van Vuuren, A.M., Krol-Kramer, F., van der Lee, R.A. and Corbijn, H. (1992) Protein digestion and intestinal amino acids in dairy cows fed fresh *Lolium perenne* with different nitrogen contents. *Journal of Dairy Science* 75, 2215–2225.

Van Vuuren, A.M. and Meijs, J.A.C. (1987) Effects of herbage composition and supplement feeding on the excretion of nitrogen in dung and urine by grazing dairy cows. In: van der Meer, H.G. *et al.* (eds) *Animal Manure on Grassland and Fodder Crops.* Martinus Nijhoff, Dordrecht, pp. 17–26.

Van Vuuren, A.M., Tamminga, S. and Ketelaar, R.S. (1991) *In sacco* degradation of organic matter and crude protein of fresh grass (*Lolium perenne*) in the rumen of grazing dairy cows. *Journal of Agricultural Science, Cambridge* 116, 429–436.

Vaughn, C.E., Center, D.M. and Jones, M.B. (1986) Seasonal fluctuations in nutrient availability in some northern California annual range soils. *Soil Science* 141, 43–51.

Velthof, G.L. and Oenema, O. (1993) Nitrous oxide flux from nitric-acid-treated cattle slurry applied to grassland under semi-controlled conditions. *Netherlands Journal of Agricultural Science* 41, 81–93.

Velthof, G.L. and Oenema, O. (1994) Effect of nitrogen fertilizer type and urine on nitrous oxide flux from grassland in early spring. *Proceedings 15th General Meeting, European Grassland Federation,* Wageningen, Netherlands, pp. 458–462.

Verbic, J., Chen, X.B., MacLeod, N.A. and Ørskov, E.R. (1990) Excretion of purine derivatives by ruminants. *Journal of Agricultural Science, Cambridge* 114, 243–248.

Vercoe, J.M. (1962) Some observations on the nitrogen and energy losses in the faeces and urine of grazing sheep. *Proceedings of the Australian Society of Animal Production* 4, 160–162.

Vertregt, N. and Rutgers, B. (1987) Ammonia volatilization from urine patches in grassland. In: Nielsen, V.C., Voorburg, J.H. and L'Hermite, P. (eds) *Volatile Emissions from Livestock Farming and Sewage Operations.* Elsevier, London, pp. 85–91.

Vertregt, N. and Rutgers, B. (1991) Ammonia emissions from grazing. In: Nielsen, V.C., Voorburg, J.H. and L'Hermite, P. (eds) *Odour and Ammonia Emissions from Livestock Farming.* Elsevier, London, pp. 177–183.

Vetter, H., Steffens, G. and Schröpel, R. (1987) The influence of different processing methods for slurry upon its fertiliser value on grassland. In: van der Meer, H.G., Unwin, R.J., van Dijk, T.A. and Ennik, G.C. (eds) *Animal Manure on Grassland and Forage Crops: Fertilizer or Waste?* Martinus Nijhoff, Dordrecht, pp. 73–86.

Vez, A. (1961) [Study of the development of white clover in relation to fluctuations in certain organic substances.] *Berichte der Schweizerischen Botanischen Gesellschaft* 71, 118–172.

Vine, A. and Bateman, D. (1981) *Organic Farming Systems in England and Wales: Practice, Performance and Implications.* Report, University College of Wales, Aberystwyth.

Vinten, A.J.A. and Smith, K.A. (1993) Nitrogen cycling in agricultural soils. In: Burt, T.P., Heathwaite, A.L. and Trudgill, S.T. (eds) *Nitrate: Processes, Patterns and Management.* John Wiley and Sons, Chichester, pp. 39–73.

Vipond, J.E., Swift, G., McClelland, T.H., Fitzsimons, J., Milne, J.A. and Hunter, E.A. (1993) A comparison of diploid and tetraploid perennial ryegrass and tetraploid

ryegrass/white clover swards under continuous sheep stocking at controlled sward heights. 2. Animal production. *Grass and Forage Science* 48, 290–300.
Virtanen, A.I. and Miettinen, J.K. (1963) Biological nitrogen fixation. In: Steward, F.C. (ed.) *Plant Physiology* Vol. 3. Academic Press, New York, pp. 539–668.
Vlek, P.L.G. and Carter, M.F. (1983) The effect of soil environment and fertilizer modifications on the rate of urea hydrolysis. *Soil Science* 136, 56–63.
Volenec, J.J. and Nelson, C.J. (1984) Carbohydrate metabolism in leaf meristem of tall fescue. II. Relationship to leaf elongation rate modified by nitrogen fertilization. *Plant Physiology* 74, 595–600.
Wacquant, J.P., Ouknider, M. and Jacquard, P. (1989) Evidence for a periodic excretion of nitrogen by roots of grass-legume associations. *Plant and Soil* 116, 57–68.
Wadman, W.P., Sluijsmans, C.M.J. and De La Lande Cremer, L.C.N. (1987) Value of animal manures: changes in perception. In: van der Meer, H.G., Unwin, R.J., van Dijk, T.A. and Ennik, G.C. (eds) *Animal Manure on Grassland and Fodder Crops: Fertilizer or Waste?* Martinus Nijhoff, Dordrecht, pp. 1–16.
Waite, R. (1970) The structural carbohydrates and the *in vitro* digestibility of a ryegrass and a cocksfoot at two levels of a nitrogenous fertilizer. *Journal of Agricultural Science, Cambridge* 74, 457–462.
Walker, T.W. (1962) Problems of soil fertility in a grass–animal regime. *Transactions, International Soil Science Conference*, New Zealand, pp. 704–714.
Walker, T.W. (1965) The significance of phosphorus in pedogenisis. In: Hallsworth, E.G. and Crawford, D.V. (eds) *Experimental Pedology*. Butterworths, London, pp. 295–316.
Walker, T.W., Adams, A.F.R. and Orchiston, H.D. (1956) Fate of labelled nitrate and ammonium nitrogen when applied to grass and clover grown separately and together. *Soil Science* 81, 339–351.
Walker, T.W., Edwards, G.H.A., Cavell, A.J. and Rose, T.H. (1952) The use of fertilizers on herbage cut for conservation. Part II. Effects on the mineral composition of herbage cut for silage. *Journal of the British Grassland Society* 7, 107–130.
Walker, T.W., Orchiston, H.D. and Adams, A.F.R. (1954) The nitrogen economy of grass legume associations. *Journal of the British Grassland Society* 9, 249–274.
Walshe, M.J. and Conway, A. (1960) Hypomagnesaemia in ruminants. *Proceedings 8th International Grassland Congress*, Reading, pp. 548–553.
Walther, W. (1989) The nitrate leaching out of soils and their significance for groundwater; results of long-term tests. In: Hansen, J.A. and Henriksen, K. (eds) *Nitrogen in Organic Wastes Applied to Soils*. Academic Press, London, pp. 346–356.
Wardle, D.A. and Greenfield, L.G. (1991) Release of mineral nitrogen from plant root nodules. *Soil Biology and Biochemistry* 23, 827–832.
Wareing, P.F. and Allen, E.J. (1977) Physiological aspects of crop choice. *Philosophical Transactions of the Royal Society* 281, 107–119.
Warembourg, F.R. and Roumet, C. (1989) Why and how to estimate the cost of symbiotic N_2 fixation? A progressive approach based on the use of ^{14}C and ^{15}N isotopes. *Plant and Soil* 115, 167–177.
Warneck, P. (1988) *Chemistry of the Natural Atmosphere*. Academic Press, San Diego.
Warren, G.P. (1988) Use of acetylene to improve the reliability of the assessment of soil nitrogen mineralized on aerobic incubation. *Soil Biology and Biochemistry* 20, 119–121.

Warren, G.P. and Whitehead, D.C. (1988) Available soil nitrogen in relation to fractions of soil nitrogen and other soil properties. *Plant and Soil* 112, 155–165.

Watson, C.J. (1986) Preferential uptake of ammonium nitrogen from soil by ryegrass under simulated spring conditions. *Journal of Agricultural Science, Cambridge* 107, 171–177.

Watson, C.J. and Adams, S.N. (1986) Effect of simulated wet spring conditions on the relative efficiency of three forms of nitrogen fertilizer on grassland. *Journal of Agricultural Science, Cambridge* 107, 219–222.

Watson, C.J., Fowler, S.M. and Wilman, D. (1993) Soil inorganic-N and nitrate leaching on organic farms. *Journal of Agricultural Science, Cambridge* 120, 361–369.

Watson, C.J., Jordan, C., Taggart, P.J., Laidlaw, A.S., Garrett, M.K. and Steen, R.W.J. (1992) The leaky N-cycle on grazed grassland. In: Archer, J.R. *et al.* (eds) *Nitrate and Farming Systems*. Association of Applied Biologists, Warwick, pp. 215–222.

Watson, C.J. and Kilpatrick, D.J. (1991) The effect of urea pellet size and rate of application on ammonia volatilization and soil nitrogen dynamics. *Fertilizer Research* 28, 163–172.

Watson, C.J., Stevens, R.J., Garrett, M.K. and McMurray, C.H. (1990a) Efficiency and future potential of urea for temperate grassland. *Fertilizer Research* 26, 341–357.

Watson, C.J., Stevens, R.J. and Laughlin, R.J. (1990b) Effectiveness of the urease inhibitor NBPT (*N*-(*n*-butyl) thiophosphoric triamide) for improving the efficiency of urea for ryegrass production. *Fertilizer Research* 24, 11–15.

Watson, E.R. (1969) The influence of subterranean clover pasture on soil fertility. III. The effect of applied phosphorus and sulphur. *Australian Journal of Agricultural Research* 20, 447–456.

Watson, E.R. and Lapins, P. (1964) The influence of subterranean clover pastures on soil fertility. II. The effects of certain management systems. *Australian Journal of Agricultural Research* 15, 885–894.

Watts, D.G., Hergert, G.W. and Nichols, J.T. (1991) Nitrogen leaching losses from irrigated orchardgrass on sandy soils. *Journal of Environmental Quality* 20, 355–362.

Wayne, R.P. (1993) Nitrogen and nitrogen compounds in the atmosphere. In: Burt, T.P., Heathwaite, A.L. and Trudgill, S.T. (eds) *Nitrate: Processes, Patterns and Management.* John Wiley and Sons, Chichester, pp. 23–38.

Webb, B.W. and Walling, D.E. (1985) Nitrate behaviour in streamflow from a grassland catchment in Devon, U.K. *Water Research* 19, 1005–1016

Webb, J. and Sylvester-Bradley, R. (1994) Effects of fertilizer nitrogen on soil nitrogen availability after a grazed grass ley and on the response of the following cereal crops to fertilizer nitrogen. *Journal of Agricultural Science, Cambridge* 122, 445–457.

Webster, C.P. and Dowdell, R.J. (1984a) Effect of drought and irrigation on the fate of nitrogen applied to cut permanent grass swards in lysimeters: leaching losses. *Journal of the Science of Food and Agriculture* 35, 1105–1111.

Webster, C.P. and Dowdell, R.J. (1984b) The fate of fertilizer applied to a grass/arable rotation: effect of sward destruction and tillage on nitrogen mineralization. *Letcombe Laboratory Report for 1983*, pp. 66–68.

Webster, C.P. and Goulding, K.W.T. (1989) Influence of soil carbon content on denitrification from fallow land during autumn. *Journal of the Science of Food and Agriculture* 49, 131–142.

Wedin, D.A. and Tilman, D. (1990) Species effects on nitrogen cycling: a test with perennial grasses. *Oecologia* 84, 433–441.

Weeda, W.C. (1967) The effect of cattle dung patches on pasture growth, botanical composition and pasture utilisation. *New Zealand Journal of Agricultural Research* 10, 150–159.

Weissbach, F. and Gordon, F.J. (1992) Grassland based animal production in Europe. *Proceedings 14th General Meeting, European Grassland Federation,* Lahti, Finland, pp. 1–18.

Wellburn, A.R. (1990) Why are atmospheric oxides of nitrogen usually phytotoxic and not alternative fertilizers? *New Phytologist* 115, 395–429.

Wells, T.C.E. and Sheail, J. (1988) The effects of agricultural change on the wildlife interest of lowland grasslands. In: Park, J.R. (ed.) *Environmental Management in Agriculture.* Bellhaven Press, London, pp. 186–201.

Wetselaar, R. and Farquhar, G.D. (1980) Nitrogen losses from tops of plants. *Advances in Agronomy* 33, 263–302.

Wheeler, J.L. (1958) The effect of sheep excreta and nitrogenous fertilizer on the botanical composition and production of a ley. *Journal of the British Grassland Society* 13, 196–202.

White, D.J., Wilkinson, J.M. and Wilkins, R.J. (1983) Support energy use in animal production from grassland. In: Corrall, A.J. (ed.) *Efficient Grassland Farming, British Grassland Society Occasional Symposium* No. 14, 33–42.

White, E.M., Krueger, C.R. and Moore, R.A. (1976) Changes in total N, organic matter, available P and bulk densities of a cultivated soil 8 years after tame pastures were established. *Agronomy Journal* 68, 581–583.

Whitehead, D.C. (1966a) Data on the mineral composition of grassland herbage from the Grassland Research Institute, Hurley, and the Welsh Plant Breeding Station, Aberystwyth. *Technical Report 4, Grassland Research Institute*, Hurley, pp. 55.

Whitehead, D.C. (1966b) *Nutrient Minerals in Grassland Herbage.* Review series 1/1966, Commonwealth Bureau of Pastures and Field Crops, Farnham Royal.

Whitehead, D.C. (1970) Carbon, nitrogen, phosphorus and sulphur in herbage plant roots. *Journal of the British Grassland Society* 25, 236–241.

Whitehead, D.C. (1982) Yield of white clover and its fixation of nitrogen as influenced by nutritional and soil factors under controlled environment conditions. *Journal of the Science of Food and Agriculture* 33, 1227–1234.

Whitehead, D.C. (1983a) The influence of frequent defoliation and of drought on nitrogen and sulphur in the roots of perennial ryegrass and white clover. *Annals of Botany* 52, 931–934.

Whitehead, D.C. (1983b) Prediction of the supply of soil nitrogen to grass. *Proceedings 14th International Grassland Congress*, Lexington, 1981, pp. 318–320.

Whitehead, D.C. (1984) Interactions between soil and fertilizer in the supply of nitrogen to ryegrass grown on 21 soils. *Journal of the Science of Food and Agriculture* 35, 1067–1075.

Whitehead, D.C. (1986) Sources and transformations of organic nitrogen in intensively managed grassland soils. In: van der Meer, H.G., Ryden, J.C. and Ennik, G.C. (eds) *Nitrogen Fluxes in Intensive Grassland Systems.* Martinus Nijhoff, Dordrecht, pp. 47–58.

Whitehead, D.C., Barnes, R.J. and Jones, L.H.P. (1983) Nitrogen, sulphur and other mineral elements in white clover and perennial ryegrass grown in mixed swards,

with and without fertilizer N, at a range of sites in the U.K. *Journal of the Science of Food and Agriculture* 34, 901–909.
Whitehead, D.C. and Bristow, A.W. (1990) Transformations of nitrogen following the application of ^{15}N-labelled cattle urine to an established grass sward. *Journal of Applied Ecology* 27, 667–678.
Whitehead, D.C., Bristow, A.W. and Lockyer, D.R. (1990) Organic matter and nitrogen in the unharvested fractions of grass swards in relation to the potential for nitrate leaching after ploughing. *Plant and Soil* 123, 39–49.
Whitehead, D.C., Bristow, A.W. and Pain, B.F. (1989a) The influence of some cattle and pig slurries on the uptake of nitrogen by ryegrass in relation to fractionation of the slurry N. *Plant and Soil* 117, 111–120.
Whitehead, D.C., Buchan, H. and Hartley, R.D. (1975) Components of soil organic matter under grass and arable cropping. *Soil Biology and Biochemistry* 7, 65–71.
Whitehead, D.C., Buchan, H. and Hartley, R.D. (1979) Composition and decomposition of roots of ryegrass and red clover. *Soil Biology and Biochemistry* 11, 619–628.
Whitehead, D.C., Goulden, K.M. and Hartley, R.D. (1986a) Fractions of nitrogen, sulphur, phosphorus, calcium and magnesium in the herbage of perennial ryegrass as influenced by fertilizer nitrogen. *Animal Feed Science and Technology* 14, 231–242.
Whitehead, D.C., Jones, L.H.P. and Barnes, R.J. (1978) The influence of fertilizer N plus K on N, S and other mineral elements in perennial ryegrass at a range of sites. *Journal of the Science of Food and Agriculture* 21, 1–11.
Whitehead, D.C. and Lockyer, D.R. (1987) The influence of the concentration of gaseous ammonia on its uptake by the leaves of Italian ryegrass, with and without an adequate supply of nitrogen to the roots. *Journal of Experimental Botany* 38, 818–827.
Whitehead, D.C. and Lockyer, D.R. (1989) Decomposing grass herbage as a source of ammonia in the atmosphere. *Atmospheric Environment* 23, 1867–1869.
Whitehead, D.C., Lockyer, D.R. and Raistrick, N. (1988) The volatilization of ammonia from perennial ryegrass during decomposition, drying and induced senescence. *Annals of Botany* 61, 567–571.
Whitehead, D.C., Lockyer, D.R. and Raistrick, N. (1989b) Volatilization of ammonia from urea applied to soil: influence of hippuric acid and other constituents of livestock urine. *Soil Biology and Biochemistry* 21, 803–808.
Whitehead, D.C., Pain, B.F. and Ryden, J.C. (1986b) Nitrogen in UK grassland agriculture. *Journal of the Royal Agricultural Society of England* 147, 190–201.
Whitehead, D.C. and Raistrick, N. (1990) Ammonia volatilization from five nitrogen compounds used as fertilizers following surface application to soils. *Journal of Soil Science* 41, 387–394.
Whitehead, D.C. and Raistrick, N. (1991) Effects of some environmental factors on ammonia volatilization from simulated livestock urine applied to soil. *Biology and Fertility of Soils* 11, 279–284.
Whitehead, D.C. and Raistrick, N. (1992) Effects of plant material on ammonia volatilization from simulated livestock urine applied to soil. *Biology and Fertility of Soils* 13, 92–95.
Whitehead, D.C. and Raistrick, N. (1993a) Nitrogen in the excreta of dairy cattle: changes during short-term storage. *Journal of Agricultural Science, Cambridge* 121, 73–81.

Whitehead, D.C. and Raistrick, N. (1993b) The volatilization of ammonia from cattle urine applied to soils as influenced by soil properties. *Plant and Soil* 148, 43–51.

Whitmore, A.P., Bradbury, N.J. and Johnson, P.A. (1992) Potential contribution of ploughed grassland to nitrate leaching. *Agriculture, Ecosystems and Environment* 39, 221–233.

Widdowson, F.V., Penny, A. and Flint, R.C. (1972) Results from an experiment comparing aqueous ammonia with nitro-chalk for grazed grass. *Journal of Agricultural Science, Cambridge* 79, 341–348.

Widdowson, F.V., Penny, A. and Williams, R.J.B. (1962) An experiment comparing urea-formaldehyde fertilizer with 'nitro-chalk' for Italian ryegrass. *Journal of Agricultural Science, Cambridge* 59, 263–268.

Widdowson, F.V., Penny, A. and Williams, R.J.B. (1965a) Experiments comparing concentrated and dilute NPK fertilizers and four nitrogen fertilizers on a range of crops. *Journal of Agricultural Science, Cambridge* 65, 45–55.

Widdowson, F.V., Penny, A. and Williams, R.J.B. (1965b) An experiment measuring effects of N, P, and K fertilizers on yield and N, P, and K contents of grass. *Journal of Agricultural Science, Cambridge* 64, 93–100.

Widdowson, F.V., Penny, A. and Williams, R.J.B. (1966) An experiment measuring effects of N, P and K fertilizers on yield and N, P and K contents of grazed grass. *Journal of Agricultural Science, Cambridge* 67, 121–128.

Widdowson, F.V. and Shaw, K. (1960) Comparisons of casein and formalized casein with ammonium sulphate, calcium nitrate and urea for Italian ryegrass. *Journal of Agricultural Science, Cambridge* 55, 53–59.

Wilcox, G.E. and Hoff, J.E. (1974) Grass tetany: an hypothesis concerning its relationship with ammonium nutrition of spring grasses. *Journal of Dairy Science* 57, 1085–1089.

Wild, A., Jones, L.H.P. and Macduff, J.H. (1987) Uptake of mineral nutrients and crop growth: the use of flowing nutrient solutions. *Advances in Agronomy* 41, 171–219.

Wilkins, R.J. (1987) Grassland into the twenty-first century. In: Norsk Hydro Fertilizers, *Farming into the Twenty-First Century*. Norsk Hydro Fertilizers, pp. 93–115.

Wilkins, R.J., Hopkins, A., Kirkham, F.W., Sargent, C., Mountford, O., Dibb, C. and Gilbey, J. (1989) Effects of changes in fertilizer use on the composition and productivity of permanent grassland in relation to agricultural production and floristic diversity. *Proceedings 16th International Grassland Congress*, Nice pp. 101–102.

Wilkinson, J.F. and Duff, D.T. (1972) Effects of fall fertilization on cold resistance, color and growth of Kentucky bluegrass. *Agronomy Journal* 64, 345–348.

Wilkinson, S.R. and Lowrey, R.W. (1973) Cycling of mineral nutrients in pasture ecosystems. In: Butler, G.W. and Bailey, R.W. (eds) *Chemistry and Biochemistry of Herbage*, Vol. 2. Academic Press, London, pp. 247–315.

Williams, C.H. and Donald, C.M. (1957) Changes in organic matter and pH in a podzolic soil as influenced by subterranean clover and superphosphate. *Australian Journal of Agricultural Research* 8, 179–189.

Williams, P.H., Hedley, M.J. and Gregg, P.E.H. (1989) Uptake of potassium and nitrogen by pasture from urine-affected soil. *New Zealand Journal of Agricultural Research* 32, 415–421.

Williams, P.H., Hedley, M.J. and Gregg, P.E.H. (1990) The effect of preferential flow of dairy cow urine and simulated rainfall on movement of potassium through undisturbed topsoil cores. *Australian Journal of Soil Research* 28, 857–868.

Williams, T.E. and Clement, C.R. (1965) Accumulation and availability of nitrogen in soils under leys. *Proceedings of 1st General Meeting of European Grassland Federation*, pp. 39–45.

Wilman, D. (1965) The effect of nitrogenous fertilizer on the rate of growth of Italian ryegrass. *Journal of the British Grassland Society* 20, 248–254.

Wilman, D., Acuña, P.G.H. and Michaud, P.J. (1994) Concentrations of N, P, K, Ca, Mg and Na in perennial ryegrass and white clover leaves of different ages. *Grass and Forage Science* 49, 422–428

Wilman, D. and Asiegbu, J.E. (1982) The effects of variety, cutting interval and nitrogen application on the morphology and development of stolons and leaves of white clover. *Grass and Forage Science* 37, 15–27.

Wilman, D. and Derrick, R.W. (1994) Concentration and availability to sheep of N, P, K, Ca, Mg and Na in chickweed, dandelion, dock, ribwort and spurrey, compared with perennial ryegrass. *Journal of Agricultural Science, Cambridge* 122, 217–223.

Wilman, D., Droushiotis, D., Koocheki, A., Lwoga, A.B. and Shim, J.S. (1976a) The effect of interval between harvests and nitrogen application on the digestibility and digestible yield and nitrogen content and yield of four ryegrass varieties in the first harvest year. *Journal of Agricultural Science, Cambridge* 86, 393–399.

Wilman, D., Droushiotis, D., Mzamane, M.N. and Shim, J.S. (1977) The effect of interval between harvests and nitrogen application on initiation, emergence and longevity of leaves, longevity of tillers and dimensions and weights of leaves and 'stems' in *Lolium*. *Journal of Agricultural Science, Cambridge* 89, 65–79.

Wilman, D. and Hollington, P.A. (1985) Effects of white clover and fertilizer nitrogen on herbage production and chemical composition and soil water. *Journal of Agricultural Science, Cambridge* 104, 453–467.

Wilman, D., Koocheki, A., Lwoga, A.B., Droushiotis, D. and Shim, J.S. (1976b) The effect of interval between harvests and nitrogen application on the numbers and weights of tillers and leaves in four ryegrass varieties. *Journal of Agricultural Science, Cambridge* 87, 45–57.

Wilman, D. and Mohamed, A.A. (1980) Early spring and late autumn response to applied nitrogen in four grasses. 2. Leaf development. *Journal of Agricultural Science, Cambridge* 94, 443–453.

Wilman, D. and Pearse, P.J. (1984) Effects of applied nitrogen on grass yield, nitrogen content, tillers and leaves in field swards. *Journal of Agricultural Science, Cambridge* 103, 201–211.

Wilman, D. and Simpson, D. (1988) The growth of white clover (*Trifolium repens*) in five sown hill swards grazed by sheep. *Journal of Applied Ecology* 25, 631–642.

Wilman, D. and Wright, P.T. (1983) Some effects of applied nitrogen on the growth and chemical composition of temperate grasses. *Herbage Abstracts* 53, 387–393.

Wilman, D. and Wright, P.T. (1986) The effect of interval between harvests and nitrogen application on the concentration of nitrate-nitrogen in the total herbage, green leaf and 'stem' of grasses. *Journal of Agricultural Science, Cambridge* 106, 467–475.

Wilson, D.B. (1964) Interactions in forage yield trials. *Canadian Journal of Plant Science* 44, 344–350.

Wilson, J.D., Catchpoole, V.R., Denmead, O.T. and Thurtell, G.W. (1983) Verification of a simple micro-meteorological method for estimating the rate of gaseous mass transfer from the ground to the atmosphere. *Agricultural Meteorology* 29, 183–189.

Wilson, R.K. (1963) An attempt to induce hypomagnesaemia in wethers by feeding high levels of urea. *Veterinary Record* 75, 698–699.

Witty, J.F. and Minchin, F.R. (1988) Measurement of nitrogen fixation by the acetylene reduction assay: myths and mysteries. In: Beck, D.P. and Materon, L.A. (eds) *Nitrogen Fixation by Legumes in Mediterranean Agriculture*. ICARDA, Syria, pp. 331–344.

Woelfel, C.G. and Poulton, B.R. (1960) The nutritive value of timothy hay as affected by nitrogen fertilization. *Journal of Animal Science* 19, 695–699.

Woledge, J. (1988) Competition between grass and clover in spring as affected by nitrogen fertilizer. *Annals of Applied Biology* 112, 175–186.

Woledge, J. and Pearse, P.J. (1985) The effect of nitrogenous fertilizer on the photosynthesis of leaves of a ryegrass sward. *Grass and Forage Science* 40, 305–309.

Woledge, J., Tewson, V. and Davidson, I.A. (1990) Growth of grass/clover mixtures during winter. *Grass and Forage Science* 45, 191–202.

Wolton, K.M. (1955) The effect of sheep excreta and fertilizer treatments on the nutrient status of pasture soil. *Journal of the British Grassland Society* 10, 240–253.

Wolton, K.M. (1979) Dung and urine as agents of sward change: a review. In: Charles, A.H. and Haggar, R.J. (eds) *Changes in Sward Composition and Productivity.* British Grassland Society Occasional Symposium No. 10. British Grassland Society, Hurley, pp. 131–135.

Wolton, K.M., Brockman, J.S., Brough, D.W.T. and Shaw, P.G. (1968) The effect of nitrogen, phosphate and potash fertilizers on three grass species. *Journal of Agricultural Science, Cambridge* 70, 195–202.

Woodmansee, R.G. (1979) Factors influencing input and output of nitrogen in grasslands. In: French, N.R. (ed.) *Perspectives in Grassland Ecology*. Springer-Verlag, New York, pp. 117–134.

Woodmansee, R.G., Vallis, I. and Mott, J.J. (1981) Grassland nitrogen. In: Clark, F.E. and Rosswall, T. (eds) *Terrestrial Nitrogen Cycles*. Swedish Natural Science Research Council, Stockholm, pp. 443–462.

Wood, M., Cooper, J.E. and Campbell, D.S. (1985) A survey of clover and *Lotus* rhizobia in Northern Ireland pasture soils. *Journal of Soil Science* 36, 357–365.

Wood, M. and McNeill, A.M. (1993) $^{15}N_2$ measurement of nitrogen fixation by legumes and actinorhizals: theory and practice. *Plant and Soil* 155/156, 329–332.

Woods, L.E., Cole, C.V., Elliott, E.T., Anderson, R.V. and Coleman, D.C. (1982) Nitrogen transformations in soil as affected by bacterial-microfaunal interactions. *Soil Biology and Biochemistry* 14, 93–98.

Woolford, M.K. (1984) *The Silage Fermentation*. Marcel Dekker, New York.

Wouters, A.P. and Verboon, M.C. (1993) Handling of slurry in relation to the environment on dairy farms in the Netherlands. In: Hopkins, A. and Younie, D. (eds) *Forward with Grass into Europe*, Occasional Syposium No. 27. British Grassland Society, pp. 85–96.

Wright, M.J. and Davison, K.L. (1964) Nitrate accumulation in crops and nitrate poisoning in animals. *Advances in Agronomy* 16, 197–247.

Yokoyama, K., Kai, H., Koga, T. and Aibe, T. (1991) Nitrogen mineralization and microbial populations in cow dung, dung balls and underlying soil affected by paracoprid dung beetles. *Soil Biology and Biochemistry* 23, 649–653.

Young, C.P. (1986) Nitrate in groundwater and the effects of ploughing on the release of nitrate. In: Solbe, J.F. (ed.) *Effects of Land Use on Freshwaters.* WRC/Ellis Horwood Ltd, Chichester, UK, pp. 221–237.

Young, D.J.B. (1958) A study of the influence of nitrogen on the root weight and nodulation of white clover in a mixed sward. *Journal of the British Grassland Society* 13, 106–114.

Younie, D. and Watson, C.A. (1992) Soil nitrate-N levels in organically and intensively managed grassland systems. In: Archer, J.R. *et al.* (eds) *Nitrate and Farming Systems.* Association of Applied Biologists, Warwick, pp. 235–238.

Glossary

Acid rain: rain whose pH is less than 5.6 (the normal equilibrium value for carbon dioxide and water), usually due to the presence of sulphuric and nitric acids (from emissions of SO_2 and NO_x)

Aerobic: conditions having a continuous supply of molecular oxygen.

Aerosol particles: solid or liquid particles in the atmosphere, other than cloud droplets, of diameter < 10 μm.

Aggregate (soil): an assemblage of individual soil particles bound together by colloidal material, and with a fairly distinct shape, such as a crumb or prism.

Amino acid: an organic compound containing both an amino group and a carboxyl group; amino acids combine to form proteins.

Ammonification: the release of ammonium/ammonia by microorganisms through the decomposition of organic matter.

Anaerobic: conditions in which there is a lack of molecular oxygen: in soils usually caused by excessive wetness.

Anion: a negatively charged atom or group of atoms, e.g. nitrate (NO_3^-) or sulphate (SO_4^{2-}).

Aquifer: an underground stratum of porous rock that retains water.

Autotrophic (organisms): organisms that are able to produce organic molecules from inorganic sources: most are photosynthetic autotrophs using the energy in sunlight to synthesize organic compounds from CO_2, but some are chemosynthetic and obtain their energy from the oxidation of inorganic compounds such as ammonia and nitrite.

Available water capacity: the amount of water (% by weight, or mm per unit depth of soil) which a soil can store in a form available to plants; it is equal to the water content at field capacity minus that at the wilting point.

Bloat: excessive accumulation of gases in the rumen of animals.

Bulk density: mass per unit volume of undisturbed soil, dried to constant weight at 105°C; typically varies from about 1.0 g cm^{-3} in topsoils just after ploughing to about 1.9 g cm^{-3} in subsoils below the plough layer.

Cation: a positively charged atom or group of atoms, e.g. potassium (K^+) or ammonium (NH_4^+).
Cation exchange: the exchange of cations in solution with those held on the negatively charged exchange sites of clays and soil organic matter.
Cation exchange capacity: the potential of a soil (or other solid material) for absorbing cations, often expressed in milligram equivalents per kg of soil.
Continuous grazing: livestock management in which animals graze a particular area continuously for long periods, though the stocking rate may be varied.
Defoliation: partial or complete removal of the above-ground parts of a sward, either by cutting or by grazing.
Denitrification: microbiological or chemical processes that convert nitrate to nitrous oxide or nitrogen gas.
Digestibility: the proportion (usually expressed as %) of the ingested herbage (either total dry matter or organic matter) that is digested and absorbed by an animal consuming the herbage.
Dry deposition: the transfer of material from the atmosphere to land or sea surfaces by the processes of gravitational settling, turbulent transport, molecular diffusion and impaction.
Ecosystem: a system of interacting biological and environmental components.
Enclosed grassland: an area of grassland bounded by fences, hedges or walls.
Eutrophication: an increase in the nutrient status of surface water, often accompanied by a massive growth of organisms such as algae, and the development of anaerobic conditions.
Evapotranspiration: the loss of water from the land by a combination of evaporation from the soil surface and transpiration from plant leaves.
Field capacity: the amount of water remaining in a soil immediately after any excess has drained into the underlying subsoil (only applicable to freely drained soils).
Fixation (of ammonium): the retention of ammonium ions by the clay fraction of the soil in a form that is non-exchangeable by cation exchange.
Fixation (of N): the conversion of gaseous nitrogen (N_2) into a form in which the N is combined with one or more other elements, e.g. NH_3 or HNO_3.
Forage crop: an arable crop (e.g. turnips, kale, oats or maize) grown specifically for consumption by livestock and usually fed to ruminants at times when the supply of fresh herbage is inadequate.
Groundwater: water occupying pores and crevices in subsoils and rocks, below the water-table and above an impermeable layer.
Heterotrophic (organisms): dependent for their source of energy on organic matter produced by other organisms.
Humus: the well decomposed, relatively stable part of the organic matter in soil.
Immobilization: the transformation of inorganic forms of N in soil into organic forms, mainly by assimilation by microorganisms.
Lamina: leaf blade.
Leaching: the removal of soluble constituents (usually from soil) by water moving downwards.
Leaf area index: the ratio of leaf area (one side of leaf) to ground area.
Leaf litter: freshly fallen plant material on the soil surface.
Ley: a sown grass or grass–clover sward intended to remain *in situ* for between 1 and 5 years, often alternating with a similar period under arable cultivation.

Mass flow: the movement of nutrient ions in the flow of water caused by the plant uptake of water.
Mineralization: the process by which organic forms of N in plant, animal and microbial residues and in humified soil organic matter, are converted into inorganic or 'mineral' forms (ammonium and nitrate).
Molar: the concentration resulting from 1 mole (see below) dissolved in 1 l of solution.
Mole: the amount of a substance that has a weight in grams numerically equal to the molecular weight of the substance.
Monolith (soil): an intact column of soil from a soil profile, usually to a depth of 1–2 m.
^{15}N: there are two stable isotopes of N, ^{14}N and ^{15}N, the former being the dominant form and the latter occurring naturally in the atmosphere at a uniform abundance of only 0.3663 atoms per 100 atoms (i.e. 0.3663%). If a plant (or soil or animal) derives its N from two main sources, and if one source is artificially enriched in (or depleted of) ^{15}N, then the relative amounts of N supplied by the two sources can be calculated from analyses of the material for total N and ^{15}N.
Nitrification: the microbiological oxidation of ammonium via nitrite to nitrate.
Nitro-chalk: a fertilizer comprising a mixture of ammonium nitrate and calcium carbonate.
Nitrogenase: an enzyme system (with two constituents) that catalyses the reduction of N_2 to ammonium in N-fixing organisms.
Oxidation: in a chemical reaction, the loss of electrons, often by transfer to an oxidizing agent, which is thus reduced.
Permanent grassland: grassland maintained as such without ploughing and re-seeding; minimum age not well defined but often taken to be > 20 years.
Permanent wilting point: the water content of a soil at which permanent irreversible wilting occurs: it usually corresponds to a suction of about 15 bar (1500 kPa).
pH: the negative logarithm of the concentration of hydrogen ions in a solution: it is a quantitative expression of acidity and alkalinity, and has a scale that ranges from 0 to 14; pH 7 is neutral, < 7 is acid and > 7 is alkaline.
Reduction: in a chemical reaction, the addition of electrons, often by the addition of hydrogen; reduction is the opposite of oxidation.
Residence time: the mean period for which an element or compound remains in the specified form.
Rhizosphere: the soil immediately surrounding plant roots, where there is usually an abundant microbiological population.
Rotational grazing: a grazing management system in which animals are moved from paddock to paddock in sequence so that each paddock is grazed quickly (1–3 days) after a period of rest (3–5 weeks).
Rumen: the first and largest compartment of the stomach of a ruminant animal.
Ruminant (animals): herbivorous mammals, such as cattle and sheep, that have complex stomachs containing microorganisms that are able to break down the cellulose in plant material.
Set stocking: a grazing management system in which the stocking rate remains constant over long periods.
Silage: grass herbage or a similar forage crop preserved in a moist condition by partial fermentation.
Slurry: a mixture of dung, urine and water from housed livestock.

Soil series: the basic unit of soil classification and mapping; soils within a series have closely similar profile characteristics.
Soil structure: the spatial distribution of aggregates and pores in soil.
Stratosphere: The portion of the atmosphere extending from the top of the troposphere (8–15 km) to about 50 km.
Stocking rate (or density): the number of animals per unit area of grazing land available to them.
Stolon: a creeping stem at, or just below, the soil surface capable of rooting and sending up new shoots at the nodes.
Stomata: pores in the leaf surface that allow the movement of water vapour and gases.
Symbiosis: an association of dissimilar organisms for their mutual advantage (e.g. clover and rhizobia).
Tiller: a shoot originating at a basal node in grasses: each tiller develops its own leaves, nodes, internodes and adventitious roots.
Troposphere: the portion of the atmosphere extending from the earth's surface to a height that ranges from 8 to 15 km.
Vegetative: term used to designate stem and leaf development in contrast to flower and seed development.
Wet deposition: the removal of gaseous and particulate matter from the atmosphere by rain or snow.

Appendix

Appendix Common and botanical names of species mentioned in the text.

Common name	Botanical name
Bahiagrass	*Paspalum notatum*
Bent	*Agrostis* spp.
Bermudagrass	*Cynodon dactylon*
Blue grama	*Bouteloula gracilis*
Bluegrass	*Poa pratensis*
Bromegrass	*Bromus* spp.
Cocksfoot	*Dactylis glomerata*
Heather	*Erica* spp.
Italian ryegrass	*Lolium multiflorum*
Ladino clover	*Trifolium repens*
Ling	*Calluna* spp.
Lucerne	*Medicago sativa*
Mat grass	*Nardus stricta*
Meadow fescue	*Festuca pratensis*
Perennial ryegrass	*Lolium perenne*
Purple moor grass	*Molinia caerulea*
Red clover	*Trifolium pratense*
Red fescue	*Festuca rubra*
Rough fescue	*Festuca campestris*
Reed canarygrass	*Phalaris arundinacea*
Smooth bromegrass	*Bromus inermis*
Subterranean clover	*Trifolium subterraneum*
Tall fescue	*Festuca arundinacea*
Timothy	*Phleum pratense*
White clover	*Trifolium repens*
Yorkshire fog	*Holcus lanatus*

Index